Molecular Microbial Ecology

Molecular Microbial Ecology

Dr A. Mark Osborn (Ed.)

Department of Animal & Plant Sciences, University of Sheffield, Sheffield, UK

Dr Cindy J. Smith (Ed.)

Department of Biological Sciences, University of Essex, Colchester, UK

Taylor & Francis
Taylor & Francis Group

Published by:
Taylor & Francis Group
In US: 270 Madison Avenue
New York, NY 10016
In UK: 4 Park Square, Milton Park
Abingdon, OX14 4RN

Transferred to Digital Printing 2005

ISBN: 1-8599-6283-1

A catalog record for this book is available from the British Library.

Library of Congress Cataloging-in-Publication Data

Molecular microbial ecology / A. Mark Osborn and Cindy J. Smith (eds).
p. ; cm
Includes bibliographical references and index.
ISBN: 1-85996-283-1
1. Molecular microbiology. 2. Microbial ecology. I. Osborn, A. Mark, Dr. II. Smith, Cindy, Dr.
[DNLM: 1. Ecology. 2. Environmental Microbiology. 3. Nucleic Acid Hybridization.
4. Polymerase Chain Reaction--methods.

QW 55 M7185 2005]
QR74.M635 2005
579'.17--dc22
2005014409

Editor: Elizabeth Owen
Editorial Assistant: Dominic Holdsworth
Production Editor: Karin Henderson
Typeset by: Phoenix Photosetting, Chatham, Kent, UK

10 9 8 7 6 5 4 3 2 1

The cover shows an uncultured filament from a drinking water bacterial community visualized using FISH (courtesy of Michael Wagner, Holger Daims and Kilian Stoecker, Department of Microbial Ecology, University of Vienna).

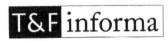

T&F informa

Taylor & Francis Group
is the Academic Division of T&F Informa plc.

Visit our web site at http://www.garlandscience.com

Contents

Contributors

Henk J. M. Aarts, RIKILT, Institute of Food Safety, Wageningen-UR, Bornsesteeg 45, P.O. Box 230, NL-6700 AE Wageningen, The Netherlands

Mark J. Bailey, Molecular Microbial Ecology Section, CEH Oxford, Mansfield Road, Oxford, OX1 3SR, UK

Holger Daims, Department of Microbial Ecology, University of Vienna, Althanstrasse 14, A-1090 Vienna, Austria

Konstantinos Damianakis, Department of Biological Sciences, University of Essex, Wivenhoe Park, Colchester, CO4 3SQ, UK

Derek Fairley, The Questor Centre, The Queens University Belfast, Belfast, BT9 5AG, UK

Andreas Felske, GBF National Research Centre for Biotechnology, Mascheroder Weg 1, D-38124 Braunschweig, Germany

Doreen E. Gillespie, Department of Plant Pathology, University of Wisconsin – Madison, 1630 Linden Dr., Madison, WI 53706, USA

N. C. M. Gomes, Federal Biological Research Centre for Agriculture & Forestry, Institute of Plant Virology, Microbiology & Biosafety, Messeweg 11–12, D-38104 Braunschweig, Germany

M. Götz, Federal Biological Research Centre for Agriculture & Forestry, Institute of Plant Virology, Microbiology & Biosafety, Messeweg 11–12, D-38104 Braunschweig, Germany

Jo Handelsman, Department of Plant Pathology, University of Wisconsin – Madison, 1630 Linden Dr., Madison, WI 53706, USA

Ian M. Head, School of Civil Engineering and Geosciences, University of Newcastle, Newcastle, UK

Renske Landeweert, Subdepartment of Soil Quality, Wageningen University, NL-6700 EC Wageningen, The Netherlands

Michael Larkin, The Questor Centre, The Queens University Belfast, Belfast, BT9 5AG, UK

Wolfgang Ludwig, Department of Microbiology, Technische Universität München, Am Hochanger 4, D-85354 Freising, Germany

Mike Manefield, Molecular Microbial Ecology Section, CEH Oxford, Mansfield Road, Oxford, OX1 3SR, UK

Annett Milling, Federal Biological Research Centre for Agriculture & Forestry, Institute of Plant Virology, Microbiology & Biosafety, Messeweg 11–12, D-38104 Braunschweig, Germany

J. Colin Murrell, Department of Biological Sciences, University of Warwick, Coventry, CV4 7AL, UK

Balbina Nogales, Departament de Biologia, Area de Microbiologia, Universitat de les Illes Balears, Crtra. Valdemosa, 07122 Palma de Mallorca, Spain

M. Oros-Sichler, Federal Biological Research Centre for Agriculture & Forestry, Institute of Plant Virology, Microbiology & Biosafety, Messeweg 11–12, D-38104 Braunschweig, Germany

A. Mark Osborn, Department of Animal & Plant Sciences, University of Sheffield, Western Bank, Sheffield, S10 2TN, UK

Vivien Prior, Department of Biological Sciences, University of Essex, Wivenhoe Park, Colchester, CO4 3SQ, UK

Stefan Radajewski, Department of Biological Sciences, University of Warwick, Coventry, CV4 7AL, UK

Jan L.W. Rademaker, NIZO food research, Department of Health & Safety, Kernhemseweg 2, P.O. Box 20, NL-6710 BA Ede, The Netherlands

Steven A. Ripp, Center for Environmental Biotechnology, University of Tennessee – Knoxville, 676 Dabney Hall, Knoxville, TN 37996, USA

Wilfred F. M. Röling, Faculty of Earth and Life Science, Vrije Universiteit, De Boelelaan 1085, NL-1081 Amsterdam, The Netherlands

Michelle R. Rondon, Department of Bacteriology, College of Agricultural & Life Sciences, 304a Fred Hall, 1550 Linden Drive, Madison, WI-53706, USA

Gary Sayler, Center for Environmental Biotechnology, University of Tennessee – Knoxville, 676 Dabney Hall, Knoxville, TN 37996, USA

K. Smalla, Federal Biological Research Centre for Agriculture & Forestry, Institute of Plant Virology, Microbiology & Biosafety, Messeweg 11–12, D-38104 Braunschweig, Germany

Eric Smit, RIVM, Antonie van Leeuwenhoeklaan 9, NL-3721 MA Bilthoven, The Netherlands

Cindy J. Smith, Department of Biological Sciences, University of Essex, Wivenhoe Park, Colchester, CO4 3SQ, UK

Francisco de Souza, Netherlands Institute of Ecology, PO Box 40, NL-6966 ZG Heteren, The Netherlands

Kilian Stoecker, Department of Microbial Ecology, University of Vienna, Althanstrasse 14, A-1090 Vienna, Austria

Sarah L. Turner, Molecular Microbial Ecology Section, CEH Oxford, Mansfield Road, Oxford, OX1 3SR, UK

Pablo Vinuesa, Centro de Ciencias Genómicas-UNAM, Programa de Ecología Genómica, Av. Universidad s/n, col. Chamilpa, Aptdo 565A, Cuernavaca, Morelos, Mexico

Michael Wagner, Department of Microbial Ecology, University of Vienna, Althanstrasse 14, A-1090 Vienna, Austria

Andrew S. Whiteley, Molecular Microbial Ecology Section, CEH Oxford, Mansfield Road, Oxford, OX1 3SR, UK

Lynn L. Williamson, Department of Plant Pathology, University of Wisconsin – Madison, 1630 Linden Dr., Madison, WI 53706, USA

Abbreviations

AFLP amplified fragment length polymorphism
ALFA automatic laser fluorescence analysis
AMF arbuscular mycorrhizal fungi
AOB ammonia-oxidizing bacteria
AP-PCR arbitrarily primed PCR
ARDRA amplified ribosomal DNA restriction analysis
ARISA automated ribosomal intergenic spacer analysis
BAC bacterial artificial chromosome
BLAST Basic Local Alignment Search Tool
BrdU bromodeoxyuridine
BSA bovine serum albumin
CAT chloramphenicol acetyl transferase
CTAB hexadecyltrimethylammonium bromide
DAF DNA amplification fingerprinting
DD differential display
ddNTP 2',3'-dideoxynucleoside triphosphate
DEPC diethylpyrocarbonate
DGGE denaturing gradient gel electrophoresis
dNTP deoxynucleoside triphosphate
EDTA ethylenediamine tetraacetic acid
ERIC enterobacterial repetitive intergenic consensus
FASTA Fast All
FERP fluorescent-enhanced rep-PCR
FISH fluorescent *in situ* hybridization
GFP green fluorescent protein
IEF isoelectric focusing
IGS inter-ribosomal gene spacer sequence
IPTG isopropyl-β-D-thiogalactopyranoside
IS insertion sequences
ISR intergenic spacer region
ITS internal transcribed spacer
ITS-PCR-RFLP inter-transfer ribosomal-PCR-RFLP
LH-PCR length heterogeneity PCR
MDH methanol dehydrogenase
MGE mobile genetic elements
MLST multi-locus sequence typing
MMO methane monooxygenase
OTU operational taxonomic unit
PBS phosphate-buffered saline
PCR polymerase chain reaction
PFA paraformaldehyde
PFGE pulsed field gel electrophoresis
PNA peptide nucleic acid
qPCR quantitative real-time PCR
RAPD random amplified polymorphic DNA
RAP-PCR RNA fingerprinting by arbitrarily primed PCR
REP repetitive extragenic palindromic sequence
rep-PCR repetitive sequence-based PCR
RFLP restriction fragment length polymorphism
RSGP reverse sample genome probing
RT reverse transcription
SDS sodium dodecyl sulfate
SIG-PCR signature PCR
SIP stable-isotope probing
SSCP single strand conformation polymorphism
TGGE temperature gradient gel electrophoresis
T-RLFP terminal restriction fragment length polymorphism
UPGMA unweighted pair group method with arithmetic mean
UV ultraviolet
VNTR variable number of tandem repeat

Preface

Microbial ecology encompasses the study of how microorganisms interact with their environment. Microoganisms are distributed across every ecosystem, and microbial transformations are fundamental to the operation of the biosphere. Yet, whilst microbial ecology arguably represents one of the most important areas of biological research, for many years our understanding of microorganisms in the environment has been based primarily on those microorganisms that we could culture on media, and which represent between only 0.1 and 10% of the total microbial flora within any given environment. However, the last 25 years have seen a revolution in microbial ecology, catalyzed by the application of molecular (largely nucleic-acid-based) approaches to enable the previously uncultured and numerically dominant microbial groups to be studied. Such research has been driven in particular by the development of methods to isolate nucleic acids directly from environmental samples and by the subsequent application of polymerase chain reaction (PCR) approaches to amplify gene sequences for further study by fingerprinting and/or sequencing, enabling investigation of microbial diversity and functional potential within any given environment.

This text aims to provide a theoretical and practical approach to molecular studies in microbial ecology and is written by some of the world's leading authorities in this area. The text begins with the isolation of nucleic acids from environmental samples and then details PCR-based approaches to study prokaryotic systematics via ribosomal RNA analysis and related methodologies allowing the study of microbial community structure, molecular typing of environmental isolates, investigation of gene expression and abundance in environmental samples, and the targeting of specific functional groups via stable isotope probing. Subsequent chapters introduce the application of DNA hybridization approaches to microbial ecology, and importantly the use of fluorescent *in situ* hybridization to enable investigation and localization of microorganisms within their environment. Chapters 10 and 11 consider the implications of, and potential for, genomic and metagenomic approaches in microbial ecology, whilst Chapters 13 and 14 discuss how molecular approaches can be utilized to investigate fungal microbial ecology, and the application of microorganisms as tools for environmental assessment of contaminated ecosystems. Chapters 12 and 15 provide a molecular toolbox for the molecular microbial ecologist with coverage of PCR primer systems for the analysis of functional genes and an introduction to key bioinformatics resources.

Whilst the scope of this text is limited to nucleic acid-based techniques, it is important to appreciate, given the complexity of the environments under study, that the most insightful discoveries in microbial ecology will be gained from polyphasic approaches combining molecular studies with microbial physiology and biochemistry and with physico-chemical characterization of the environment. With the additional development of microarray-, proteomic- and metabolomic-based methods and the potential for miniaturization in lab-on-a-chip systems, we have unparalleled opportunities to advance our understanding of microbial ecology over the next decade.

Finally we would like to take this opportunity to thank all of the contributors to this text for their enthusiasm and also their patience during the book's development. We would also like to thank all of the staff at Taylor & Francis, and in particular Dominic Holdsworth, for their continued support during this project. We also thank Nigel Farrar for his input into the early development of the text.

Dr A. Mark Osborn, Dr Cindy J. Smith

Nucleic acid extraction from environmental samples

1

Annett Milling, Newton C. M. Gomes, Miruna Oros-Sichler, Monika Götz and Kornelia Smalla

1.1 Introduction

Over 20 years ago, the first protocol titled 'DNA extraction from soil' was published by Torsvik (1). However, only in the late 1980s/early 1990s, when molecular tools such as nucleic acid hybridization, the polymerase chain reaction (PCR) and DNA cloning and sequencing became increasingly available, more attention was focused on the analysis of DNA extracted from environmental bacteria without prior cultivation. Obviously, the analysis of nucleic acids extracted directly from environmental samples allows the researcher to investigate microbial communities by obviating the limitations of cultivation techniques. The phenomenon that only a small proportion of bacteria can form colonies when traditional plating techniques are used (2) was first described by Staley and Konopka (3) as the great plate anomaly. A further limitation of the cultivation-based studies of microbial communities is that under environmental stress bacteria can enter a state termed 'viable but non-culturable' (vbnc), and again these bacteria would not be accessible to traditional cultivation techniques (4,5). Consequently researchers were attracted by the opportunities afforded by analyzing nucleic acids recovered directly from environmental samples that should be representative of the microbial genomes present in such samples. The analysis of DNA can provide information on the structural diversity of environmental samples, or on the presence or absence of certain functional genes (e.g. genes conferring xenobiotic biodegradative capabilities, antibiotic resistance or plasmid-borne sequences), or to monitor the fate of bacteria (including genetically modified organisms) released into an environment. However, in general the analysis of DNA does not allow conclusions to be drawn on the metabolic activity of members of the bacterial or fungal community or on gene expression. This information might be obtained from analysis of RNA (rRNA or mRNA) (see also Chapter 5).

A large and diverse suite of protocols has been published on nucleic acid extraction from environmental matrices (for review see 6). Two principal approaches exist, each with their own advantages and limitations. The first approach pioneered by Ogram *et al.* (7) is based on direct or *in situ* lysis of microbial cells in the presence of the environmental matrix (e.g. soil or sediments), followed by separation of the nucleic acids from matrix components and cell debris. This is by far the most frequently utilized method. The advantage of the direct nucleic acid extraction approach is that it is less time-consuming and that a much higher DNA yield is achieved. However, directly extracted DNA often contains considerable amounts of co-extracted substances such as humic acids that interfere with subsequent molecular analysis (8). Furthermore, a considerable proportion of directly extracted DNA might originate from non-bacterial sources or from free DNA.

In the second approach, the microbial fraction is recovered from the environmental matrix prior to cell lysis and subsequent DNA extraction and purification. The major concern with the so-called indirect or *ex situ* DNA extraction approach is a differential recovery efficiency of surface-bound cells. Dissociation of cells from surfaces is generally achieved by repeated blending/homogenization steps and differential centrifugation. Thus the indirect method is more time-consuming and prone to contamination. A clear advantage of the indirect approach is that the nucleic acids recovered are less contaminated with co-extracted humic acids and DNA of non-bacterial origin. During the last 10 years considerable progress has been made in developing faster and more efficient nucleic acid extraction procedures. While the first protocols for both approaches all required CsCl/ethidium bromide density gradient ultracentrifugation to purify the DNA and thus were rather tedious and time-consuming (7,9,10), the use of DNA purification kits based on different kinds of resins considerably improved the purification efficiency and reduced the time needed to obtain DNA suitable for molecular analysis (8,11,12). However, none of the protocols was suitable for all soil types, in particular for soils and sediments originating from contaminated sites. Only recently have commercial kits for DNA extraction from soils become available and these represent a major breakthrough in view of the simplification and miniaturization of this crucial method for many cultivation-independent analysis methods. Commercial soil DNA extraction kits can be used to extract DNA in the presence of the environmental matrix and/or from the microbial pellet obtained after efficient dislodgement and centrifugation. Despite the current ease of using commercial kits for nucleic acid extraction, a number of critical factors remain that influence the quantity and quality of nucleic acid extracts, and will be discussed later in this chapter. Whilst DNA extraction seems to work reliably for different matrices, efficient RNA extraction is often still problematic. Thus this chapter will also focus on protocols that enable assessment of the metabolically active microbial fraction within an environmental sample. Furthermore, we demonstrate the utility of 16S/18S rDNA-based molecular fingerprints for comparing different protocols and for illustrating how different methodologies affect the composition of the microbial community recovered.

1.2 Recovery of cells from environmental matrices

The indirect DNA extraction approach might preferably be used when problematic environmental matrices are to be analyzed, or when cloning large DNA fragments [e.g. to generate bacterial artificial chromosome (BAC) libraries; see Chapter 11] from soil or sediment DNA where a high proportion of DNA of bacterial origin is crucial. Different protocols aiming at the representative dislodgement and extraction of surface-attached cells have been published (9,13–16). All these methods have in common that they use repeated homogenization and differential centrifugation as originally suggested by Faegri *et al.* (17). However, the protocols differ considerably with respect to the solutions used to break up soil colloids and dislodge surface-attached cells that adhere to surfaces by various bonding mechanisms such as polymers (e.g. exopolysaccharides or fimbriae), electrostatic forces and water bridging, and that act to differing degrees. Homogenization is usually achieved by shaking suspensions with gravel or blending in Stomacher or Waring blenders. In particular, for soils with a high clay content a further purification can be achieved by sucrose/Percoll density gradient centrifugation or flotation of the bacterial fraction on a Nycodenz cushion (14). Although a complete dislodgement of cells seems to be impossible, it is important that cells that are bound to the surface with different degrees of strength are released with similar efficiency. This can easily be evaluated by using DNA fingerprinting (see Chapter 3), e.g. denaturing gradient gel electrophoresis to analyze 16S or 18S rDNA fragment profiles amplified from the DNA extracted from the microbial pellet in comparison to profiles generated from directly extracted DNA.

1.3 Cell lysis and DNA extraction protocols

The efficient disruption of the bacterial and fungal cell walls is crucial for the recovery of representative DNA which reflects the genomes of microbes present in an environmental sample and their relative abundance. Cell lysis can be achieved by mechanical cell disruption and/or by enzymatic or chemical disintegration of cell walls. Most of the published protocols include a combination of these steps. The efficiency of cell lysis protocols might differ considerably depending on the kind of environmental matrix, since compounds within the matrix might have adverse effects, e.g. reduced enzyme activity due to non-optimal pH or ionic conditions, or simply due to a high adsorption capacity. A number of studies compared the efficiency of lysis protocols with respect to DNA yield and fragmentation. Whereas, in most studies, bead beating was reported to yield the highest amounts of DNA, albeit the DNA produced showed some degree of shearing, some authors favored grinding in the presence of liquid nitrogen (18–20). According to our experience, bead beating is important as a first step before treatment with other cell lysis methods. The length of time and intensity of bead beating, the size of the beads as well as the ratio of beads to soil suspension, and also the content of clay minerals in the soil matrix have all been reported to influence the degree of DNA shearing (12,21–24). The efficiency of the cell lysis is usually estimated by microscopic examination (18–19,21,24). A correlation of lysis efficiency and clay content was demonstrated by Zhou *et al.* (18). We have compared common lysis approaches such as bead beating (cell homogenizer, Braun-Melsungen), Retsch mill, freeze-boiling, grinding in the presence of liquid nitrogen, vortexing (M. Oros-Sichler *et al.*, unpublished data) and also three different extraction and purification protocols (see *Figure 1.1*). While the quantity and degree of

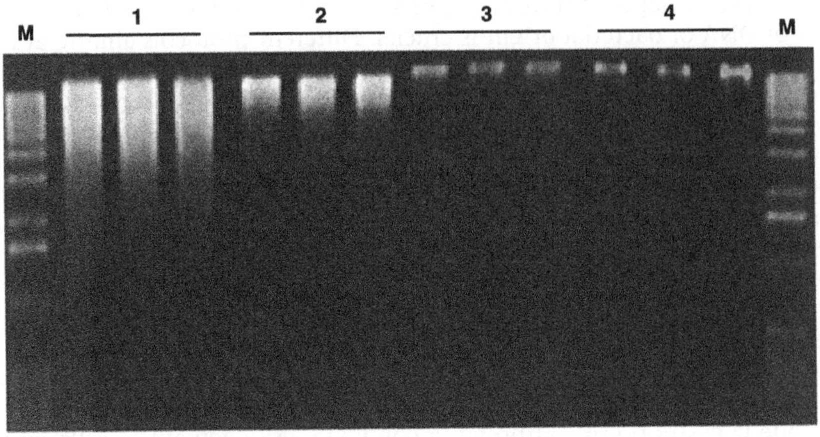

Figure 1.1

Comparison of four common lysis procedures with respect to the yield and fragmentation of the DNA (0.8% agarose gel). Lane M: 1 kb ladder; lane 1: bead beating; lane 2: Retsch mill; lane 3: vortex; lane 4: freeze-boiling cycles (20 min at 63°C followed by 20 min at –20°C).

nucleic acid fragmentation was assessed by agarose gel electrophoresis, the influence of the different protocols on the bacterial and fungal diversity was evaluated by denaturing gradient gel electrophoresis (DGGE) analysis of 16S/18S rDNA fragments amplified from the different kinds of DNA. In contrast to our traditional DNA extraction protocol (12,25), the use of commercial kits is clearly less time-consuming and avoids extraction with phenol and chloroform. In addition, the DNA extracted with the commercial kits less frequently contained PCR-inhibiting substances. A disadvantage may be that the use of rather small amounts of soil (0.25–0.5 g) for DNA extraction, from some environmental matrices, may represent a serious limitation for the recovery of a sufficient quantity of DNA to be representative of that environment (26). Like other authors (27) we found that the yield of DNA recovered per gram of soil depends on the lysis method and on the extraction protocol used. In addition, the soil type strongly affected the quality (degree of shearing, PCR-inhibiting substances) and the quantity of the DNA. Although the quantification of DNA based on agarose gels stained with DNA-staining dyes is more complicated than fluorimetric measurements, this approach also provides insights into the degree of DNA shearing. High molecular weight DNA is an important criterion when evaluating and comparing different protocols because sheared DNA can cause PCR artifacts and is not suitable for direct cloning of large DNA fragments. DNA yields reported in the literature on soil range from 5 to 250 $\mu g\ g^{-1}$ of soil (12,20). The complete removal of co-extracted humic acids is critical for the subsequent molecular analysis. Humic acids were shown to interfere with DNA hybridization, restriction enzyme digestions and PCR amplification (8,12). The degree of co-extracted humic substances strongly depends on the environmental matrix, i.e. soil type. DNA extracted by means of commercial kits was often suitable for use in PCR amplifications without additional purification steps, except when applied to sandy soils. A range of methods has been used to remove co-extracted substances, e.g. silica-based or ion exchange resins or Sephadex-G200 gel filtration. Often these additional purification steps lead to a reduced DNA yield. Some authors have also used low melting agarose gel electrophoresis to remove co-extracted humic substances (28,29). However, this purification approach should not be used when the recovered DNA is analyzed for the presence of plasmid-borne genes, e.g. antibiotic resistance genes.

1.4 Immunochemical isolation of DNA from metabolically active cells

Uptake of [^{3}H]thymidine has routinely been used for measuring the *in situ* growth of bacteria in different environments. Bromodeoxyuridine (BrdU) is structurally similar to thymidine, and since it can be incorporated into newly synthesized DNA it is widely used in medical research. Recently, two groups have shown the potential use of BrdU to detect metabolically active bacteria in microbial communities from lake water (30), bacterioplankton (31) or in soil (32,33). The procedure consists of four steps:

(i) incubation of environmental bacteria (soil or microbial fraction) with BrdU;
(ii) extraction of DNA directly from the environmental sample or the microbial fraction;
(iii) immunocapture of DNA containing incorporated BrdU by using magnetic beads covered with anti-BrdU antibody;
(iv) immunoprecipitation.

Borneman (32) used the BrdU approach to study soil bacterial communities responding to certain environmental stimuli such as the addition of glucose. Soils with or without glucose amendment were mixed with BrdU and incubated at room temperature. We have adapted the protocol to study bacterial rhizosphere communities from tomato plants which were inoculated with a *Ralstonia solanacearum* antagonist strain 24_4 (A. Milling *et al.*, unpublished data). The incorporation time and the amount of BrdU were optimized. The DNA recovered was analyzed by DGGE of PCR-amplified 16S rDNA fragments. We showed that only a part of the bacterial populations were able to incorporate BrdU. Furthermore, the bacterial control strain remained metabolically active throughout the greenhouse experiment. One problem of the procedure is that after prior amplification with group-specific primers, a DGGE pattern was also generated from samples that were not incubated with BrdU after immunocapture. This is probably due to aspecific binding of DNA and BrdU antibody-coated magnetic beads. In contrast, following direct amplification of 16S rDNA fragments from the BrdU DNA after immunocapture, only weak or no background was observed.

A second potential limitation of the BrdU method might be the proportion of bacteria that are capable of incorporating BrdU. Whilst the majority of bacteria are thought to take up and incorporate [^{3}H]thymidine into DNA, this has not been fully investigated for its analogue BrdU.

A further evaluation of the method is therefore necessary concerning the efficiency of BrdU uptake and incorporation into genomic DNA of different species and also the aspecific binding of unlabeled DNA to magnetic beads.

1.5 RNA extraction from environmental matrices

Whereas direct and indirect DNA extraction protocols are widely used, methods for RNA extraction have been less frequently used and are less well established. Due to the short half-life of bacterial messenger RNA as well as the high abundance and persistence of RNases, an unbiased recovery of total RNA still represents a methodological challenge (but see also Chapter 5). Considerable effort is required to ensure the absence of RNases. Thus all working solutions need to be treated with 0.1% diethylpyrocarbonate (DEPC), and glassware is baked at temperatures >200°C. Different protocols aimed at the simultaneous extraction of RNA and DNA have recently been published either using cell lysis in presence of the environmental matrix (20,34,35) or after prior cell lysis (22). As with DNA extractions, important criteria for the quality and suitability of RNA extraction protocols are the yield, the integrity, and the purity of the RNA. Co-extracted humic substances and DNA have been reported to affect RNA hybridization results with oligonucleotide probes. Probe hybridization decreased as the concentration of DNA or humic substances increased (36). The presence of humic substances and/or high concentrations of co-extracted DNA were shown to result in membrane saturation and consequently reduced amounts of bound RNA. Even the presence of low concentrations of DNA seemed to cause a reduced accessibility of the rRNA target. Another major concern is that some DNase preparations might contain residual RNase activity and cause partial degradation of rRNA molecules. rRNA is judged as intact when distinct bands of the small and large subunit RNA can be visualized after electrophoresis (*Figure 1.2*, page 15, lane A1). While the efficient recovery of RNA was the focus of most of the published RNA extraction protocols, Alm *et al.* (36) and also Alm and Stahl (37) emphasize undegraded RNA to be an equally important parameter. Different regions of the ribosomal RNA are differentially susceptible to the attack of RNases that might result in a partial loss of probe or primer target sites. Such partial RNA degradation is particularly critical when the RNA is used for quantitative hybridization. Thus methods for efficiently removing humic substances and residual DNA without partial degradation of the RNA are required for the reliable use of RNA for microbial community analysis. While the size and amount of rRNA can be assessed by polyacrylamide or agarose gels, the presence of mRNA is usually confirmed by PCR amplification or by membrane hybridization with probes. The turnover rate of mRNA is supposed to be related to the growth rate of a cell. Since most of the environmental bacteria grow much more slowly than *E. coli* (on which much of our knowledge of turnover is based), which is reported to have mRNA half-lives of between 0.5 and 20 min, the half-life of mRNA in bacteria in the environment should be correspondingly longer (38). However, so far no appropriate protocol is available which permits the integrity and degradation of mRNA to be investigated.

Acknowledgments

This work was funded by the EU project QLK3-2000-01598 (A.M.), BMBF projects 0311295A (M.G.) and 0312629A (M.O.-S.) and a DAAD postdoctoral fellowship to N.C.M.G.

References

1. Torsvik V (1980) Isolation of bacterial DNA from soil. *Soil Biol Biochem* **12**: 15–21.
2. Amann RI, Ludwig W, Schleifer K-H (1995) Phylogenetic identification and *in situ* detection of individual microbial cells without cultivation. *Microbiol Rev* **59**: 143–169.
3. Staley JT, Konopka A (1985) Measurement of *in situ* activities of nonphotosynthetic microorganisms in aquatic and terrestrial habitats. *Annu Rev Microbiol* **39**: 321–346.
4. Roszak DB, Colwell RR (1987) Survival strategies of bacteria in the natural environment. *Microbiol Rev* **51**: 365–379.
5. Oliver JD (2000) Problems in detecting dormant (VBNC) cells and the role of DNA elements in this response. In: Jansson JK, van Elsas JD, Bailey MJ (eds), *Tracking Genetically-engineered Microorganisms*, pp. 1–15. Eurekah, Austin, TX.
6. Van Elsas JD, Smalla K, Tebbe CC (2000) Extraction and analysis of microbial community nucleic acids from environmental matrices. In: Jansson JK, van Elsas JD, Bailey MJ (eds), *Tracking Genetically-engineered Microorganisms*, pp. 29–51. Eurekah, Austin, Texas.
7. Ogram A, Sayler GS, Barkay TJ (1987) DNA extraction and purification from sediments. *J Microbiol Meth* **7**: 57–66.
8. Tebbe CC, Vahjen W (1993) Interference of humic acids and DNA extracted directly from soil in detection and transformation of recombinant DNA from bacteria and a yeast. *Appl Environ Microbiol* **59**: 2657–2665.
9. Holben WE, Jansson JK, Chelm BK, Tiedje JM (1988) DNA probe method for the detection of specific microorganisms in the soil bacterial community. *Appl Environ Microbiol* **54**: 703–711.
10. Jacobsen CS, Rasmussen OF (1992) Development and application of a new method to extract bacterial DNA from soil based on separation of bacteria from soil with cation-exchange resin. *Appl Environ Microbiol* **58**: 2458–2462.
11. Tsai Y-L, Olson BH (1992) Rapid method for separation of bacterial DNA from humic substances in sediments for polymerase chain reaction. *Appl Environ Microbiol* **58**: 2292–2295.
12. Smalla K, Cresswell N, Mendonca-Hagler LC, Wolters A, van Elsas JD (1993) Rapid DNA extraction protocol from soil for polymerase chain reaction-mediated amplification. *J Appl Bacteriol* **74**: 78–85.
13. Hopkins DW, Macnaughton SJ, O'Donnell AG (1991) A dispersion and differential centrifugation technique for representatively sampling microorganisms from soil. *Soil Biol Biochem* **23**: 217–225.
14. Bakken LR, Lindahl V (1995) Recovery of bacterial cells from soil. In: Trevors JT, van Elsas JD (eds), *Nucleic Acids in the Environment*, pp. 13–27. Springer-Verlag, Berlin.
15. Herron PR, Wellington EMH (1990) New method for extraction of streptomycete spores from soil and application to the study of lysogeny in sterile amended and nonsterile soil. *Appl Environ Microbiol* **56**: 1406–1412.
16. Gebhard F, Smalla K (1999) Monitoring field releases of genetically modified sugar beets for persistence of transgenic plant DNA and horizontal gene transfer. *FEMS Microbiol Ecol* **28**: 261–272.

17. Faegri A, Torsvik VL, Goksoyr J (1977) Bacterial and fungal activities in soil: separation of bacteria and fungi by a rapid fractionated centrifugation technique. *Soil Biol Biochem* **9**: 105–112.
18. Zhou J, Bruns MA, Tiedje JM (1996) DNA recovery from soils of diverse composition. *Appl Environ Microbiol* **62**: 316–322.
19. Frostegård Å, Courtois S, Ramisse V, Clerc S, Bernillon D, Le Gall F, Jeannin P, Nesme X, Simonet P (1999) Quantification of bias related to the extraction of DNA directly from soils. *Appl Environ Microbiol* **65**: 5409–5420.
20. Hurt RA, Qiu X, Wu L, Roh Y, Palumbo AV, Tiedje JM, Zhou J (2001) Simultaneous recovery of RNA and DNA from soils and sediments. *Appl Environ Microbiol* **67**: 4495–4503.
21. Duarte GF, Rosado AS, Seldin L, Keijzer-Wolters AC, van Elsas JD (1998) Extraction of ribosomal RNA and genomic DNA from soil for studying the diversity of the indigenous microbial community. *J Microbiol Meth* **32**: 21–29.
22. Moré MI, Herrick JB, Silva MC, Ghiorse WC, Madsen EL (1994) Quantitative cell lysis of indigenous microorganisms and rapid extraction of microbial DNA from sediment. *Appl Environ Microbiol* **60**: 1572–1580.
23. Bürgmann H, Pesaro M, Widmer F, Zeyer J (2001) A strategy for optimizing quality and quantity of DNA extracted from soil. *J Microbiol Meth* **45**: 7–20.
24. Miller DN, Bryant JE, Madsen EL, Ghiorse WC (1999) Evaluation and optimization of DNA extraction and purification procedures for soil and sediment samples. *Appl Environ Microbiol* **65**: 4715–4724.
25. Van Elsas JD, Smalla K (1995) Extraction of microbial community DNA from soils. In: Akkermans ADL, van Elsas JD, de Bruijn FJ (eds), *Molecular Microbial Ecology Manual*, section 1.3.3, pp. 1–11. Kluwer Academic Publishers, Dordrecht.
26. Webster G, Newbury CJ, Fry JC, Weightman AJ (2003) Assessment of bacterial community structure in the deep sub-seafloor biosphere by 16S rDNA based techniques: a cautionary tale. *J Microbiol Meth* **55**: 155–164.
27. Martin-Laurent F, Philippot L, Hallet S, Chaussod R, Germon JC, Soulas G, Catroux G (2001) DNA extraction from soils: Old bias for new microbial diversity analysis methods. *Appl Environ Microbiol* **67**: 2354–2359.
28. Porteous LA, Armstrong JL (1993) A simple method to extract DNA directly from soil for use with polymerase chain reaction amplification. *Curr Microbiol* **27**: 115–118.
29. Young CC, Burghoff RL, Keim LG, Minak-Bernero V, Lute JR, Hinton SM (1993) Polyvinylpyrrolidone-Agarose gel electrophoresis purification of polymerase chain reaction-amplifiable DNA from soil. *Appl Environ Microbiol* **59**: 1972–1974.
30. Urbach E, Vergin KL, Giovannoni SJ (1999) Immunochemical detection and isolation of DNA from metabolically active bacteria. *Appl Environ Microbiol* **65**: 1207–1213.
31. Pernthaler A, Pernthaler J, Schattenhofer M, Amann R (2002) Identification of DNA-synthesizing bacterial cells in coastal North sea plankton. *Appl Environ Microbiol* **68**: 5728–5736.
32. Borneman J (1999) Culture-independent identification of microorganisms that respond to specified stimuli. *Appl Environ Microbiol* **65**: 3398–3400.
33. Yin B, Crowley D, Sparovek G, De Melo WJ, Borneman J (2000) Bacterial functional redundancy along a soil reclamation gradient. *Appl Environ Microbiol* **66**: 4361–4365.
34. Griffiths RI, Whiteley AS, O'Donnell AG, Bailey MJ (2000) Rapid method for coextraction of DNA and RNA from natural environments for analysis of ribosomal DNA- and rRNA-based microbial community composition. *Appl Environ Microbiol* **66**: 5488–5491.
35. Weinbauer MG, Fritz I, Wenderoth DF, Höfle MG (2002) Simultaneous

extraction from bacterioplankton of total RNA and DNA suitable for quantitative structure and function analyses. *Appl Environ Microbiol* **68**: 1082–1087.

36. Alm EW, Zheng D, Raskin L (2000) The presence of humic substances and DNA in RNA extracts affects hybridization results. *Appl Environ Microbiol* **66**: 4547–4554.
37. Alm EW, Stahl DA (2000) Critical factors influencing the recovery and integrity of rRNA extracted from environmental samples: use of an optimized protocol to measure depth-related biomass distribution in freshwater sediments. *J Microbiol Meth* **40**: 153–162.
38. Fleming JT, Yao Wen-Hsiang, Sayler GS (1998) Optimization of differential display of prokaryotic mRNA: application to pure culture and soil microcosms. *Appl Environ Microbiol* **64**: 3698–3706.
39. Vainio EJ, Hantula J (2000) Direct analysis of wood-inhabiting fungi using denaturing gradient gel electrophoresis of amplified ribosomal DNA. *Mycol Res* **104**: 927–936.
40. Cremonesi L, Firpo S, Ferrari M, Righetti PG, Gelfi C (1997) Double-gradient DGGE for optimized detection of DNA point mutations. *Biotechniques* **22**: 326–330.
41. Heuer H, Wieland G, Schönfeld J, Schönwälder A, Gomes NCM, Smalla K (2001) Bacterial community profiling using DGGE or TGGE analysis. In: Rouchelle P (ed.), *Environmental Molecular Microbiology: Protocols and Applications*, pp. 177–190. Horizon Scientific Press, Wymondham, UK.

Protocol 1.1: DNA extraction from bulk soil

OVERVIEW

Three protocols for DNA extraction directly from bulk soil are compared in terms of DNA yield, quality and PCR amplifiability:

(i) a classical DNA extraction method, described below (referred to as DNA sample A in the Figures);

(ii) two commercial DNA extraction kits:

- Ultra Clean™ Soil DNA Kit, MoBio Laboratories Inc., Solana Beach, CA 92075, Art. Nr. 6560-200, www.mobio.com (referred to as DNA sample B in the Figures);
- Fast DNA® Spin® Kit for Soil, BIO101, Carlsbad, CA 92008, Art. No. 12800-100, www.bio101.com (referred to as DNA sample C in the Figures).

The soil samples used were taken from strawberry and rape fields located in three different sites in Germany. All three DNA extraction methods used 0.5 g (dry weight) of an aliquot from two composite samples as starting material. Each composite sample was obtained by pooling 10 cores of bulk soil (15 cm from the top) from each plot at the same site. Aliquots for extraction were taken after mixing the composite sample by sieving (2 mm diameter pores). For the DNA extraction based on kits, the cell lysis was only performed mechanically, using the bead beating procedure (see below under 'Equipment'). Otherwise, the manufacturers' instructions were followed.

MATERIALS

Reagents

Reagents for cell lysis:

100 mM sodium phosphate buffer (pH 8.0), containing freshly added lysozyme (5 mg/ml)

Reagents for DNA extraction and precipitation:

20% sodium dodecyl sulphate (SDS)

Tris-buffered phenol (pH 8.0)

Chloroform:isoamyl alcohol (24:1)

Isopropanol

70% ethanol

TE buffer: 10 mM Tris–HCl (pH 8.0), 1 mM ethylenediamine tetraacetic acid (EDTA) (pH 8.0)

Deionized water

Reagents for DNA purification:

Cesium chloride (CsCl)

8 M potassium acetate (KAc) (pH not adjusted)

Additional reagents and materials:

Glass beads (0.17–0.18 mm diameter; sterilized)

2 ml screw cap Eppendorf tubes

2 ml and 1.5 ml Eppendorf tubes

Ultra Clean™ 15 DNA Purification Kit, MoBio Laboratories, Solana Beach, CA 12100-300

Geneclean Spin® Kit, BIO101, 2251 Rutherford Rd., Carlsbad, CA 1101-600,

Ice

Equipment

Bead beater (MSK cell homogenizer, B. Braun Diessel Biotech, Melsungen, Germany)

Microcentrifuge

Speed vacuum concentrator

Vortex

Water bath

METHODS

Cell lysis

1. Resuspend 0.5 g (dry weight) of soil in 750 µl 100 mM sodium phosphate buffer (pH 8.0) (containing 5 mg/ml lysozyme) in a 2 ml Eppendorf tube.
2. Homogenize the soil suspension and incubate in a water bath at 37°C for 15 min.
3. Place the samples immediately on ice.
4. Add to each sample 750 mg glass beads, keep samples on ice.
5. Disrupt the cells for 90 s in the bead beater.
6. Add to the disrupted cell suspension 45 µl 20% SDS.

7. Incubate the samples on ice for 1 h.

DNA extraction

8. Add 700 µl Tris-buffered phenol (pH 8.0) to the cell lysate.
9. Mix well (e.g. by gently vortexing) and centrifuge for 5 min at 10 000 ***g*** at room temperature.
10. Carefully transfer the aqueous (upper) phase to a new Eppendorf tube.
11. Re-extract the phenol phase by adding 200 µl 100 mM sodium phosphate buffer (pH 8.0) and repeat the Step 9, in order to remove possible rests of phenol.
12. Combine both of the aqueous phases from Steps 9 and 11.
13. Add to the pooled aqueous phases an equal volume of chloroform/isoamyl alcohol (24:1).
14. Mix well and centrifuge the samples for 10 min at 10 000 ***g*** at room temperature.
15. Transfer the aqueous phase to a new Eppendorf tube.

DNA precipitation

16. Add to each sample 0.7 volumes of isopropanol and incubate at room temperature for ≥1 h.
17. Centrifuge for 30 min at 16 000 ***g*** at room temperature.
18. Discard the supernatant and add 1 ml 70% ethanol to the pellet.
19. Centrifuge again for 20 min at 16 000 ***g*** at 4°C.
20. Repeat Steps 18 and 19.
21. Dry the pellet in a speed vacuum concentrator.
22. Dissolve the pellet in 100 µl TE (pH 8.0). The DNA obtained after this step is further referred to as crude DNA (sample A1 in the Figures).

Troubleshooting

Sometimes the phase separation in Step 10 might not be evident, which can lead to contamination of the crude extracts with phenols. In this case a second repeat of the purification with chloroform/isoamyl alcohol (Steps 13–15) might be helpful.

Two-step purification by CsCl and KAc precipitation

(Perform all steps at room temperature, unless stated otherwise.)

CsCl precipitation

1. Add 50 mg CsCl to 50 µl crude DNA extract from Step 22 and incubate for 2 h.
2. Centrifuge at 20 000 ***g*** for 30 min at room temperature.
3. Transfer the supernatant to a fresh Eppendorf tube and add 4 volumes deionized water and 0.6 volumes isopropanol and incubate for 30 min.
4. Centrifuge at 20 000 ***g*** for 20 min at room temperature.
5. Discard the supernatant and resuspend the pellet in 50 µl TE (pH 8.0).

KAc precipitation

6. Add to the dissolved pellet 0.2 volumes of 8 M potassium acetate (pH not adjusted).
7. Mix well and incubate the samples for 45 min.
8. Centrifuge at 20 000 ***g*** for 15 min at room temperature.
9. Transfer the supernatant to a new Eppendorf tube and add 0.6 volumes isopropanol.
10. Mix well and incubate for 30 min at room temperature to precipitate the DNA.
11. Centrifuge again at 20 000 ***g*** for 20 min at room temperature.
12. Discard the supernatant and wash the pellet with 70% ethanol, centrifuge again, discard the supernatant and dry the pellet in a speed vacuum concentrator.
13. Dissolve the pellet in 50 µl TE buffer. The DNA obtained after this step is further referred to as two-step purified DNA (sample A2 in the Figures).

Hints and troubleshooting

Often the two-step purified DNA is already suitable for PCR amplification. However, depending on the soil type additional purification might be required.

For additional purification we used the kits Ultra Clean™ 15 DNA Purification Kit (MoBio Laboratories, Solana Beach, CA), resulting in the

DNA referred to as sample A3 in the figures and Geneclean Spin® Kit (BIO101, Vista, CA), resulting in the DNA referred to as sample A4 in the Figures. A comparison of the DNA yield is shown in *Figure 1.2*, whilst *Figure 1.3* shows a comparison between DGGE fingerprints (see Chapter 3 for description of DGGE) of 18S rDNA fragments amplified from total community DNA with primers described by Vainio and Hantula (39) after the extraction and purification methods used.

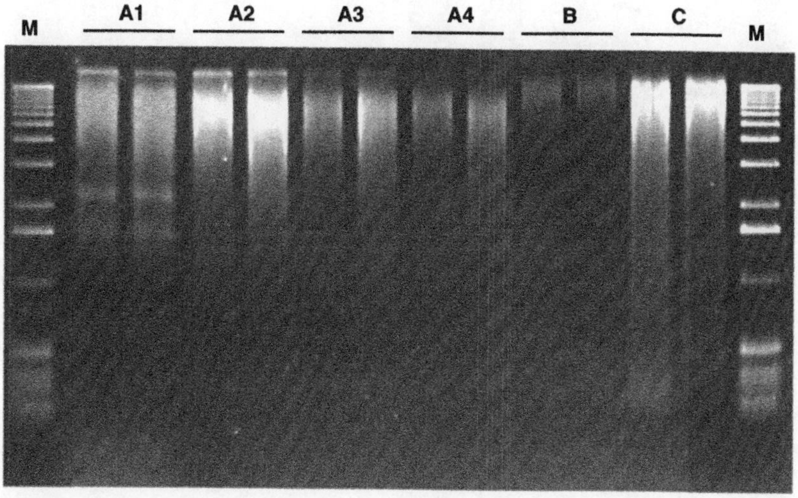

Figure 1.2

Agarose gel electrophoresis of nucleic acids recovered from bulk soil. Comparison of a classical extraction method (12) and commercial kits for extraction and purification of soil DNA (0.8% agarose gel). M: 1 kb DNA marker; A1: crude DNA; A2: two-step purified DNA; A3: DNA purified with the Ultra Clean™ 15 DNA Purification Kit; A4: DNA purified with the Geneclean Spin® Kit; B: DNA extracted with the Ultra Clean™ Soil DNA Kit; C: DNA extracted with the Fast DNA Spin® Kit for Soil. Each extraction and/or purification approach is presented for two replicate samples taken from different plots at the same site.

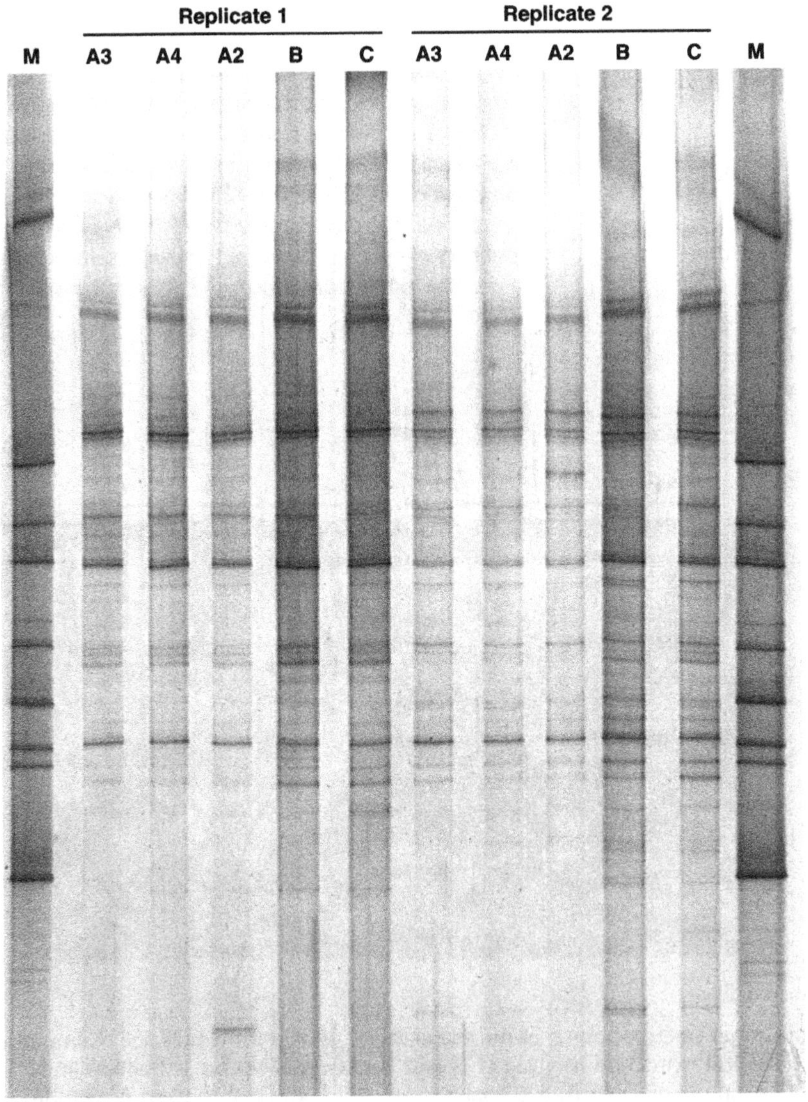

Figure 1.3

DGGE fingerprints of 18S rDNA fragments (1650 bp) amplified from total community DNA with primers described by Vainio and Hantula (39). Comparison of a classical extraction and purification method (12) and commercial kits for extraction and purification of soil DNA (18–38% denaturing gradient). M: Mixture of 18S rRNA gene fragments amplified from genomic DNA of different fungal isolates; A3: DNA purified with the Ultra Clean™ 15 DNA Purification Kit; A2: two-step purified DNA; A4: DNA purified with the Geneclean Spin® Kit; B: DNA extracted with the Ultra Clean™ Soil DNA Kit; C: DNA extracted with the Fast DNA Spin® Kit for Soil. Data are shown for two replicate samples taken from different plots at the same site.

Protocol 1.2: BrdU immunocapture

OVERVIEW

To provide information on the metabolically active bacterial fraction the bromodeoxyuridine (BrdU) method described by Borneman (32) was established and optimized for the study of rhizosphere bacteria. Briefly, the rhizosphere sample (microbial rhizosphere pellet, roots with adhering soil) was incubated with BrdU. DNA was extracted using the Ultra Clean™ DNA Purification Kit (MoBio Laboratories) followed by an immunocapture step with anti-BrdU-coated magnetic beads and recovery of the DNA after immunoprecipitation.

Samples were supplemented with 0.5, 1.0, 2.5 and 5 µmol BrdU per gram of the rhizosphere pellet and incubated for 2, 6, 14, 24, 48 and 120 h at room temperature. The incorporation of BrdU into the bacterial cells was shown to be dependent upon the concentration of BrdU tested. In addition, the incubation time and the application of BrdU (onto a bacterial pellet or direct application to roots with adhering soil) influenced the incorporation of BrdU. After 24 h incorporation of BrdU concentration, no changes in the DGGE profiles of 16S rDNA fragments amplified from BrdU DNA were observed.

MATERIALS

Reagents

Reagents for immunocapture:

5-Bromo-2′-deoxyuridine (BrdU) (Roche Diagnostics GmbH, Mannheim, Germany)

Dynabeads (M-450) Sheep anti-mouse IgG (Dynal Biotech ASA, Oslo, Norway)

Anti-BrdU-antibody (0.1 µg/µl) (Roche Diagnostics GmbH)

Bovine serum albumin (BSA IgG-free, Protease-free) (Dianova, Hamburg, Germany)

Herring sperm DNA (10 mg/ml) (Promega, Mannheim, Germany)

Phosphate-buffered saline (PBS): $NaH_2PO_4 \times H_2O$ (0.16 g), $Na_2HPO_4 \times 2H_2O$ (0.98 g), NaCl (8.1 g) made up to 1 l with distilled H_2O (pH 7.4) and autoclaved.

10% (w/v) bovine serum albumin (BSA) solution

PBS-BSA containing 0.1% (w/v) BSA

PBS-BSA herring sperm DNA: containing 0.1% (w/v) BSA and 5 mg ml^{-1} herring sperm DNA

Anti-BrdU antibody: working concentration = 0.01 µg/µl (stock solution 1:10 diluted in PBS-BSA)

50 ml Centrifuge tubes (Sorval)

1.5 ml and 2 ml Eppendorf tubes

Equipment

Dynal MX1 Sample Mixer (Dynal Biotech, ASA, Oslo, Norway)

Dynal Magnetic Particle Concentrator (Dynal, MPC)

Centrifuge and microcentrifuge

Stomacher 400 blender (Seward)

Heating block (95°C)

METHODS

Incubating soil or rhizosphere with BrdU

Soil:

1. Add 100 µl of 50 mM BrdU to 1 gram of soil (corresponds to 5 µmol BrdU/g).
2. Vortex carefully and incubate at room temperature for 24–48 h in the dark.

Rhizosphere:

1. Place 3 g of plant roots with firmly adhering soil into sterile plastic bags, resuspend in 9 ml of distilled water and treat in a Stomacher 400 blender (Seward) at high speed for 1 min.
2. After this treatment collect the supernatant without roots in a centrifuge tube and resuspend the remaining roots in 9 ml of distilled water.
3. Repeat the Stomacher blending step twice more.
4. Combine the supernatants of the three blending steps and centrifuge at high speed (20 000 *g*) for 30 min to collect the microbial pellet.
5. Resuspend the pellet in 1 ml sterile water and transfer the suspension to a 2 ml Eppendorf tube.
6. Centrifuge again at 10 000 *g* for 20 min and discard the supernatant.

7. Add 100 µl of 50 mM BrdU to 1 g of the rhizosphere pellet (corresponds to 5 µmol BrdU g^{-1} soil).
8. Vortex carefully and incubate at room temperature for 48 h in the dark.

DNA extraction

1. In order to extract the total community DNA the UltraClean™ Soil DNA Isolation Kit, MoBio Laboratories (Solana Beach, CA) was used. Add 0.25–0.5 g of rhizosphere pellet or soil to the 2 ml bead solution tube. The extractions were done according to the MoBio protocol and instructions, including a bead-beating step (2 × 30 s at 4000 rpm) for cell lysis.

BrdU immunocapture

1. Mix the Dynabeads very well before use.
2. Add 25 µl Dynabeads into a 1.5 ml Eppendorf tube.
3. Wash the beads three times with 200 µl PBS-BSA. Collect them by magnetic capture and discard the supernatant.
4. Resuspend the beads in 84 µl PBS-BSA herring sperm DNA.
5. Add 16 µl anti-BrdU antibody.
6. Incubate the beads at room temperature for 60 min while rotating the tubes.
7. Wash three times with 200 µl PBS-BSA, collecting the beads by magnetic capture and discard the supernatant.
8. Resuspend the beads in 80 µl PBS-BSA herring sperm DNA.
9. Add 20 µl freshly denatured BrdU labeled DNA (DNA is denatured by heating at 95°C for 5 min, prior to transfer to ice for 5 min).
10. Incubate at room temperature for 120 min while rotating the tubes.
11. Wash very carefully five times with 200 µl PBS-BSA, collecting the beads by magnetic capture and discard the supernatant.
12. Resuspend the beads in 40 µl H_2O.
13. Heat this suspension at 95°C for 5 min and cool on ice for 5 min.

14. Centrifuge for 5 s at 10 000 ***g***.
15. Remove the Dynabeads by magnetic capture and transfer the non-bead fraction to a new tube.
16. Store the supernatant (non-bead fraction) at –20°C for further analysis.

This method was used to follow population dynamics in the rhizosphere of tomato plants grown in a soil with a high *Ralstonia solanacearum* titre during greenhouse testing of *Pseudomonas chlororaphis* strain 24-4 with *in vitro* antagonistic activity towards *R. solanacearum*. The development of the bacterial community during the greenhouse experiment was followed by PCR-DGGE analysis of 16S rDNA fragments (see Chapter 3) amplified from total community DNA obtained from the microbial pellet with and without BrdU incubation. DGGE profiles of non-BrdU-incorporating and BrdU-incorporating bacterial communities extracted from rhizosphere samples taken 14 days post planting are shown in *Figure 1.4*.

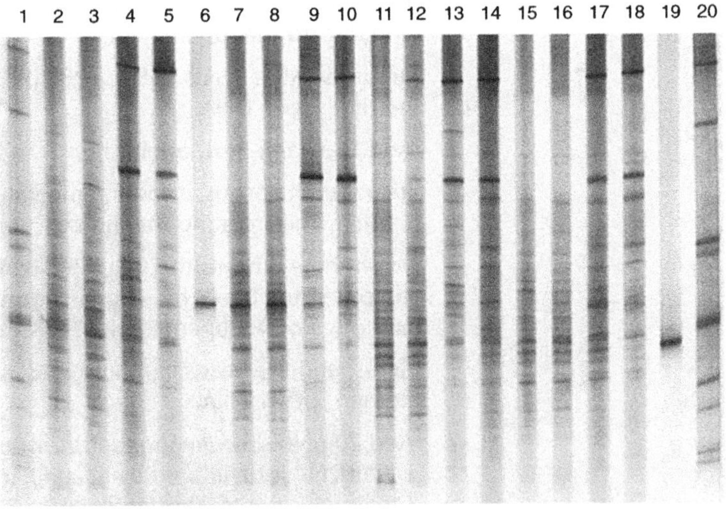

Figure 1.4

PCR-DGGE analysis of 16S rDNA fragments [*E. coli*: 968-1401 according to Heuer *et al.* (41)] amplified from rhizosphere total community DNA without and with BrdU incorporation, and in the presence or absence of *Ralstonia solanacearum* (*Rs*) and/or its antagonist *Pseudomonas chlororaphis* 24-4 (*Pc*). Denaturing gradient: 27.5–58%, 6% acrylamide. Lanes 1, 20: standard; lanes 2, 3: tomato; lanes 4, 5: tomato + BrdU; lane 6: *Rs*; lanes 7, 8: tomato + *Rs*; lanes 9, 10: tomato + *Rs* + BrdU; lanes 11, 12: tomato + *Rs* + antagonist (*Pc*); lanes 13, 14: tomato + *Rs* + antagonist (*Pc*) + BrdU; lanes 15, 16: tomato + antagonist (*Pc*); lanes 17, 18: tomato + antagonist (*Pc*) + BrdU; lane 19: antagonist (*Pc*).

Protocol 1.3: Simultaneous DNA/RNA extraction

OVERVIEW

The conditions for the extraction of total DNA/RNA from soil were developed based on a protocol published by Hurt *et al.* (20). The modifications made were required after we included a bead beating step instead of grinding with liquid nitrogen.

MATERIALS

Reagents for DNA/RNA extraction

Inactivate RNase by diethylpyrocarbonate (DEPC) treatment. Treat all solutions (except for solutions containing Tris) and water for reagent preparation with 0.1% DEPC. Autoclave for 20 min at 120°C

Absolute ethanol

Extraction buffer (pH 7.0): 100 mM sodium phosphate (pH 7.0), 10 mM Tris–HCl (pH 7.0), 1 mM EDTA (pH 8.0), 1.5 M NaCl, 1% (w/v) hexadecyltrimethylammonium bromide (CTAB) and 2% (w/v) SDS

Chloroform/isoamyl alcohol (24:1)

Isopropanol

TE buffer: 10 mM Tris–HCl, 1 mM EDTA (pH 8.0)

Additional reagents and materials

All non-disposable materials used for DNA/RNA extraction must be treated to inactivate RNase. Glassware: baked at 200°C overnight. Plasticware: treated with 0.1% DEPC or cleaned with RNase Away (Molecular Bio-Products, San Diego, CA).

2 ml Eppendorf tubes

QIAGEN RNA/DNA Mini Kit (QIAGEN, Hilden, Germany)

Glass beads (0.17–0.18 mm diameter)

Equipment

Bead beater (MSK cell homogenizer, B. Braun Diessel Biotech, Melsungen, Germany)

Water bath

Microcentrifuge

METHODS

DNA/RNA extraction

1. Add 0.5 g of soil into an Eppendorf tube containing 0.2 g of glass beads and keep the sample on ice.
2. Add 0.5 ml of absolute ethanol on the sample and homogenize twice for 30 s by bead beating (4000 rpm). Samples should be kept on ice (30 s) between the first and second bead-beating steps.
3. Centrifuge for 5 min at 16 000 ***g*** and discard the supernatant.
4. Add 1.0 ml of extraction buffer and incubate the sample for 30 min at 65°C mixing carefully every 10 min and centrifuge at 16 000 ***g*** for 5 min.
5. Transfer the supernatant (~1 ml) into pre-chilled 2 ml Eppendorf tubes on ice containing 1 ml aliquots of 24:1 chloroform-isoamyl alcohol, mix and centrifuge at 16 000 ***g*** for 5 min.
6. Transfer the aqueous phase to an Eppendorf tube and precipitate the nucleic acid by addition of 0.6 volumes of isopropanol for ≥30 min at room temperature, and centrifuge at 16 000 ***g*** for 20 min at room temperature. Pelleted nucleic acids can then be washed in ice cold 70% (vol/vol) ethanol and if necessary be kept at –70°C overnight before starting the DNA/RNA separation procedure.
7. Discard supernatant and resuspend the pellet in 0.5 ml of DEPC-treated water and start to perform the separation of RNA from DNA with QIAGEN RNA/DNA Mini Kit according to the manufacturers' protocol.
8. To improve the RNA protection we recommend to store the RNA as aliquots of isopropanol precipitates.

The DGGE profiles resulting from either DNA or RNA (cDNA) templates showed a complex bacterial soil community (*Figure 1.5*). Although differences in band intensities were detected most of the bands were found in both profiles.

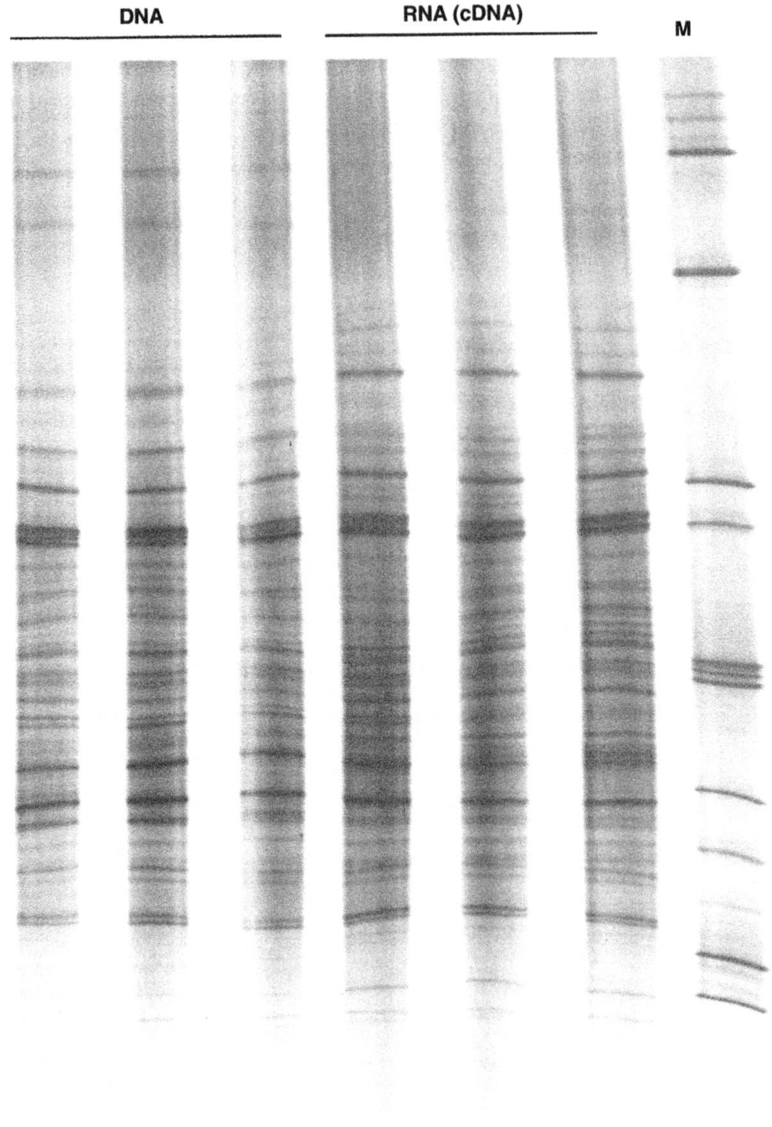

Figure 1.5

Scanned images of standard silver-stained DGGE [6–9% acrylamide gradient (40), 26–58% denaturant: 6 h run at 220 V and 58°C]: DGGE profiles from bacterial soil community DNA and RNA (reverse-transcribed by Random Primers, Invitrogen). The fingerprints were generated by separation of PCR fragments amplified with eubacterial primers F984 (GC-clamp) and R1378 (41). M: Marker DNA containing, from top to bottom, PCR products amplified using F984 and R1378 from the following bacterial species: *Clostridium pasteurianum* DSM 525, *Pectobacterium carotovora*, *Agrobacterium tumefaciens* DSM 30205, *Pseudomonas fluorescens* R2f, *Pantoea agglomerans*, *Nocardia asteroides* N3, *Rhizobium leguminosarum* DSM 30132, *Actinomadura viridis* DSM 43462, *Kineosporia aurantiaca* JCM 3230, *Nocardiopsis atra* ATCC 31511 and *Actinoplanes philippiensis* JCM.

Prokaryotic systematics: PCR and sequence analysis of amplified 16S rRNA genes

2

Wilfred F. M. Röling and Ian M. Head

2.1 Introduction

Until the 1980s, the determination of microbial community structure and the identification of microorganisms in environmental samples depended on culture-based studies. These can be both time-consuming and cumbersome and were already known to be selective, as only a small part of a microbial community is accessed (1). Two major discoveries have revolutionized microbial ecology and have permitted culture-independent characterization of microbial communities: the recognition that phylogenetic relationships between microorganisms can be inferred from molecular sequences (2) and the ability to selectively amplify minute amounts of nucleic acids extracted from environmental samples by the polymerase chain reaction [PCR; 3).

After more than 15 years of nucleic-acid-based analysis of natural microbial communities, the 16S rRNA gene remains central in contemporary molecular microbial ecology. This gene is a very suitable molecular marker (2). The gene is present in all prokaryotes and its product shows functional constancy: 16S rRNA is part of the ribosomes that are required by all organisms to synthesize protein. The gene is sufficiently long (~1.5 kb) to be used as a document of evolutionary history, and evidence for horizontal transfer of rRNA genes is limited. Due to the functional necessity of rRNAs, their primary and secondary structure is constrained. 16S rRNA consists of several sequence domains that have evolved at different rates; some domains have remained almost universally conserved and are interspersed by more variable regions, specific for phylum up to subspecies level. This permits unambiguous alignment of homologous positions in a sequence and the identification of near-universally conserved and taxon-specific 'signature' sequence motifs.

In this chapter an outline of the basic approaches used in the 16S rRNA sequence-based analysis of natural microbial communities is given: PCR, construction screening and sequencing of clone libraries, and the determination of phylogenetic relationships (*Figure 2.1*). The abilities and limitations of these techniques will be discussed and related to other molecular approaches to study microbial communities (*Figure 2.1*).

2.2 Application of PCR in microbial ecology

PCR allows the selective amplification of small amounts of DNA extracted from natural samples (*Figure 2.1*). It is nowadays generally used as the first step in phylogenetic analysis of microbial communities and for many other molecular ecological approaches (*Figure 2.1*).

A PCR reaction consists of a buffered mixture containing at least a thermostable DNA polymerase, oligonucleotide primers, free deoxynucleoside triphosphates (dNTPs: dATP, dCTP, dGTP and dTTP), magnesium ions and template DNA. The oligonucleotide primers are designed to hybridize to regions of DNA flanking the desired gene sequence. Primer selection is a

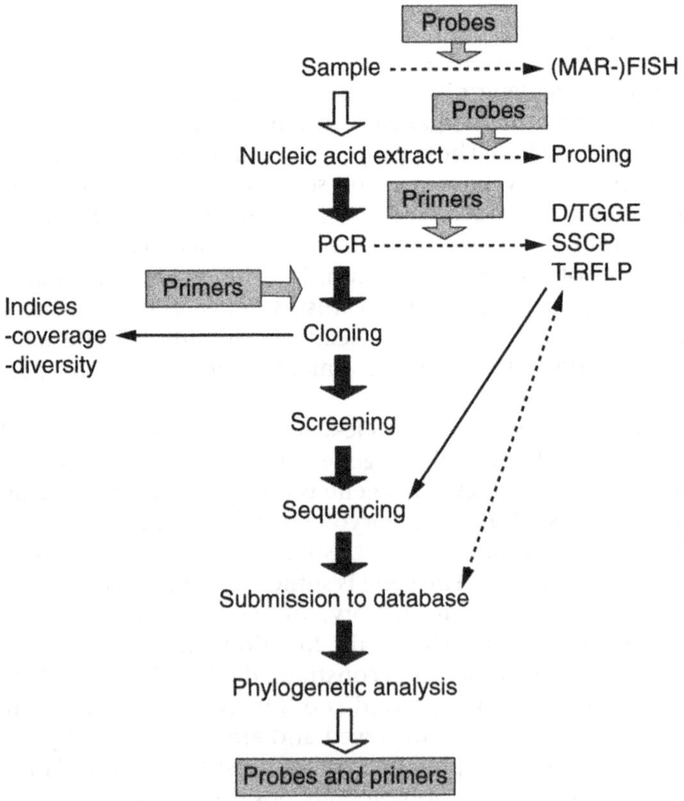

Figure 2.1

Flow scheme of the different steps in phylogenetic analysis (connected by black arrows) and the interaction of phylogenetic analysis with other molecular techniques (indicated by white arrows).
MAR = microautoradiography

crucial consideration and will depend on the particular application and whether rRNA genes from a wide range of organisms (e.g. all members of the domain Bacteria) or specific group of organisms are being targeted. Their design is based on sequence alignments and should be done carefully (4). As the database of 16S rRNA sequences is continually expanding, it is worthwhile to check 'old' primers routinely to assess their specificity. A number of software packages are now available that permit such an analysis to be more readily accomplished [e.g. ARB (www.arb-home.de) and PRIMROSE (www.cf.ac.uk/biosi/research/biosoft/)]. Probes designed on the basis of rRNA sequence alignments that were available in the past, and which contained many fewer sequences than current databases, should routinely be checked before their application in new experiments because target sequences that appeared to be specific for a particular taxon at the time they were designed may prove to be present in a wider range of organisms when screened against a more complete database.

The starting point for PCR-based analyses of microbial populations is a nucleic acid preparation extracted from an environmental sample (see Chapter 1) which is used as a template in the PCR reaction. The PCR process involves three stages: the DNA is denatured to convert double-stranded DNA into single-stranded DNA; oligonucleotide primers are annealed to complementary priming sites in the target DNA; finally the DNA is extended from the primers by the addition of nucleotides through DNA polymerase activity, resulting in double-stranded products. Repetitive cycling through these three steps results in an exponential increase in the DNA fragments of interest.

The three different steps in PCR are performed at different temperatures in an automated programmable thermal cycler. DNA is denatured at 94–96°C. The annealing temperature depends on the melting behavior of the primers; the melting temperature (T_m) can be calculated crudely from the primer base composition:

$$T_m = 2*(A + T) + 4*(G + C).$$

In general, a temperature 5°C below T_m is used for annealing. Higher temperature results in higher specificity and the specificity of the PCR reaction is generally tested empirically by conducting the PCR reaction at different annealing temperatures. A number of thermal cyclers are now available that allow a temperature gradient to be maintained across the heating block, allowing rapid optimization of the annealing temperature. The primer extension is performed at the temperature at which the chosen DNA polymerase show maximal activity: 68–72°C.

The performance of PCR is influenced by the concentrations of the individual components including the magnesium ion concentration, the type of DNA polymerase, concentration and properties of primers and template in the PCR mixture, as well as the length of the product (4). When a new type of reaction is set up, PCR conditions must be optimized to obtain high yields of only the product of interest. Standard PCR conditions are generally adequate for the amplification of rRNA genes, though techniques such as hot start (5) and touchdown PCR (6) can reduce mispriming and increase specificity of amplification. Though in some instances high specificity may not be desired.

2.2.1 Good working practices in PCR: avoiding PCR inhibition and contamination

A major issue regarding amplification of rRNA genes, and any other gene, from environmental samples is overcoming inhibition of the PCR by substances co-extracted with the nucleic acids (7). These are often assumed to be humic acids, though the actual nature of PCR inhibitory material is not usually rigorously established. In each PCR experiment a positive control, comprising a DNA template containing the primer target sequences and known to yield a product in PCR reactions, should be included to distinguish between errors in preparation of the reaction mix and sample-related problems. If the positive control yields a product but the sample of interest does not, one possibility is that inhibitory material is present in the DNA preparation. If no PCR product is obtained and it is suspected that very low levels of template DNA are present, an additional test should be performed by spiking the sample with positive control DNA to establish if the absence of PCR product is due to inhibitory material in the PCR reaction. Details of procedures designed to remove inhibitory substances from nucleic acid extracts are given in Chapter 1. Furthermore, the PCR reaction can be optimized to reduce the effects of inhibitors. A simple method that often yields successful PCR is dilution of the nucleic acid preparation to reduce the levels of the inhibitory contaminants to below the level at which the PCR is inhibited. The addition of non-acetylated bovine serum albumin (400 ng μl^{-1}) or *T4* gene 32 protein (150 ng μl^{-1}) can lead to much higher tolerance to contaminants (8). Also, DNA polymerases show different minimum inhibitory concentrations of 'humic acids', ranging from 0.08 to 8.33 $\mu g\ ml^{-1}$ (9), and DNA polymerases that are more resistant to inhibitors should help to diminish the problem. Besides humic acids, there are many other factors that are known to inhibit PCR (10). Some are related to the conditions under which PCR is performed. Therefore, good practice in the laboratory and aseptic conditions are advised.

Good working practices in the laboratory are especially important with respect to another major problem associated with 16S rDNA-based PCR: false-positive reactions due to contamination of the PCR with DNA fragments from a source other than the added template (*Figure 2.2*). This is especially likely to occur when universal primers (i.e. those whose target sequence is present in the vast majority of small subunit rRNA characterized to date) or primers targeting very broad phylogenetic groupings are used. Therefore, in each PCR experiment a negative control, to which no template DNA is added, should be included to check for contaminating DNA. Sometimes, contamination is difficult to avoid, as DNA can be present in commercial DNA polymerase preparations due to co-extraction during the isolation of DNA polymerases from the microorganisms used to produce the enzyme. Also reagents used in DNA extraction can contain low amounts of contaminating DNA (11). Contaminating DNA can be removed by UV treatment or DNase I treatment (12,13) (*Figure 2.2*). Contamination is a particular problem if low levels of template DNA are used (i.e. template DNA levels are less than the level of contamination) or if nested PCR is used (*Figure 2.2*). Nested PCR of negative control reactions will often reveal contamination in the second round of amplification even when the first

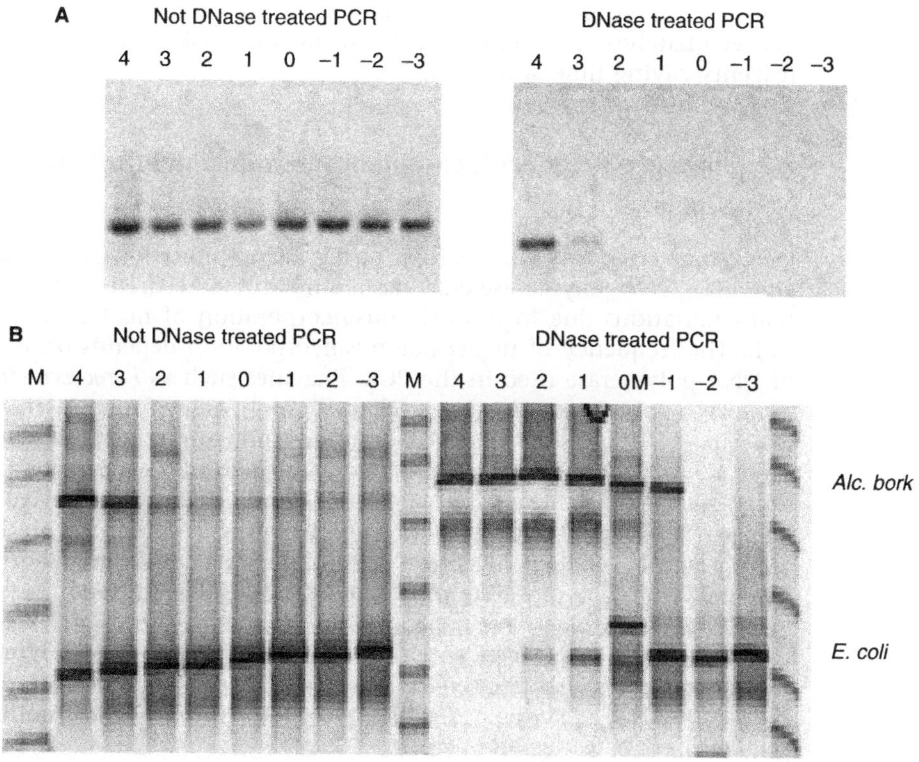

Figure 2.2

Effects of template concentration and DNase treatment on contamination levels. *Alcanivorax borkumensis* DNA was used as template in PCR, with concentrations ranging from 10^4 to 10^{-3} cell equivalents per PCR (indicated by log-transformed numbers above the lanes). Primers pA and pH′ (72) were applied in the first round of PCR. DNase treatment was performed by adding 1 IU of DNase I per PCR and a preincubation for 30 min at 37°C and DNase inactivation for 10 min at 95°C, before template was added and 35 PCR cycles performed. Results are shown in (**A**) and indicate a large degree of contamination in the untreated PCR. A nested PCR was performed, and the products separated by DGGE (73). This revealed severe contamination by *E. coli*, the bacterium used to produce the Taq polymerase. Quantitative analysis (74) indicated the presence of ~5000 *E. coli* 16S rRNA genes per reaction (~5×10^2 cells), DNase treatment lowered the amount to <10 per reaction: ~1 cell (**B**) It is possible to detect a single copy by this nested PCR.

round of PCR shows no PCR product detectable on an ethidium bromide-stained agarose gel. It is therefore important to conduct PCR amplification using an aliquot from the first-round negative control as template when conducting nested PCR. Comparison of denaturing gradient gel electrophoresis (DGGE) profiles may also be useful in assessing the nature of contamination.

To reduce the risk of contamination, it is desirable to have a dedicated work area for PCR, with instruments (pipettes etc.) that are used only for PCR. Reagents should be prepared with high quality water and stored remote from PCR products and template DNA preparations. Purchase of high quality reagents is recommended as this ensures reproducibility between batches of reagents and also reduces the risk of contamination of reagents, saving time and effort.

2.2.2 Pitfalls of PCR: artifacts and differential amplification

2.2.2.1 PCR artifacts

PCR can give rise to a number of artifacts, such as mutations, deletions and chimeras. DNA polymerase enzymes are not 100% accurate and introduce point mutations due to intrinsic misincorporation of nucleotides during PCR. The frequency of nucleotide misincorporation depends on the type of DNA polymerase used in the PCR. Enzymes such as *Pyrococcus furiosus* (*Pfu*) DNA polymerase, that have a 3' to 5' proofreading function, have very low rates of nucleotide misincorporation. More commonly used enymes, such as *Taq* DNA polymerase from *Thermus aquaticus*, have a higher processivity but lack a proofreading exonuclease activity and hence have lower fidelity. However, the degree of error resulting from misincorporation during PCR is generally relatively small compared to differences in rRNA sequences, often considered indicative of species level differences [~2–3% (14)]. The percentage of sequences with polymerase errors increases with the number of PCR cycles and the length of the amplified fragment. A second artifact is also frequently observed. Due to stable secondary structures in rRNA, deletions can easily occur during PCR, resulting in smaller-sized PCR fragments (15).

A third well-known PCR artifact is the formation of chimeric PCR products (16). Chimeric genes result from the incomplete synthesis of a rRNA gene fragment during amplification. If the incomplete fragment anneals to a homologous rRNA gene fragment forming a heteroduplex, it can be extended to full length. This results in a rRNA gene fragment that has been replicated from different templates and thus represents a complete rRNA sequence that does not exist naturally in a living organism. The frequency of chimeras can be up to 30% and is enhanced by:

(i) the availability of partial rDNA fragments in a nucleic acid extract (often as a consequence of using sheared DNA as a template, or from premature termination of elongation during PCR); and
(ii) the result of a sequence possessing a high percentage of highly conserved stretches along the primary rDNA structure where, after denaturation, single strands originating from different rDNAs can anneal in complementary regions.

Chimera formation can be diminished by increasing the elongation time and decreasing the number of cycles (16).

A number of computer algorithms have been developed to identify chimeric sequences [e.g. CHIMERA_CHECK (http://rdp.cme.msu.edu)]. These programs have difficulties in recognizing chimeric molecules when the 'parent' sequences have >85% homology. Due to higher sequence identity, chimeras are readily generated between very closely related sequences and these chimeras are far more difficult to detect. Such programs should therefore only be used as a guide and the occurrence of chimeras should be confirmed by careful analysis of secondary structure interactions and independent phylogenetic analyses with different regions of the molecule. The occurrence of chimeric molecules can best be detected by conducting phylogenetic analyses on opposite ends of the rRNA sequence. If the sequence is chimeric, the trees generated from the two fragments will be incongruent. If the sequence is genuine, the trees generated independently from different regions of the rRNA molecule should be identical or at least very similar.

Both misincorporation of nucleotides and chimera formation lead to an overestimation of diversity (see Section 2.4).

2.2.2.2 Differential amplification

Ideally a PCR product amplified from environmental nucleic acids, and any clone library produced from it, should reflect the composition of the microbial community from which DNA was extracted. Quantitative abundance of species from PCR can only be inferred when all molecules are equally accessible to primer hybridization, when primer–template hybrids are formed with the same efficiency, and DNA polymerase extends at equal efficiency with all templates, throughout the whole PCR process. Unfortunately, these assumptions often do not hold and several biases compromise the ability to draw quantitative conclusions on species abundance from abundances of PCR-generated fragments. These biases are related to the template and the PCR approach used.

Problems relating to template are rRNA gene copy number (17), rRNA operon heterogeneity (18), differences in G + C content (19), presence of sequences outside the amplified sequence that inhibit amplification (20), modified template (e.g. DNA methylation) (9) and template concentration (21). Microorganisms differ in the number of rRNA operons in their genome; the number can vary from one to more than ten (http://rrndb.cme.msu.edu). This will lead to overrepresentation of sequences from organisms with higher numbers of rRNA operons during PCR (17). Also heterogeneity in rRNA operons within a single organism (22) will complicate quantification and lead to an overestimation of diversity. The 16S rRNA genes of different species differ considerably in G + C content. Genes with a lower G + C content denature with a higher efficiency and may therefore be preferentially amplified (19). This selectivity can be reduced by the addition of 5% acetamide as a denaturant (19), which also minimizes non-specific primer annealing. Furthermore, DNA-associated molecules that remain DNA-bound during the preparation of the DNA extraction can result in a lower efficiency of amplification for Gram-positive bacteria, as these molecules can form loops in template strands that interfere with elongation (9). From some environments, such as subsurface sediment samples, it is difficult to isolate large quantities of DNA due to the low amount of biomass present. Chandler *et al.* (21) suspected that very low concentrations of DNA (tens of picograms) generate random fluctuations in amplification efficiency, leading to differences in the composition of clone libraries. They noted significant differences in the composition of 16S rDNA clone libraries when different dilutions of environmental DNA (tens of picograms) were used in PCR. Conversely excess concentrations of non-target DNA may inhibit PCR, possibly by preventing specific interaction between polymerase and target DNA (7).

Differential amplification as a consequence of PCR has been found to relate to primer efficiency and selectivity and competition between primer annealing and template reannealing. Sub-optimal binding of the primer will result in less efficient amplification of the respective DNA. When 'universal' primers are used, different levels of mismatch between the primer and target sequences can result in preferential amplification of certain rRNA gene sequences. Introduction of degeneracy into primer sequences can minimize this but degenerate primers, essentially a mixture of similar but not identical primers, also have the potential to cause biases in PCR amplification. This can result from differences in the annealing temperatures of oligonucleotides in a degenerate mixture. Thus, the choice of primers may influence the recovery of target sequences. To reduce problems relating to primer efficiency and specificity, it is advisable to use computer algorithms such as CHECK_PROBE in the RDP package (http://rdp.cme.msu.edu), or PROBE_MATCH in the ARB software (www.arb-home.de) to analyze probes.

Competition between primer annealing and template reannealing has been recognized as another source of bias in the PCR amplification of rRNA genes (23). Using defined mixtures of rRNA gene templates, it was shown that some primer pairs gave a strong correlation between the ratio of genes in the starting mix and the ratio in the final PCR product (23). However, for other primer pairs the ratio in the final PCR product was close to 1:1, independent of the starting ratio of the two genes. Increasing the number of cycles in PCR enhanced this effect. The explanation for a tendency towards a 1:1 ratio of products regardless of the initial ratio of genes present was that in a mixture of two rRNA genes with one present in excess, the concentration of the most abundant template reaches a critical concentration as the PCR proceeds. Once this concentration of template is reached, template reannealing is favored over primer annealing and amplification of this template decreases. Thus, the originally less dominant template becomes more effectively amplified in the later cycles of the PCR until it too reaches a concentration at which template reannealing outcompetes primer annealing. However, this phenomenon is not likely to be a serious problem when amplifying from environmental DNA since it consists of a mixture of different species, all at relatively low concentrations. Therefore, any single species would be unlikely to reach a concentration where reannealing would be favored over primer annealing (23).

2.2.3 rRNA versus rDNA: indicative of active versus total microbial communities?

Microbial ecologists are especially interested in the metabolically active members of microbial communities. Characterization of microbial communities on the basis of amplification of 16S rRNA genes (rDNA) has the severe drawback that the detected sequences do not necessarily represent microorganisms that are presently active. The DNA might even be derived from naked DNA or from microorganisms transported to an environment with conditions that are unfavorable for their activity or present in an environment where conditions have changed such that the microbes are still intact but in an inactive state or dead.

Several authors have tried to circumvent this problem by analysing 16S rRNA, which is converted to cDNA using reverse transcriptase, prior to PCR (24–26). The rationale behind this approach is that a positive relationship exists between growth rate and cellular rRNA content in cultured organisms, since ribosome abundance influences the rate of protein synthesis, which in turn is essential to growth. Thus, the rRNA approach seems to allow a better analysis of active communities. Still, this approach needs to be treated with great caution and there is no way to conclude with certainty that a sequence detected by this approach really represents an active microorganism. Also, inactive but viable microorganisms will contain ribosomes, in order to be able to react to changing environmental conditions. The amount of ribosomes in starved microorganisms can be considerable, when compared to the ribosome content found during growth at maximal growth rates, and high levels can be maintained over long periods of time (27) and it has been shown that autotrophic ammonia-oxidizing bacteria maintain very high levels of rRNA even in the absence of detectable ammonia oxidation activity (28). A general relationship of RNA/DNA (with the RNA mainly being rRNA) versus growth rate has been established (29,30) and is shown in *Figure 2.3*. In most natural environments, growth rates will be low, with doubling times estimated at <1 day (31); at such low growth rates rRNA mainly indicates only the presence of microbes and only a small part of the rRNA population corresponds to activity (~20% at a doubling time of 1 day). Some knowledge of overall growth rates or activities in the samples of interest is therefore desirable before employing the rRNA approach, since RNA isolation is more complicated than DNA isolation and

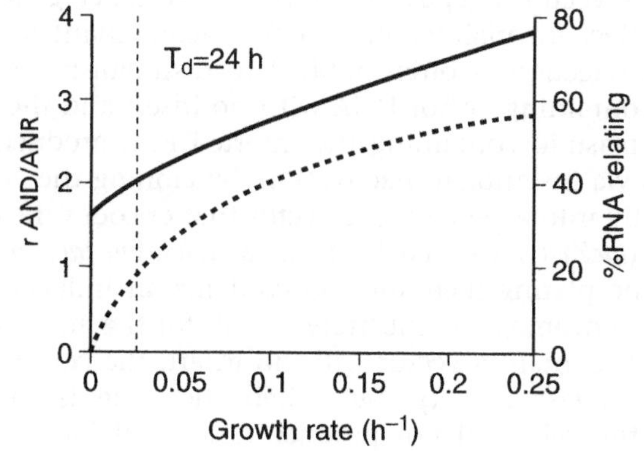

Figure 2.3

Dependence of rRNA content on growth rate. The solid line shows RNA/DNA ratio versus growth rate (μ), based on the relationship determined by Kemp (30); RNA/DNA = $1.65 + 6.01*\mu^{0.73}$. The dotted line indicates the percentage of RNA relating to activity at various growth rates, as determined from [RNA/DNA(μ) – 1.65]/[RNA/DNA(μ)] in which 1.65 refers to the RNA/DNA ratio at $\mu = 0$. The RNA/DNA ratio coincident with a doubling time of 1 day is indicated specifically.

requires careful controls such as reactions that omit the reverse transcription (RT) step to ensure that the RNA preparations are free from DNA (see Chapters 1 and 5). The additional effort required is only warranted if meaningful results regarding the activity of the organisms detected can be inferred.

Comparing results of the rRNA approach to the rDNA approach is not necessarily helpful in determining which sequences represent (more) active microorganisms, as the two approaches have different, approach-specific biases that also contribute to differences. The rDNA approach is influenced by differences in the number of rRNA gene copies (17). Analyses of ribosome contents and RNA/DNA ratios of different strains that were grown at similar growth rates has also revealed significant differences (30,32). The rRNA of some microorganisms has been found to be very labile [e.g. the 23S rRNA of *P. denitrificans* (N. Saunders, personal communication)] and thus might not be detected by RT-PCR. Modification of nucleotides in DNA or RNA, or proteins remaining bound to nucleic acids might also interfere with comparative analysis of rRNA- and rDNA-based clone libraries or community fingerprints. Furthermore, not all primers are suitable for the RT stage, due to secondary structure complications (33).

2.3 Generation and analysis of clone libraries

2.3.1 Generation of clone libraries

Following amplification of rRNA/rDNA, the next step towards the culture-independent phylogenetic analysis of a natural microbial community is cloning of the PCR fragments into a cloning vector (*Figure 2.1*). Competent *Escherichia coli* cells are transformed with the recombinant plasmid DNA (vector and inserted PCR product) and the cells are plated on a medium that selects for a trait conferred on the recipient *E. coli* by the uptake of the vector. Usually an antibiotic resistance gene on the vector is used to select for maintenance of the recombinant plasmid in the host. A second selection is often applied to distinguish between *E. coli* transformants containing vector DNA with no insert and those that carry a recombinant plasmid containing the inserted PCR product. In general, this is achieved via insertional inactivation, by cloning the PCR-amplified 16S rDNA directly into a gene on the vector that encodes the α subunit of β-galactosidase (*lacZ*) (the *E. coli* host has a defective *lacZ* gene). Besides an antibiotic, the plating medium also contains an inducer for the *lacZ* gene as well as a chromogenic substrate (X-gal) for the enzyme. When the transformed cells contain a vector with no insert, the cells will turn blue due to β-galactosidase activity, but when the gene is interrupted by inserted DNA, the cells will not produce functional β-galactosidase and the colonies will remain white.

Cloning of the heterogeneous population of rRNA sequences separates single sequences into individual *E. coli* clones in a clone library. Nowadays complete cloning kits are commercially available, e.g. pGEM-T (Promega), TOPO cloning (Invitrogen). The current most popular kits exploit T-vectors (34), which are plasmids that when linearized have single deoxythymidine residues at their 3' end. This can be conveniently achieved by cleavage with restriction endonucleases that produce blunt ends. 3' dT overhangs can then be generated by incubation with *Taq* DNA polymerase and dTTP. These vectors allow sticky-end ligation of PCR products generated by non-proofreading thermostable DNA polymerases without the need for restriction digestion (TA-overhang cloning). Non-proofreading DNA polymerases that lack a 3' to 5' proofreading function (e.g. *Taq* DNA polymerase) have terminal deoxynucleotide transferase activity and add a template-independent deoxyadenosine residue to the 3' ends of the PCR product (35). For PCR fragments generated with proofreading DNA polymerases (e.g *Pfu* DNA polymerase) a 3' dA overhang can be created in a similar way to that used to generate T-vectors.

Other methods of cloning are blunt-end (36) and sticky-end ligation (37). Blunt-end ligation procedures are less efficient than sticky-end ligation, and PCR products generated with non-proofreading polymerases must be modified to produce blunt ends. This is normally achieved using a DNA polymerase that has a 3' to 5' proofreading function (e.g. T4 DNA polymerase or *Pfu* DNA polymerase). Sticky-end cloning requires the addition of restriction sites to the 5' end of the amplification primers. It has the advantage that if different restriction sites are incorporated in each primer then double digestion can be carried out, preventing recircularization of the cloning vector, which improves cloning efficiency. Moreover, it also permits directional cloning of PCR products. However, restriction endonuclease cleavage at sites within amplified rRNA gene products can result in the recovery of truncated rRNA sequences in clone libraries.

2.3.1.1 Biases in cloning

Like PCR, cloning will not necessarily lead to a correct representation of the original microbial community. Rainey *et al.* (38) noticed that the same batch of PCR product cloned using either blunt end or sticky end cloning procedures gave different results. However, it is not clear how internal restriction enzyme cleavage affected the results since the clone libraries were screened by dot-blot hybridization procedures and the size of the insert DNA in the screened clones was not reported (38). Moreover, replicate cloning experiments using the same cloning procedure were not evaluated to assess variation inherent in a single cloning procedure.

Furthermore, heteroduplex molecules might be cloned; such molecules have strands from two different PCR products as the result of reannealing of denatured PCR fragments. During vector multiplication, this results in two different cloned sequences within a single cell, complicating screening. Also, cloned heteroduplex molecules may be subjected to *E. coli* DNA repair mechanisms, resulting in hybrid plasmid inserts. The problem can be overcome by treatment with single-strand specific nucleases (e.g S1 nuclease) prior to cloning.

2.3.2 Screening of clone libraries

Plasmid isolation and subsequent restriction analysis can be used to confirm the presence of correctly sized DNA inserts. Rapid screening can also be performed by colony PCR with primers that target plasmid-encoded priming sites flanking the cloned DNA.

The cloned, correctly sized rRNA gene fragments can be sequenced from all, or a selection, of the clones and a detailed picture of the sequence types present in a particular environment can be gained. However, phylogenetic analysis of complex communities is laborious, time-consuming and costly. The most costly element of the analysis is sequencing and it is often desirable to cut down the number of clones to be sequenced. Several screening methods allow the detection of similar or identical rRNA sequences.

Some methods are based on detecting signature sequences in the cloned 16S rDNA, such as colony hybridization procedures with oligonucleotide probes of defined phylogenetic resolution (37). The specificity of the probe used is critical. If a probe is too specific, clones containing sequences of interest can be overlooked. Conversely, it is possible to discount unique clones if they contain the target site for the oligonucleotide probe, but are otherwise quite different. Recently, a PCR approach called 'signature PCR' (SIG-PCR) has been described to classify 16S rDNA sequences into main taxa (39). SIG-PCR employs a mixture of nine oligonucleotide primers and yields PCR products of taxon-specific lengths.

The methods above offer either coarse phylogenetic resolution or require a large number of specific probes or primers. A more common means of screening clones to diminish the number of clones selected for sequencing is profiling of clones by one-dimensional electrophoresis methods, comparing the patterns, and grouping clones that produce similar electrophoretic profiles, followed by sequencing of representatives of each group. Suitable profiling methods are amplified ribosomal DNA restriction analysis (ARDRA) (see Chapter 4), terminal restriction fragment length polymorphism (T-RLFP), denaturing/thermal gradient gel electrophoresis (D/TGGE) and single strand conformation polymorphism (SSCP) (see Chapter 3). These methods detect sequence variation within cloned genes by digestion with restriction endonucleases that recognize tetrameric sequences (ARDRA, T-RFLP) or determination of melting behavior (D/TGGE, SSCP). ARDRA is most frequently used, since it offers the ability of high-speed screening of large numbers of clones in a simple, reproducible way, at low cost. As an additional advantage, with respect to screening of clones, several bands per clone will be generated by ARDRA (*Figure 2.4*), yielding more diagnostic information per clone. However, the low resolution of agarose gels means that many similar but not identically sized bands can be seen as a single band, thus potentially the number of clone types and hence diversity can be underestimated. This is exacerbated by the fact that digestion with at least three restriction endonucleases is required to give maximal taxonomic resolution (40). In screening clones by restriction analysis of PCR fragments amplified from vectors in which 16S rDNA has been cloned by TA-cloning or blunt end ligation, with primers targeting priming sites in the vector that flank the insert DNA, it should be realized that the amplified vector DNA may contain a restriction site and the cloned PCR fragment can be cloned in two different orientations, potentially generating two slightly different patterns for the same sequence (41; *Figure 2.4*).

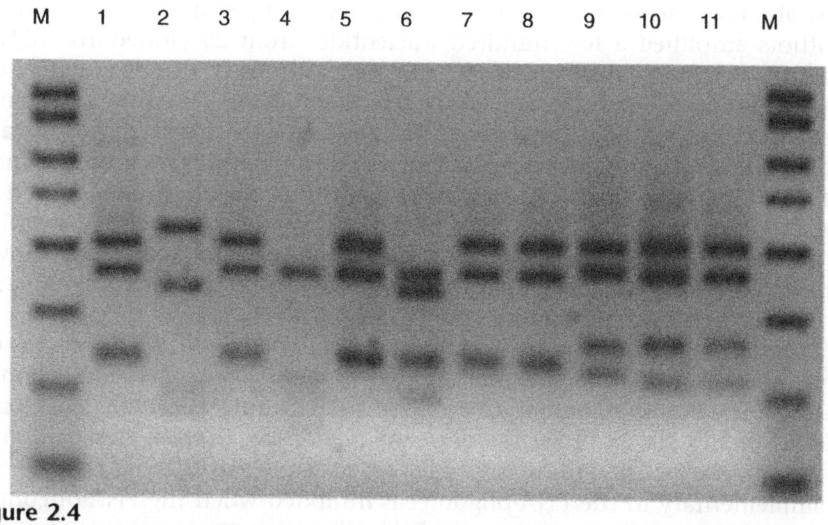

Figure 2.4

Amplified ribosomal DNA restriction analysis (ARDRA) of 11 clones, with restriction enzyme *RsaI*. Clones were reamplified with primers targeting sequences in the vector, flanking the insert. Sequences of clones 8 and 9 were identical, but due to differences in orientation of the inserts in the vector, and the presence of an *RsaI* site in the part of the vector which was included in the PCR product, differences in ARDRA profiles are observed. M = 100 bp PCR marker (Sigma; top band at 800 bp, lowest band at 100 bp).

The most time-consuming aspect of ARDRA screening is the comparison of patterns. This process can be accelerated considerably by the use of software that allows databasing of profiles, automatic pattern analysis and pattern recognition (42). A large range of commercially available software exists, with Gelcompar (Applied Maths, Kortrijk, Belgium), Molecular Analyst (Biorad, Hercules, CA, USA) and Phoretix 1D (Phoretix Int., Newcastle-upon-Tyne, UK) most commonly used. Screening should be done under well-defined, reproducible conditions, with the inclusion of suitable markers to allow within- and between-gel comparison. Again, this is most easily achieved using ARDRA.

A drawback of ARDRA is that it is often very difficult to make direct comparisons with ARDRA-generated community profiles, as several bands per clone are present and community profiles are often very complex. T-RFLP and D/TGGE yield only one diagnostic band per clone, allowing better comparison with community profiles (e.g. 24).

2.3.2.1 Selecting for less abundant sequences

Clone libraries may be dominated by a few clones, even when generated with 'universal' primers. Covering a larger part of the diversity, and the less abundant sequences, with the above-described methods requires the screening of a large number of clones. Several approaches allow for more rapid selection for less abundant sequences. A screening method to avoid sequencing of identical or closely related clone sequences and detection of less abundant sequences was described by Wintzingerode *et al.* (43). These authors amplified a few hundred nucleotides from 29 cloned 16S rDNA inserts, with digoxigenin-labeled primers, and used the fragments for hybridization to filters on which purified plasmid from 308 clones was blotted. All clones not identified by hybridization were sequenced. In contrast to the design of insert-specific oligonucleotide probes, this method does not require prior sequencing and washing conditions are almost the same for all 29 16S rDNA fragments, reducing optimization requirements. Recently, a subtractive hybridization protocol has also been described which allows the rapid generation of habitat-specific probes to screen for subdominant members in a clone library by colony hybridization (44).

Another approach to enrich for rare sequences is to repeat the process of PCR and clone library construction after obtaining phylogenetic information on the dominant clones from a particular sample. Peptide nucleic acid (PNA) probes specific for the dominant clones can then be designed and added to the PCR reaction. Amplification of sequences that are completely complementary to the PNA-oligomer is inhibited when the PNA-oligomer is included in a PCR reaction, as the PNA-oligomer acts as a competitor with one of the primers for a binding site or binds to an internal target sequence (45). PNA-mediated PCR clamping relies on the thermal stability of PNA–DNA duplexes that is greater than for corresponding DNA–DNA duplexes, and the inability of DNA polymerases to recognize PNA oligomers. Consequently, PNA oligomers cannot serve as primers in PCR. Thus, PCR clamping introduces a preferential bias to selectively enrich low-abundance sequences from a mixed template (45).

2.4 Quantitative analysis of clone libraries: coverage and diversity indices

2.4.1 Coverage and sampling

An important question in community analysis using clone libraries is how far the actual species composition (richness) in a natural sample is captured in a clone library. A very simple estimate can be obtained by calculating the coverage [*C*; (46)]:

$$C_x = 1 - (n_x/N) \qquad \text{(eq. 1)}$$

where n_x is the number of clone types (e.g. ARDRA types, sequence types) that are encountered only once in library x and *N* is the total number of clones analyzed. Hence if there is a large proportion of unique sequences recovered in a clone library, n_x/N tends towards unity and coverage is small. Alternatively, these calculations are often done on sequence types. To simplify the derivation of a value for coverage, it has been suggested that sequences that are >97% similar should be considered identical (47). This is based on the observation that organisms with 16S rRNA sequence homologies <97% are unlikely to exhibit genomic DNA homology >80% [indicative of a relationship at the species level (14)]. This may underestimate species diversity since a number of organisms that are known to have rRNA sequence homology of >99% are clearly distinct species based on DNA–DNA reassociation experiments and phenotypic data (48). However, this conservative approach is justified since heterogeneity of different rRNA operons within a single organism can be significant (18).

Recently, coverage calculations were included in a statistical approach to determine the significance of differences between clone libraries (49). Differences between homologous coverage curves [coverages calculated from eq. 1, using different sequence similarities/evolutionary distances (*D*) as cut-off criteria for uniqueness of sequences] and heterologous coverage curves (in which n_x in eq. 1 is replaced by n_{xy}, indicating the number of sequences in library x that are not found in library y) were calculated by a Cramer–von Mises statistic $[\Sigma(C_x - C_{xy})^2]$ with comparison by a Monte Carlo permutation test procedure.

Other ways to determine how well a community is sampled are also available. Plotting accumulation curves and rank–abundance curves (*Figure 2.5*), for example, can provide useful information about the composition of a

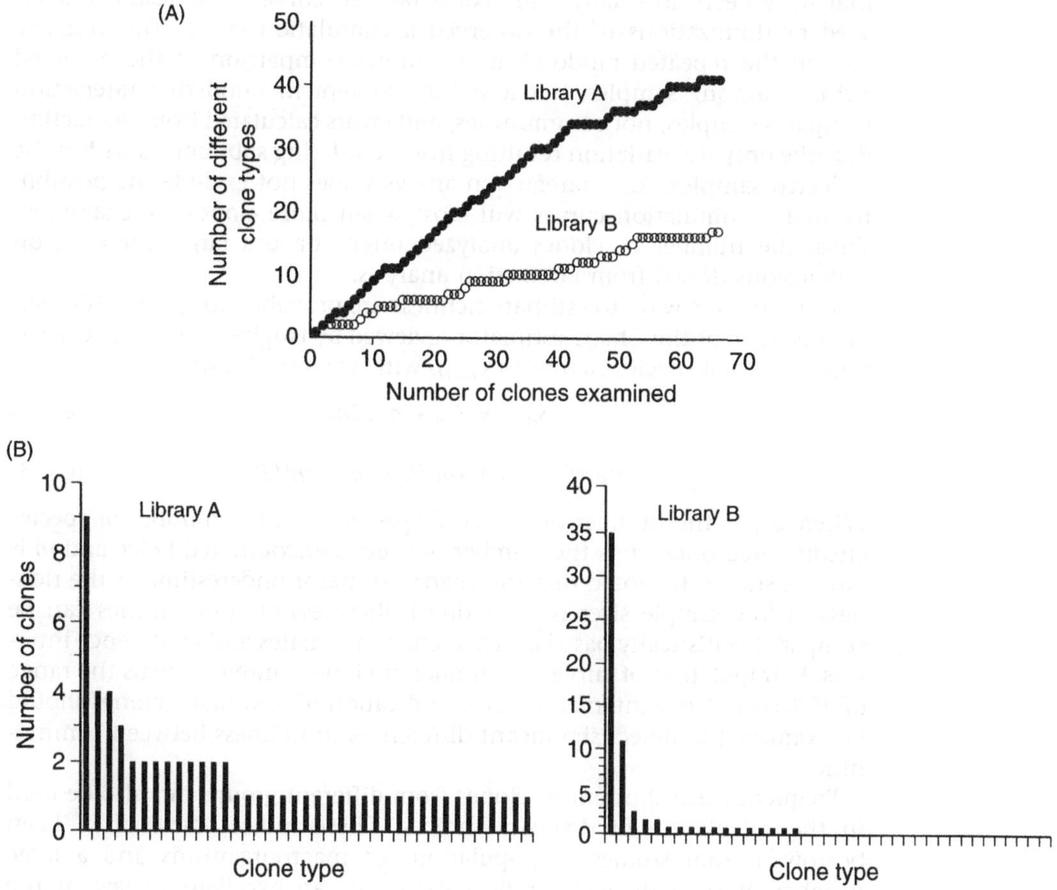

Figure 2.5

Accumulation curves (**A**) and rank abundance plots (**B**) for clone libraries derived from beach sediment before (library A) and 6 days after oil pollution and fertilization with nitrogen and phosphate (library B). Coverages (eq. 1) were 58.1 and 83.6% for library A and B, respectively. Chao1-estimator for richness (eq. 2) with standard deviation (root eq. 3) were 69.2 ± 20.3 and 47.2 ± 33.4, respectively.

clone library. An accumulation curve is a plot of the cumulative number of different clone types observed versus total number of individuals (the clones). As all communities contain a finite number of species, upon continued sampling an asymptote will be reached that represents the number of types present: the richness. Thus, the shape of the curve provides information on how well communities have been sampled. For rank–abundance curves, the different clones are ordered from the most to the least abundant on the *x*-axis, and the abundance of each clone type is plotted on the *y*-axis.

2.4.2 Diversity indices

Richness is an important parameter in the determination of diversity. Statistical approaches to estimate microbial diversity have recently been reviewed (50). Rarefaction compares the observed richness among habitats that have been unequally sampled. A rarefied curve is the result of averaged randomizations of the observed accumulation curve. The variance around the repeated randomizations allows comparison of the observed richness among samples, but it should be kept in mind that rarefaction compares samples, not communities, and errors calculated from rarefaction describe only the variation resulting from reordering subsamples within the collected samples. Also, rarefaction analysis does not exclude the possibility that accumulation curves will cross when more clones are examined. Thus, the number of clones analyzed might have a large influence on conclusions drawn from rarefaction analysis.

More suitable ways to estimate richness are probably nonparametric estimators, such as the Chao1 estimator reviewed by Hughes *et al.* (50). 'Chao1' estimates total species richness (S_{chao1}), with variance [Var(S_{chao1})]:

$$S_{chao1} = S_{obs} + n_1^2/2n_2 \qquad \text{(eq. 2)}$$

$$\text{Var}(S_{chao1}) = n_2(m^4/4 + m^3 + m^2/2) \qquad \text{(eq. 3)}$$

Where S_{obs} is the number of observed species, n_1 is the number of species encountered once, n_2 is the number of species encountered twice and m is n_1/n_2. It should be noted that the chao1 estimator underestimates the richness at low sample sizes (<100 clones). Richness of communities can be compared statistically based on their chao1 estimates and confidence intervals. Extrapolation of curves of number of clones sampled versus the range of 95% confidence intervals gives an indication of how many clones should be examined to detect significant differences in richness between communities.

Frequency distributions of clones from different groups can also be used in the calculation of diversity indices. These indices have often been borrowed from studies of populations of macroorganisms and a large number of such diversity indices exist. For an excellent review of the strengths and weaknesses of different diversity measures, the book by Magurran is highly recommended (51). A frequently used index is the Shannon index of diversity (H'), which is calculated as $-\Sigma p_i{*}\log(p_i)$ in which p_i is the proportion of clones contributed by group i to the whole clone library. The larger H', the higher the diversity. It takes into account the number of different groups [often called richness (*S*)] and evenness of clones distributed over the clone library [$E = H'/\log(S)$]. One should be careful of overinterpreting Shannon index data because they give a gross descriptor of diversity and do not take into account the specific nature of the organisms present. For example, a community consisting of four different Gram-positive *Bacillus* species, each comprising 25% of the community, would have the same Shannon diversity index as a community comprised of equal amounts of a Gram-positive bacterium, a Cytophaga, a gammaproteobacterium and a planctomycete. It is obvious that from a phylogenetic point of view the latter community appears far more diverse and it has therefore been suggested that similarities in DNA sequences should be included in diversity calculations (52).

Due to the biases in DNA extraction, PCR and cloning, actual diversity cannot reliably be estimated from clone distributions. However, if similar methods are used throughout a study, the relative diversities can still be compared. As with coverage measures, diversity indices are strongly dependent on the definition of the operational taxonomic unit (OTU). As long as the definition of OTU is kept constant in a study, diversity can be compared. However, use of diversity indices in a context other than comparative analysis should be considered with caution, as, at least for microbial communities, values of diversity indices are unlikely to relate to any absolute description of diversity (50).

2.5 Phylogenetic inference

2.5.1 Sequencing

The primary goal of generating 16S rRNA gene clone libraries is to determine the phylogenetic relationships of the organisms present, based on sequence comparisons with cultured organisms or of sequences recovered from environmental samples. Determination of nucleotide sequences was considerably advanced by the introduction of the chain-termination DNA sequencing method (53). The reaction closely resembles a PCR reaction, except for two key differences. First, only one oligonucleotide primer is used, thus resulting in linear as opposed to exponential sequence amplification. Second, in addition to dNTPs, 2',3'-dideoxynucleoside triphosphate (ddNTP) nucleotide analogs are included in the reaction. Elongation of a DNA fragment will be terminated when a ddNTP is incorporated, since DNA polymerase requires a free 3' hydroxyl group for the enzymatic formation of a phosphodiester bond with new dNTPs and this is not present in ddNTPs. The ddNTP incorporation occurs randomly, resulting in a mixture of fragments of different length. In the past, radiolabeled primers or deoxynucleotides were used and four different reactions had to be performed with each different ddNTP incorporated in separate reactions to determine the position of A, C, G and T residues. The reactions were run separately in adjacent lanes on a high-resolution polyacrylamide gel and the nucleotide sequences were read from autoradiograms. Nowadays, a sequencing reaction can be performed by cycle sequencing with the four different ddNTPs each conjugated to a different fluorescent reporter molecule included in a single reaction. This sequencing reaction is then analyzed in a single lane on a polyacrylamide denaturing gel or by capillary electrophoresis in an automated system. During electrophoresis the labeled DNA fragments are separated by size and a laser beam scans the gel thousands of times per hour. Fluorescence emissions specific for each of the different ddNTP labels are detected and sent to data collection software for storage. When data collection is complete, automated analysis is initiated resulting in a chromatogram/electropherogram of the fluorescence peaks detected and the corresponding nucleotide sequence.

2.5.2 Phylogenetic analysis

The sequence obtained must be checked carefully for reading errors, after which information on its identity can be obtained by comparing the sequence to one of the online sequence databases (EMBL, GenBank, RDP) using BLASTN or FASTA (54, 55). 16S rRNA sequence identity of <97% is generally regarded as indicating that the sequences belong to different species (14). However, if sequence identities >97% are obtained, interpretation must be considered with caution since many well-defined species have >97% 16S rRNA sequence identity (56). Hugenholtz *et al.* (57) state that 500 nucleotides of sequence is sufficient for reliable phylogenetic placement of a novel sequence if a longer sequence present in the database is closely related (>90% identity in aligned sequence) to the sequence in question. However, in the case of novel sequences that share <85% identity with any known sequence, sequences of <500 nucelotides are usually insufficient for a meaningful phylogenetic placement. It should also be noted that variability is not equally spread over the 16S rRNA gene, the 5' end shows greater variability than the 3' end due to the presence of more hypervariable sequence domains (58). Amann *et al.* (1) suggest that ≥1000 nucleotides should be sequenced for meaningful phylogenetic analysis. While this is almost certainly true when sequences have low identity to previously characterized sequences, if sequences are closely related to known sequences it is likely that analysis of longer tracts of sequence will give results similar to those obtained with shorter sequences, at least with regard to the identification of the closest phylogenetic neighbor. Furthermore, if specific primers are used to obtain sequence data from particular organisms it may not be possible to obtain more than four or five hundred nucleotides of sequence. In this case, the sequences will often be closely related to known sequences.

The first step in detailed phylogenetic analysis of macromolecular sequences is the careful alignment of the sequences. This is a crucial step in phylogenetic analyses. The fundamental basis of phylogenetic analysis is the comparison of homologous characters shared between organisms. For example it is valid to compare the wing of a bird with the wing of a bat or the limb of a primate since these are evolutionarily homologous structures derived from a common vertebrate ancestor. By contrast, comparison of the wing of a bird with the wing of an insect is not phylogenetically valid. Thus, in molecular phylogenetics it is essential that only nucleotides (or amino acids) present at homologous positions in different macromolecular sequences be compared. This is critical because molecular phylogenetic analysis essentially examines differences in the primary sequence of macromolecules that are the result of evolutionary events where the sequence is changed by the mutation of a nucleotide at a specific position. For highly conserved regions the alignment is straightforward, but it is more problematic for regions with greater sequence variability. As secondary structure interactions for 16S rRNA molecules are known for microorganisms belonging to many phyla, comparing the sequences to secondary structure models will often resolve alignment problems and should be used to optimize sequence homologies (2). Regions that even then cannot unambiguously be aligned should not be included in subsequent analyses for inference of phylogenetic relationships. Rigorous alignment criteria excluding sequence positions that cannot be unambiguously aligned and insertions/deletions reduce the amount of sequence data used in phylogenetic analysis, but are essential to reduce the possibility of erroneously inferring close relationships between sequences. It is also important to note that a phylogeny inferred from macromolecular sequence data represents the evolutionary history of the gene (or protein) and this does not necessarily reflect the evolutionary history of the organism that carries the gene. Nevertheless, in the case of 16S rRNA genes, there is relatively little evidence of horizontal transfer and it is often considered that 16S rRNA and organismal phylogenies are congruent.

The primary output from a phylogenetic sequence analysis is a phylogenetic tree representing the evolutionary relationship between different sequences. A phylogenetic tree consists of terminal nodes/tips which correspond to the taxa. The branch points within a tree are called internal nodes. Branches that end at a tip are named peripheral branches, all other branches are interior branches. If just three branches connect to an internal node then the node is said to represent a bifurcation or dichotomy.

Most of the procedures used to infer a phylogenetic tree result in an unrooted tree; a phylogeny in which the earliest point in time (the common ancestor) is not known. However, they may be drawn as rooted trees by locating the root of the tree, using added information, most commonly by adding sequences related to taxa that are assumed to lie outside of the presumed group. These taxa are named the outgroup taxa. It should be noted that the rooting does not usually have any evolutionary significance regarding the outgroup sequences, being genuinely ancestral to the other sequences represented in a tree.

Inferring a phylogeny is an estimation procedure—a 'best guess' of the evolutionary history, based on incomplete information as generally direct information about the past is lacking. All methods to infer phylogenetic trees from aligned sequences are imperfect, especially when dealing with lineages that evolve at different rates. Even within a single macromolecule it is clear that different positions evolve at different rates and saturation of the evolutionary signal in highly variable regions can occur (i.e. many mutational events are superimposed: multiple substitutions), resulting in underestimation of evolutionary divergence. Some attempts have been made to take account of these factors in phylogenetic analyses (e.g. 59) but they remain important limitations in phylogenetic analysis. These limitations are primarily an issue if the questions asked fundamentally relate to the detailed and exact evolutionary history of particular lineages. This may not be the primary requirement of microbial ecologists where inference of the identity of an uncultured organism may be the main concern, at least in the first instance. This is particulary true where phylogenetic position relates to physiological properties of an organism (e.g. autotrophic ammonia-oxidizing bacteria and sulfate-reducing bacteria).

Confident inference of the phylogenetic affiliation of an organism identified from 16S rRNA sequences is best judged from the use of several methods, in combination with some measure of the confidence that can be placed on the phylogenetic groupings identified (e.g. bootstrap analysis; 11). Many methods have been developed to infer phylogenetic relationships from molecular sequences; three of these are widely used for analysis of 16S rRNA sequences recovered from uncultivated microorganisms: parsimony, corrected distance and maximum likelihood analyses. For detailed information on these methods we refer the reader to Swofford *et al.* (60), who give an excellent account, with examples, on exactly how phylogenetic trees are generated from sequence data, with useful examples that explain the mathematics behind the phylogenetic inference methods. A summary of the fundamental principles that underlie these methods is given below.

2.5.3 Parsimony

Parsimony methods select trees that minimize the tree length (i.e. the smallest number of changes required to convert one sequence to another) required to explain a set of data. Parsimony methods do not require a specific model of evolutionary change, whereas the other two methods do.

In parsimony analysis each position in a set of aligned sequences is a character, with state A, C, G or T/U, that can be compared between different sequences. However, not all positions in a macromolecular sequence are informative and if the character state is identical at a particular position in all, or all but one, sequences in an alignment that position is excluded from the analysis. When all non-informative sites have been removed from the analysis the minimum number of steps required to explain the data (pattern of character states at each sequence position) is determined for each informative site. For each site there are several possible trees that can explain the data and the number of possible trees depends on the number of taxa that are being compared. For unrooted trees this can be calculated from eq. (4) and for rooted trees the number of possible trees can be calculated from eq. (5):

$$N = (2s - 5)!/2^{s-3}(s - 3)! \qquad \text{(eq. 4)}$$

$$N = (2s - 3)!/2^{s-2}(s - 2)! \qquad \text{(eq. 5)}$$

Where N is the number of possible trees and s is the number of taxa (note that '!' is the standard notation for 'factorial' which is the product of a series of positive integers from 1 to a given number, e.g. $5! = 5 \times 4 \times 3 \times 2 \times 1 = 120$). Thus, for a tree containing four taxa there are three possible unrooted trees and 15 possible rooted trees, and as the number of taxa in a tree increases the number of possible trees expands dramatically. For example, in a tree with 10 taxa there are 2 027 025 possible unrooted trees and 34 459 425 possible rooted trees.

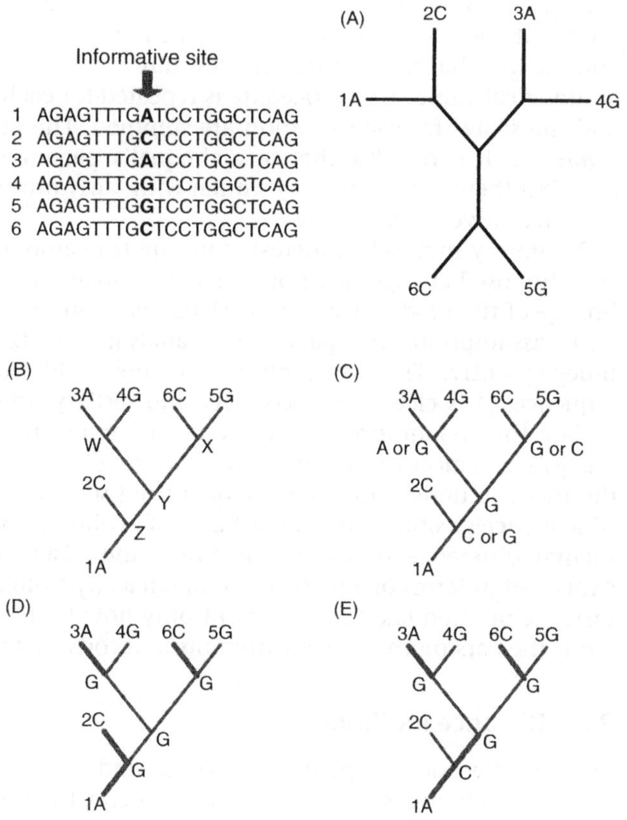

Figure 2.6

Parsimony analysis. Parsimony analysis evaluates all possible trees at each site in a sequence alignment. In this example there are six sequences with one informative site. At this site the character states are A, C, A, G, G and C for sequences 1 to 6 respectively. With six sequences there are 105 possible unrooted trees one of which is shown (**A**). The process involved in parsimony analysis is illustrated for this tree but in a real analysis the process would be conducted for all possible trees and the most parsimonius tree would be selected. By rooting the tree with sequence 1, the rooted tree (**B**) would be obtained. The number of steps required to obtain a tree with this topology is estimated by starting at the tips of the tree and moving to the internal nodes. If two tips have the same character state then the node that joins them is assigned the same character state. If the character state at two tips is different then the character state at the node is assigned the two alternative states. Thus node W = A or G, X = G or C, Y = G (since both W and X include G) and Z is assigned the state C or G (**C**). When this process reaches the root the process is reversed. Since node Z (C or G) does not include the character state of the ancestral node (A in sequence 1), it is assigned arbitrarily as C or G (see **D** and **E**). Node Y is already assigned (as G) and the process proceeds to nodes W and X. Node W = G since it does not require a change from the character state at the ancestral node (i.e. this is the most parsimonius explanation of the data). Using the same reasoning, node X is assigned as G. Branches along which the character state changes (shown in bold in **D** and **E**) are counted and in this case the number is the same whether node Z = C or G. All other possible rooted trees are evaluated in the same manner and the tree which gives the smallest number of branches along which the character state changes is selected and this number of branches is the score assigned to this site. In parsimony analysis this is done for all informative sites, the number of changes is summed and the tree(s) with the lowest overall score is the most parsimonious tree. The illustration of the basis behind parsimony analysis is based on Hall (69).

In parsimony analysis each of the possible trees that can be drawn to represent the character states at each informative position is determined and the tree that requires the smallest number of steps is selected and given a numerical value. This procedure is repeated for each informative position and the scores for each position are summed. The process is illustrated in *Figure 2.6*. It is possible that several equally parsimonious trees are generated, but these tend to be very similar with fairly minor differences between the alternative trees.

Parsimony methods underestimate the true amount of change, because superimposed changes are not assessed (changes at a particular site along a lineage of the phylogeny may mask earlier changes at that site). An important assumption in parsimony analysis is that all sites evolve independently. This assumption may not hold true and in 16S rRNA sequences it is clear that secondary and tertiary interactions are likely to violate this assumption for many positions in an alignment. Parsimony analysis uses more of the information present in a sequence alignment than distance methods, which reduce data from several hundreds or thousands of sequence positions to a single value of evolutionary distance. The fundamental difference between parsimony and distance methods has been explained in terms of a person arriving in a city from another place (2). The distance method takes into account only how far the person has travelled, while the parsimony analysis attempts to reconstruct the actual route taken.

2.5.4 Distance methods

Distance methods are perhaps easiest to understand. Pairwise comparisons of a set of aligned sequences are used to construct a distance matrix. The distance matrix expresses the divergence between pairs of sequences in terms of the fraction of sites that are different. It is apparent that sequences that differ in 2% of the positions in an alignment are more closely related than sequences that differ in 5% of the positions. It is also logical to infer that, assuming constant rates of nucleotide substitution, more time has passed from the point of divergence in a sequence that is 5% divergent from a given sequence than the divergence time of a sequence which is 2% divergent from the same sequence. The calculation of distances often includes a model of base substitution to account for multiple substitutions at a single nucleotide position. Even so, distance methods tend to underestimate evolutionary distances. Several different base substitution models exist and within the limitations of this chapter it is only possible to briefly consider a small number of the models. The Jukes and Cantor model (61) is most commonly used. This model assumes that there is independent change at all sites, with equal probability. Whether a nucleotide changes is independent of its identity, and when it changes there is an equal probability that it may mutate to any of the other three nucleotides. Many of these assumptions are not valid; for example, simply by looking at the variable and conserved regions characteristic of 16S rRNA molecules, it is clear that not all sites on the molecule evolve at the same rate. A number of more complicated models have been developed including the Kimura 2 parameter model (62) that distinguishes between transitions [mutation from a purine (A, G) to purine, or a pyrimidine (C, T) to a pyrimidine] and transversion (mutation from a purine to a pyrimidine or *vice versa*). Distance models that attempt to estimate different evolutionary rates at different positions in a molecule have also been developed (63). The different models often, though not always, produce similar results.

The matrix of pairwise distances between sequences in an alignment is used to construct a tree by one of a number of methods. UPGMA (unweighted pair group method with arithmetic mean) and neighbor-joining are most commonly used. UPGMA is a clustering method and the first step in UPGMA is identification of the sequence pair with the smallest distance between them. The branch point is then estimated as half the distance between the two sequences. The two taxa are then defined as a cluster and the matrix is recalculated with the first two taxa combined. This process is repeated with the number of entries in the matrix reduced by one each time until finally the matrix contains a single value. The tree is then built from the root by adding clusters defined in each of the matrices. There are a number of assumptions inherent in the construction of a UPGMA tree. One is that the tree is additive (i.e. the distance between any two nodes is equal to the sum of the branch lengths between them) and a second is that the tree is ultrametric (all taxa are equally distant from the root). The second assumption is not likely to be sustainable and consequently nowadays UPGMA is rarely used in phylogenetic sequence analyses.

The neighbor-joining method (64) is the most widely used algorithmic method to generate trees from distance matrices. The neighbor-joining method is related to cluster analysis in that it involves a sequential reduction in the size of a distance matrix, but with the difference that it does not assume that all lineages have diverged an equal amount from the common ancestor. Instead of producing clusters, neighbor-joining involves calculation of distances from each taxon to internal nodes. The first step in constructing a neighbor-joining tree is the calculation of the overall divergence of each sequence from all other sequences. This is the sum of the distances from a sequence to all other sequences. The overall divergence is then used to recalculate the distance matrix. The distances between the least divergent pair of sequences and the node that connects them is then calculated. Unlike UPGMA trees the sequences need not be equidistant from the node that joins them. A new matrix is then calculated where the first pair of sequences is replaced with the node that joins them. The process is repeated until the tree is constructed.

Although clearly imperfect from an evolutionary perspective, distance methods are popular because of their conceptual simplicity, they are not computationally expensive and for the needs of most microbial ecologists they provide results that fulfill their primary requirement of determining the closest phylogenetic neighbor of an organism identified from a nucleic acid sequence recovered from an environmental sample.

2.5.5 Maximum likelihood

Maximum likelihood methods are a bridge between parsimony and distance methods. They require a model of evolutionary change that accounts for the conversion of one sequence into another, as in distance

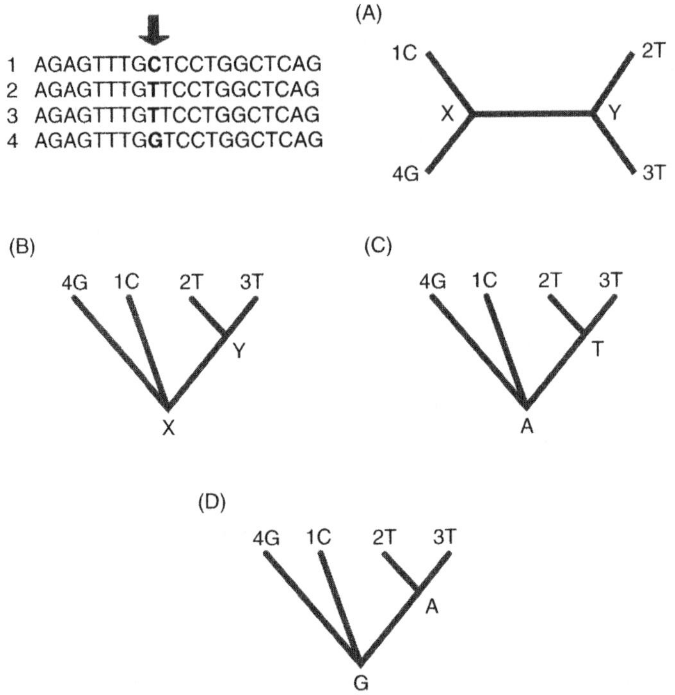

Figure 2.7

Maximum likelihood analysis. Maximum likelihood analysis tests the probability of a phylogenetic tree given a sequence alignment and a model of evolutionary change. Based on this, the tree most likely to explain the observed sequence data is selected. The evolutionary model provides an estimate of the rate of evolutionary change at a particular site in the alignment and a tree that may be explained by the sequence data. Four sequences can give rise to three possible unrooted trees. One of the possible unrooted trees that can be derived from the highlighted position in the sequence data is shown **(A)**. If we do not assume a particular direction of divergence the tree can be rooted at any node, e.g. X **(B)**. The character state at nodes X and Y are unknown but at each node this may be either A, C, G or T. Thus there are 2^4 (= 16) possible permutations of nucleotides at node X and Y that may explain the tree. An example of one of the possible trees is illustrated **(C)**. The probability of this particular permutation is the probability of having an A at the root (P_A) and depends on the particular evolutionary model used and could for example be 0.25 or the frequency of A at this alignment position. The probability of the A at the root changing to the G at the tip (P_{AG}) can be calculated from the chosen evolutionary model and the length of the branch length from A to G, as can the probability of the A at the root changing to the C at the tip (P_{AC}). It follows that the probability of the complete tree shown in **B** is the product of the probabilities of the nucleotide changes between each of the nodes, i.e.

$$P_{\text{tree C}} = P_A \times P_{AG} \times P_{AC} \times P_{AT} \times P_{TT} \times P_{TT}$$

The probability of the situation shown in **(D)** can likewise be calculated from

$$P_{\text{tree D}} = P_G \times P_{GG} \times P_{GC} \times P_{GA} \times P_{AT} \times P_{AT}$$

When all 16 possible permutations are considered then the probability of the tree shown in **(B)** based on the position highlighted in the alignment can be calculated.

$$P_{\text{tree B}} = P_{\text{tree C}} + P_{\text{tree D}} + P_{\text{tree E}} + \ldots + P_{\text{tree 16}}$$

continued

methods, and like parsimony analysis, a nucleotide position by nucleotide position comparison is involved. The fundamental basis of maximum likelihood methods is the inference of the 'best' evolutionary tree by finding the one tree from many alternative trees that maximize the probability of observing the sequence data in an alignment. In general, maximum likelihood methods are robust in violations of the assumptions in the evolutionary model used. Alternative trees are tested against a goodness-of-fit statistic to assess which is the most likely tree given the data. The log likelihood ratio test is often used.

Maximum likelihood methods start with a sequence alignment, an evolutionary model and an initial tree (which can for example be a neighbor-joining tree). The likelihood of the initial tree given the particular data set is computed and in principle the likelihood of all possible trees is calculated and compared. In practice, this is not possible since the number of trees generated and needing to be compared in alignments of more than a small number of sequences makes the process extremely time-consuming and demanding of computer processor time. For this reason, maximum likelihood methods begin with an initial tree and compare it with a similar tree. The most likely tree in the comparison is then retained and any less likely tree is rejected. This means that the final result is highly dependent on the initial tree, and the optimal tree that is found may only be optimal relative to the initial tree, but may not be the best possible tree. For this reason the analysis is usually conducted several times with different initial trees. Often the 'best' tree is arrived at by stepwise addition of sequences to a tree. The initial tree contains three sequences and there is one possible unrooted tree. New sequences are added in a stepwise fashion and in the case of a four sequence tree there are three possible places where

Figure 2.7 – continued

The probability of the tree based on all positions in the alignment is the probability of observing the data at all of the sites. This is the product of the probabilities for each of the sites (i) from 1 to *N*.

$$P_{tree} = \prod_{i=1}^{N} P_i$$

The values for the probabilities are typically very small (<<<1); most computers are unable to deal with the calculations and therefore the probability/likelihood of a tree for each site is usually expressed as the natural logarithm of the likelihood (ln L_i) and consequently the likelihood of the tree for all sites is the sum of the log likelihood for each of the sites.

$$\ln L_{tree} = \sum_{i=1}^{N} \ln L_i$$

ln L_{tree} is the likelihood that the alignment can be explained, given the particular evolutionary model used and the topology of the tree (including branch lengths). In principle, maximum likelihood methods determine the log likelihood for all possible trees and the one which gives the highest log likelihood value is considered the optimum tree. For four taxa, the log likelihood for 15 possible rooted trees must be calculated in the manner illustrated above. For 10 taxa, >34 million possible rooted trees must be evaluated. For this reason, methods for identifying the 'best' tree based on heuristic approaches are normally used (see main text).

The illustration of the basis behind maximum likelihood analysis is based on Hall (69).

the fourth sequence can be added to the tree (i.e. in each of the existing branches in the three sequence tree). The likelihood of each of the three alternative four taxon trees is then evaluated and only the best tree is used for the subsequent addition of a fifth sequence that generates 15 possble trees rather than the 45 (3 × 15) that would need to be compared if all possible trees were to be evaluated. Again the best five-sequence tree is selected and a sixth sequence is added to each of the seven branches on the optimal tree and so on. This is known as a heuristic approach to tree optimization. The process involved in maximum likelihood analysis is outlined in *Figure 2.7*.

2.5.6 Bootstrap analysis

The statistical validity of a tree generated with the above methods can be tested by bootstrapping, though this is rarely computationally feasible for maximum likelihood analysis. Bootstrapping (65) creates a new data set by sampling *N* characters from the alignment randomly with replacement, so that the resulting data set has the same size as the original, but some characters have been left out while others are duplicated. The random variation of the results from analyzing these bootstrapped data sets can be shown statistically to be typical of the variation that would be obtained from collecting new data sets. All new data sets are subjected to phylogenetic analysis. Comparing the resulting trees allows a majority-rule consensus tree to be constructed with bootstrap values for the different nodes. High bootstrap values are indicative of the significance of the grouping to the right of the node.

The description of phylogenetic analysis methods above is far from exhaustive, and new approaches are being developed continually, e.g. log-det methods (66) and Bayesian methods (67). Readers wishing to pursue the details of phylogenetic analysis are directed to the primary literature or the excellent treatise on molecular systematics provided by Hillis *et al.* (68). An excellent practical manual for those entering the world of molecular systematics that gives extremely accessible explanations of some phylogenetic inference methods has been written by Hall (69). A comprehensive list (with web links) of phylogenetic analysis software is available at http://evolution.genetics.washington.edu/phylip/software.html.

2.6 Phylogenetic analysis compared to other molecular techniques

16S rDNA-based phylogenetic analysis allows the most detailed insight into microbial community composition but suffers from the facts that:

(i) it is time-consuming and expensive, which can preclude detailed comparisons between large numbers of samples for all but the wealthiest laboratories;
(ii) it is not reliably quantitative;
(iii) it is not necessarily informative regarding community function.

However, phylogenetic analysis forms the starting point for overcoming these problems (see *Figure 2.1*).

2.6.1 Comparing large numbers of samples

Comprehensive phylogenetic analysis of complex microbial communities would require screening of hundreds, if not thousands, of clones. Detailed and comparative analysis of microbial communities can be more readily achieved by applying one of the fingerprinting methods described in Chapter 3, such as DGGE and T-RFLP. These techniques are best combined with phylogenetic analysis. For example, prior phylogenetic analysis can indicate the presence of interesting groups of microorganisms for which specific primers can be designed to examine their dynamics using general fingerprinting techniques. Although the fingerprints obtained by DGGE with general bacterial primers are not informative on phylogenetic composition, bands can be excised from gels and sequenced to allow a more detailed phylogenetic analysis. However, the PCR fragments used in DGGE are relatively short (<500 bp), which can compromise subsequent phylogenetic analysis. In complex environments where DGGE can generate large numbers of bands, it can be problematic to obtain single bands, free from contamination with other rRNA gene fragments, which can be sequenced directly. Clones from 16S rRNA libraries can also be profiled by DGGE and their positions compared to the community fingerprint to identify bands in DGGE, although this should best be confirmed by probing (24). T-RFLP in principle allows direct phylogenetic inference based on fragment size, but this depends strongly on a good 16S rRNA database. While the database at the moment contains >100 000 sequences, a large proportion consists of partial sequences and the database is biased towards Proteobacteria (70). Furthermore, even under the most ideal circumstances (i.e. single nucleotide resolution and ability to analyze fragments ranging from ~50 to 1500 bp in length), it is theoretically only possible to identify ~1400 T-RFLP types if only one labeled primer that amplifies almost the entire rRNA gene is used, and group-specific primers that amplify smaller fragments will reduce the number of sequences that can potentially be discriminated for each group. Considering the conserved nature of much of the 16S rRNA gene and the fact that most restriction endonucleases recognize palindromic sequences, thus reducing the number of possible sequence motifs that can be recognized in T-RFLP analysis, it seems likely that the phylogenetic resolution of T-RFLP may be relatively limited when using 16S rRNA fragments amplified with universal primers.

2.6.2 Quantification

Biases in PCR and cloning allow only a semi-quantitative assessment of abundance, which, however, forms a suitable starting point for more comprehensive quantification with other techniques. Specific PCR approaches allow for more quantitative analysis but the best approach is to count cells specifically visualized by either *in situ* PCR or fluorescent *in situ* hybridization (FISH) (Chapter 8). The decision on which probes or primers to use can be based on previous phylogenetic studies. Detailed studies of microbial communities with FISH are at present either directed at specific phylogenetic groups or a few broadly defined taxa. Detailed fine-scale analysis (e.g. enumeration of the majority of cells present at the species-level) is at present very labor intensive.

2.6.3 Relationship between phylogenetic information and function

Often phylogenetic information cannot be unambiguously related to specific processes. For example, current knowledge suggests the involvement of a single phylogenetic group for some important subsurface processes, such as anaerobic degradation of BTEX by *Azoarcus/Thauera* spp, *Geobacteraceae* and *Desulfobacteria* under denitrifying, iron-reducing and sulfate-reducing conditions, respectively (71). However, these phylogenetic groups also include non-BTEX degraders. For many 16S rRNA gene sequences, no closely related cultured microorganisms are known. Until recently, attempts to relate 16S rDNA information to function was dependent on culturing studies, but new nucleic-acid-based techniques allow specific microbial processes and functions to be related to individual members of microbial communities in a cultivation-independent manner (see Chapters 6 and 10). The particular activities of a microorganism can be identified by probing (FISH microautoradiography; Chapter 8) or cloning and sequencing of 16S rRNA or 16S rRNA genes recovered from heavy-isotope-labeled nucleic acids (stable isotope probing; Chapter 6).

2.7 Concluding remarks

In conclusion, recovery and analysis of rRNA sequences from environmental samples is not a definitive cataloging exercise but the first step towards identifying relatively abundant, uncultured members of the microbial population. It provides the information required to attempt the targeted isolation of taxa that have not yet been cultivated but potentially catalyze important biogeochemical functions. Only by isolation of the organisms that the sequences represent, or the direct combination of 16S rRNA-based techniques with functional assays, can their true role in the environment be discovered. Analysis of rRNA sequences also provides the framework for subsequent studies using methods more amenable to comprehensive sampling and rapid analysis.

References

1. Amann RI, Ludwig W, Schleifer KH (1995) Phylogenetic identification and in situ detection of individual microbial cells without cultivation. *Microbiol Rev* **59**: 143–169.
2. Woese CR (1987) Bacterial evolution. *Microbiol Rev* **51**: 221–271.
3. Saiki RK, Gelfand DH, Stoffel S, Scharf SJ, Higuchi R, Horn GT, Mullis KB, Erlich HA (1988) Primer-directed enzymatic amplification of DNA with a thermostable DNA-polymerase. *Science* **239**: 487–491.
4. Steffan RJ, Atlas RM (1991) Polymerase chain reaction: applications in environmental microbiology. *Annu Rev Microbiol* **45**: 137–161.
5. D'aquila RT, Bechtel LJ, Videler JA, Eron JJ, Gorczyca P, Kaplan JC (1991) Maximizing sensitivity and specificity of PCR by preamplification heating. *Nucleic Acids Res* **19**: 3749–3749.
6. Don RH, Cox PT, Wainwright BJ, Baker K, Mattick JS (1991) Touchdown PCR to circumvent spurious priming during gene amplification. *Nucleic Acids Res* **19**: 4008–4008.

7. Tebbe CC, Vahjen W (1993) Interference of humic acids and DNA extracted directly from soil in detection and transformation of recombinant-DNA from bacteria and a yeast. *Appl Environ Microbiol* **59**: 2657–2665.
8. Kreader CA (1996) Relief of amplification inhibition in PCR with bovine serum albumin or T4 gene 32 protein. *Appl Environ Microbiol* **62**: 1102–1106.
9. von Wintzingerode F, Gobel UB, Stackebrandt E (1997) Determination of microbial diversity in environmental samples: pitfalls of PCR-based rRNA analysis. *FEMS Microbiol Rev* **21**: 213–229.
10. Wilson IG (1997) Inhibition and facilitation of nucleic acid amplification. *Appl Environ Microbiol* **63**: 3741–3751.
11. Tanner MA, Goebel BM, Dojka MA, Pace NR (1998) Specific ribosomal DNA sequences from diverse environmental settings correlate with experimental contaminants. *Appl Environ Microbiol* **64**: 3110–3113.
12. Rochelle PA, Weightman AJ, Fry JC (1992) DNase-I treatment of Taq DNA-polymerase for complete PCR decontamination. *Biotechniques* **13**: 520–520.
13. Furrer B, Candrian U, Wieland P, Luthy J (1990) Improving PCR efficiency. *Nature* **346**: 324–324.
14. Stackebrandt E, Goebel BM (1994) A place for DNA-DNA reassociation and 16S ribosomal-RNA sequence-analysis in the present species definition in bacteriology. *Int J Syst Bacteriol* **44**: 846–849.
15. Cariello NF, Thily WG, Swenberg JA, Skopek TR (1991) Deletion mutagenesis during polymerase chain reaction: dependence on DNA polymerase. *Gene* **99**: 105–108.
16. Wang GCY, Wang Y (1997) Frequency of formation of chimeric molecules is a consequence of PCR coamplification of 16S rRNA genes from mixed bacterial genomes. *Appl Environ Microbiol* **63**: 4645–4650.
17. Farrelly V, Rainey FA, Stackebrandt E (1995) Effect of genome size and rrn gene copy number on PCR amplification of 16S rRNA genes from a mixture of bacterial species. *Appl Environ Microbiol* **61**: 2798–2801.
18. Clayton RA, Sutton G, Hinkle PS, Bult C, Fields C (1995) Intraspecific variation in small-subunit rRNA sequences in GenBank: why single sequences may not adequately represent prokaryotic taxa. *Int J Syst Bacteriol* **45**: 595–599.
19. Reysenbach AL, Giver LJ, Wickham GS, Pace NR (1992) Differential amplification of rRNA genes by polymerase chain reaction. *Appl Environ Microbiol* **58**: 3417–3418.
20. Hansen MC, TolkerNielsen T, Givskov M, Molin S (1998) Biased 16S rDNA PCR amplification caused by interference from DNA flanking the template region. *FEMS Microbiol Ecol* **26**: 141–149.
21. Chandler DP, Fredrickson JK, Brockman FJ (1997) Effect of PCR template concentration on the composition and distribution of total community 16S rDNA clone libraries. *Mol Ecol* **6**: 475–482.
22. Nübel U, Engelen B, Felske A, Snaidr J, Wieshuber A, Amann RI, Ludwig W, Backhaus H (1996) Sequence heterogeneities of genes encoding 16S rRNAs in *Paenibacillus polymyxa* detected by temperature gradient gel electrophoresis. *J Bacteriol* **178**: 5636–5643.
23. Suzuki MT, Giovannoni SJ (1996) Bias caused by template annealing in the amplification of mixtures of 16S rRNA genes by PCR. *Appl Environ Microbiol* **62**: 625–630.
24. Felske A, Wolterink A, VanLis R, Akkermans ADL (1998) Phylogeny of the main bacterial 16S rRNA sequences in Drentse A grassland soils (The Netherlands). *Appl Environ Microbiol* **64**, 871–879.
25. Felske A (1999) Reviewing the DA001-files: a 16S rRNA chase on suspect # X99967, a *Bacillus* and Dutch underground activist. *J Microbiol Meth* **36**: 77–93.
26. Miskin IP, Farrimond P, Head IM (1999) Identification of novel bacterial lineages

as active members of microbial populations in a freshwater sediment using a rapid RNA extraction procedure and RT-PCR. *Microbiology-UK* **145**: 1977–1987.

27. Fukui M, Suwa Y, Urushigawa Y (1996) High survival efficiency and ribosomal RNA decaying pattern of *Desulfobacter latus*, a highly specific acetate-utilizing organism, during starvation. *FEMS Microbiol Ecol* **19**: 17–25.
28. Wagner M, Rath G, Amann R, Koops HP, Schleifer KH (1995) In-situ identification of ammonia-oxidizing bacteria. *Syst Appl Microbiol* **18**: 251–264.
29. Kemp PF, Lee S, Laroche J (1993) Estimating the growth rate of slowly growing marine bacteria from RNA content. *Appl Environ Microbiol* **59**: 2594–2601.
30. Kemp PF (1995) Can we estimate bacterial growth rates from ribosomal RNA content? In: *Molecular Ecology of Aquatic Microbes*. Springer-Verlag, Berlin, pp. 279–302.
31. Whitman WB, Coleman DC, Wiebe WJ (1998) Prokaryotes: the unseen majority. *Proc Natl Acad Sci USA* **95**: 6578–6583.
32. Kerkhof L, Kemp P (1999) Small ribosomal RNA content in marine Proteobacteria during non-steady-state growth. *FEMS Microbiol Ecol* **30**: 253–260.
33. Weller R, Ward DM (1989) Selective recovery of 16S ribosomal-RNA sequences from natural microbial communities in the form of cDNA. *Appl Environ Microbiol* **55**: 1818–1822.
34. Marchuk D, Drumm M, Saulino A, Collins FS (1991) Construction of T-vectors, a rapid and general system for direct cloning of unmodified PCR products. *Nucleic Acids Res* **19**: 1154–1154.
35. Clark JM (1988) Novel non-templated nucleotide addition-reactions catalyzed by procaryotic and eukaryotic DNA-polymerases. *Nucleic Acids Res* **16**: 9677–9686.
36. Weisburg WG, Barns SM, Pelletier DA, Lane DJ (1991) 16S ribosomal DNA amplification for phylogenetic study. *J Bacteriol* **173**: 697–703.
37. Rheims H, Sproer C, Rainey FA, Stackebrandt E (1996) Molecular biological evidence for the occurrence of uncultured members of the actinomycete line of descent in different environments and geographical locations. *Microbiology – UK* **142**: 2863–2870.
38. Rainey FA, Ward N, Sly LI, Stackebrandt E (1994) Dependence on the taxon composition of clone libraries for PCR amplified, naturally occurring 16S rDNA on the primer pair and the cloning system used. *Experientia* **50**: 796–797.
39. Uphoff HU, Felske A, Fehr W, Wagner-Dobler I (2001) The microbial diversity in picoplankton enrichment cultures: a molecular screening of marine isolates. *FEMS Microbiol Ecol* **35**: 249–258.
40. Moyer CL, Tiedje JM, Dobbs FC, Karl DM (1996) A computer-simulated restriction fragment length polymorphism analysis of bacterial small-subunit rRNA genes: efficacy of selected tetrameric restriction enzymes for studies of microbial diversity in nature? *Appl Environ Microbiol* **62**: 2501–2507.
41. Marchesi JR, Weightman AJ (2000) Modified primers facilitate rapid screening of 16S rRNA gene libraries. *Biotechniques* **29**: 48–50.
42. Rademaker JLW, Louws FJ, Rossbach U, Vinuesa P, de Bruijn FJ (1999) Computer-assisted pattern analysis of molecular fingerprints and database construction. In: Akkermans ADL, van Elsas JD, de Bruijn FJ (eds) *Molecular Microbial Ecology Manual*, pp. 713/1–713/33. Kluwer Academic Publishers, Dordrecht.
43. von Wintzingerode F, Selent B, Hegemann W, Gobel UB (1999) Phylogenetic analysis of an anaerobic, trichlorobenzene transforming microbial consortium. *Appl Environ Microbiol* **65**: 283–286.
44. Mau M, Timmis KN (1998) Use of subtractive hybridization to design habitat-based oligonucleotide probes for investigation of natural bacterial communities. *Appl Environ Microbiol* **64**: 185–191.

45. von Wintzingerode F, Landt O, Ehrlich A, Gobel UB (2000) Peptide nucleic acid-mediated PCR clamping as a useful supplement in the determination of microbial diversity. *Appl Environ Microbiol* **66**: 549–557.
46. Good IJ (1953) The population frequencies of species and the estimation of the population parameters. *Biometrika* **40**: 237–264.
47. Giovannoni SJ, Mullins TD, Field KG (1995) Microbial diversity in oceanic systems: rRNA approaches to the study of unculturable microbes. In: Joint I (ed.) *Molecular Ecology of Aquatic Microbes*, NATO ASI Series G38. Springer-Verlag, Berlin, pp. 217–248.
48. Nakagawa Y, Sakane T, Yokota A (1996) Emendation of the genus *Planococcus* and transfer of *Flavobacterium okeanokoites* Zobell and Upham 1944 to the genus *Planococcus* as *Planococcus okeanokoites* comb nov. *Int J Syst Bacteriol* **46**: 866–870.
49. Singleton DR, Furlong MA, Rathbun SL, Whitman WB (2001) Quantitative comparisons of 16S rRNA gene sequence libraries from environmental samples. *Appl Environ Microbiol* **67**: 4374–4376.
50. Hughes JB, Hellmann JJ, Ricketts TH, Bohannan BJM (2001) Counting the uncountable: Statistical approaches to estimating microbial diversity. *Appl Environ Microbiol* **67**: 4399–4406.
51. Magurran AE (1988) *Ecological Diversity and its Measurement*. Chapman & Hall, London.
52. Watve MG, Gangal RM (1996) Problems in measuring bacterial diversity and possible solution. *Appl Environ Microbiol* **62**: 4299–4301.
53. Sanger F, Nicklen S, Coulson AR (1977) DNA sequencing with chain-terminating inhibitors. *Proc Nat Acad Sci USA* **74**: 5463–5467.
54. Altschul SF, Gish W, Miller W, Myers EW, Lipman DJ (1990) Basic local alignment search tool. *J Mol Biol* **215**: 403–410.
55. Pearson WR, Lipman DJ (1988) Improved tools for biological sequence comparison. *Proc Natl Acad Sci USA* **85**: 2444–2448.
56. Fox GE, Wisotzkey JD, Jurtshuk P (1992) How close is close—16S ribosomal-RNA sequence identity may not be sufficient to guarantee species identity. *Int J Syst Bacteriol* **42**: 166–170.
57. Hugenholtz P, Pitulle C, Hershberger KL, Pace NR (1998) Novel division level bacterial diversity in a Yellowstone hot spring. *J Bacteriol* **180**: 366–376.
58. Van de Peer Y, Chapelle S, de Wachter R (1996) A quantitative map of nucleotide substitution rates in bacterial ribosomal subunit RNA. *Nucleic Acids Res* **24**: 3381–3391.
59. Olsen GJ (1987) Earliest phylogenetic branchings: comparing rRNA-based evolutionary trees inferred with various techniques. *Cold Spring Harbor Symp Quant Biol* **LII**: 825–837.
60. Swofford DL, Olsen GJ, Waddell PJ, Hillis DM (1996) Phylogenetic inference. In: Hillis DM, Moritz GG, Mable BK (ed.) *Molecular Systematics*. Sinauer, Sunderland, MA, pp. 407–514.
61. Jukes TH, Cantor CR (1969) Evolution of protein molecules. In: Munro HN (ed.) *Mammalian Protein Metabolism*, Vol. 3, pp. 21–132. Academic Press, New York, NY.
62. Kimura T (1980) A simple method for estimating evolutionary rates of base substitutions through comparative studies of nucleotide sequences. *J Mol Evol* **16**: 111–120.
63. Van de Peer Y, Neefs J-M, De Rijk P, De Wachter R (1993) Reconstructing evolution from eukaryotic small ribosomal subunit RNA sequences: calibration of the molecular clock. *J Mol Evol* **37**: 221–232.
64. Saitou N, Nei M (1987) The neighbor-joining method: a new method for reconstructing phylogenetic trees. *Mol Biol Evol* **4**: 406–425.

65. Felsenstein J (1985) Confidence limits on phylogenies: an approach using the bootstrap. *Evolution* **39**: 783–791.
66. Lockhart PL, Steel MA, Hendy MD, Penny D (1993) Recovering evolutionary trees under a more realistic model of sequence evolution. *Mol Biol Evol* **11**: 605–612.
67. Mau B, Newton M, Larget B (1999) Bayesian phylogenetic inference via Markov chain Monte Carlo methods. *Biometrics* **55**: 1–12.
68. Hillis DM, Moritz GG, Mable BK (1996) *Molecular Systematics*. Sinauer, Sunderland, MA.
69. Hall BG (2001) *Phylogenetic Trees Made Easy: A How-to Manual for Molecular Biologists*. Sinauer, Sunderland, MA.
70. Dunbar J, Ticknor LO, Kuske CR (2001) Phylogenetic specificity and reproducibility and new method for analysis of terminal restriction fragment profiles of 16S rRNA genes from bacterial communities. *Appl Environ Microbiol* **67**: 190–197.
71. Spormann AM, Widdel F (2000) Metabolism of alkylbenzenes, alkanes and other hydrocarbons in anaerobic bacteria. *Biodegradation* **11**: 85–105.
72. Edwards U, Rogall T, Blöcker H, Emde M, Böttger EC (1988) Isolation and complete nucleotide determination of entire genes Characterisation of a gene coding for 16S ribosomal RNA. *Nucleic Acids Res* **17**: 7843–7853.
73. Muyzer G, de Waal EC, Uitterlinden AG (1993) Profiling of complex microbial populations by denaturing gradient gel-electrophoresis analysis of polymerase chain reaction-amplified genes coding for 16S ribosomal RNA. *Appl Environ Microbiol* **59**: 695–700.
74. Felske A, Akkermans ADL, DeVos WM (1998) Quantification of 16S rRNAs in complex bacterial communities by multiple competitive reverse transcription PCR in temperature gradient gel electrophoresis fingerprints. *Appl Environ Microbiol* **64**: 4581–4587.

Protocol 2.1

MATERIALS

Reagents

PCR reagents:	PCR buffer (appropriate for thermostable DNA polymerase use), store at −20°C
	Thermostable DNA polymerase (commercial type, non-proofreading), store at −20°C
	Deoxynucleoside triphosphate (dNTP) solution (10 mM of dATP, dCTP, dGTP and dTTP each), store at −20°C in small aliquots (100 μl) to prevent excessive degradation due to repeated cycles of freezing and thawing
	Oligonucleotide primers (10 μM), store at −20°C in small aliquots (100 μl) to prevent excessive degradation due to repeated cycles of freezing and thawing
	Deionized water, filter sterile (0.2 μm filter)
	Mineral oil (needed for some thermal cyclers only)
Reagents for agarose gel electrophoresis:	TAE buffer (40 mM Tris, 20 mM acetic acid, 1 mM EDTA, pH 8.3)
	Agarose
	Ethidium bromide (10 mg/ml); mutagenic, use gloves
	Loading buffer [0.05% (w/v) bromophenol blue, 40% (w/v) sucrose, 0.1 M EDTA 0.5% (w/v) sodium dodecyl sulphate, pH 8.0]
	DNA marker (e.g. 100 bp ladder)
TA overhang cloning:	PCR clean-up kit (optional)
	Commercial cloning kit, for high numbers of clones, containing: 1. Vector DNA, with associated enzymes (topoisomerase or T4 ligase) and buffers, store at −20°C, 2. *E. coli* cells, store at −70°C
	IPTG (isopropyl-β-D-thiogalactopyranoside), 0.1 M, filter sterilize and store at −20°C.

X-gal (5-bromo-4-chloro-3-indolyl-β-D-galactopyranoside); 50 mg/ml X-gal in *N,N*-dimethyl formamide, store at –20°C

Ampicillin (50 mg/ml), filter-sterilize and store at –20°C

LB/Ap/IPTG/X-gal agar (2.5 g of yeast extract, 5 g tryptone, 2.5 g NaCl and 7.5 g agar are dissolved in distilled water (500 ml) and adjusted to pH 7.5 with NaOH. Sterilized by autoclaving and cooled to 50°C before 0.5 ml of ampicillin solution, 0.5 ml of IPTG solution (0.1 M) and 0.4 ml of X-gal solution are added prior to pouring). Solidified plates (20 ml) are stored at 4°C and dried at 37°C before use.

Reagents for sequencing:

Dye Terminator Ready Reaction Cycle Sequencing Kit (ABI)

Loading dye [5 vol. deionized formamide + 1 vol. 25 mM EDTA (pH 8.0) with blue dextran (50 mg/ml)]

10 × TBE preparation (108 g Tris base, 55 g boric acid, 7.44 g $Na_2EDTA \cdot 2H_2O$, per litre), filter over 0.45 μm filter

Urea

40% acrylamide

Ammonium persulfate

TEMED (N, N, N', N',-tetramethylethylenediamine)

Equipment

Thermal cycler

Agarose gel electrophoresis apparatus with power supply

UV transilluminator, connected to gel documentation system

Autoclave

Balance

Waterbath

Incubator

Sequencer, with sequencing plates and combs, and connected to data acquisition software. N.B. Many systems now use capillary electrophoresis.

Computer with internet connection

METHODS

PCR amplification of 16S rRNA genes

1. To ensure reproducibility between PCRs and to minimize the number of pipetting steps necessary, prepare a bulk reaction mix. To ensure that there is sufficient reagent for each PCR (including controls) prepare enough master mix to perform one more PCR than is actually being run. The following protocol provides sufficient reaction mix for 20 × 50 µl reactions, and uses primers pA and pH′ (9) to amplify nearly entire bacterial 16S rRNA genes:

   100 µl of 10 × PCR buffer

   30 µl $MgCl_2$ (50 mM)

   20 µl dNTP mix (10 mM)

   20 µl forward primer pA (5′-AGAGTTTGATCCTGGCTCAG-3′), 10 µM

   20 µl reverse primer pH′ (5′-AAGGAGGTGATCCAGCCGCA-3′), 10 µM

   393 µl of sterile distilled water

   4 µl of thermostable DNA polymerase (Bioline; 5 IU μl^{-1})

   Dispense 49 µl into microcentrifuge tubes. Overlay all of the reactions with a few drops of mineral oil (if the thermal cycler is fitted with a heated lid no oil is required).

2. Add 1 µl of DNA template to each of the tubes. Sterile distilled water (1 µl) is added to one of the tubes as a negative control. Note: The negative control should be the last PCR set up. As a positive control, to one tube a template is added which is known to yield a PCR product.

3. Place the tubes in a thermal cycler. Subject the samples to the following PCR cycling program: denaturation at 95°C for 3 min followed by 30 cycles of denaturation at 95°C for 30 s, primer annealing at 55°C for 60 s and primer extension at 72°C for 90 s. Final extension 72°C for 10 min.

Agarose gel electrophoresis

1. Clean and assemble the agarose gel mold (15 × 15 cm) with combs (20 wells per comb, 10–20 μl capacity each. Dissolve 1.2 g agarose in 100 ml TAE in a beaker by heating in a microwave oven until completely molten. Let it cool (80°C), add 1 μl ethidium bromide solution* to the molten agarose and pour the gel. Let the gel solidify for 30 min. Fill the reservoir of the electrophoresis tank with TAE buffer. Remove the combs and put it in the reservoir and top up the tank with TAE buffer until it completely covers the gel. Since ethidium bromide is mutagenic, the gel should always be handled with gloves.

   *Alternatively the gel can be stained with ethidium bromide (0.5 μg ml^{-1}) after electrophoresis.

2. Mix 5 μl of each of the PCRs with 1 μl loading buffer. Load into the gel. Put a DNA marker in the outside wells
3. Connect the power supply and run the gel for 50 min at 100 V.
4. Visualize the DNA in the gel using a UV transilluminator (302 nm) and photograph or document the gel using a camera or image capture system.

TA cloning

1. Purify the PCR product (optional), using a commercially available kit (e.g. QIAquick spin columns, Qiagen, Crawley, UK), according to the manufacturer's instructions. Gel purification of the PCR product may be required particularly if the PCR reaction does not produce a single distinct DNA band of the correct size.
2. Recheck a fraction (5 μl) of the purified PCR product on gel. Run a standard in the gel for which the sizes and amount per band are known. Quantify the amount of PCR product, in comparison to the standard, by using suitable software (e.g. Quantity One) or make a rough estimate by visual comparison with a dilution series of DNA marker of known concentration.

3. Clone the PCR product using commercially available TA cloning kits (e.g. TOPO), according to the instructions of the manufacturer. The steps consist of covalently linking the PCR fragment into the vector by topoisomerase or DNA ligase activity, transformation into competent *E. coli* cells (requiring a water bath at 42°C), plating on selective medium and incubation overnight at 37°C, and takes in total 2 (topoisomerase method) to 3 (ligase method) days. Plates and liquid medium have to be prepared in advance.
4. Pick white colonies with sterile tooth picks and transfer them to a fresh LB plate containing ampicillin, X-gal and IPTG. Grow the transformants overnight.

Checking and cataloging of transformants

1. Pick up a small aliquot of the cells grown on the agar plates, transfer into 200 μl water dispersed into microcentrifuge tubes and boil for 3 min.
2. Prepare a PCR mixture as above, but with primers targeting sequences in the vector flanking the insert [e.g. pUC-f (5′-AAACAGCTATGACCATG-3′) and pUC-r (5′-TAATACGACTCACTATAGGG-3′)] for pUC-based vectors. Add 1 μl template and perform PCR.
3. Check the size of the PCR fragments by agarose gel electrophoresis, as above.
4. For correctly sized fragments, perform a double digestion as follows (per reaction):

   12 μl PCR product

   5 IU *Hae*III

   5 IU *Rsa*I

   2 μl 10 × restriction buffer

   0.2 μl BSA

   Add water to a total volume of 20 μl

   Incubate for 3 h at 37°C.
5. Add 4 μl loading buffer and run the samples on a 3% agarose gel for 1 h at 100 V, as above. Add markers (mixture of 20 and

100 bp ladder) every six or seven lanes. Document the gels.

6. Compare the ARDRA profiles observed for the clone library, either visually or using gel analysis software with database options (e.g. GelCompar, Phoretix, Diversity Database), according to manufacturer's instructions. Categorize all ARDRA profiles.

Sequencing

1. PCR amplify the cloned fragment with primers targeting the cloning vector and flanking the cloned fragment.
2. Check the results on a 1% agarose gel, and remove the primers and free dNTPs by the use of a commercial PCR purification kit. Quantify the amount of product.
3. Mix 4 μl of cocktail mix from the Dye Terminator Ready Reaction Cycle Sequencing Kit (ABI) with a single sequencing primer (20 pmol) and 20-200 ng PCR product. To obtain an almost complete 16S rRNA gene sequence use primers pA, pD, pD′, pF, pF′ and pH′ (72) as sequencing primers.
4. Perform a PCR: 25 cycles of 96°C for 0.5 minute, 50°C for 15 s, and 60°C for 4 min.
5. Remove unincorporated dye terminators by centrifugation of reaction products through spin columns.
6. Mix PCR product with 2 μl loading dye. Denature the samples at 100°C for 2 min and transfer immediately to ice before loading.

   Run the samples on either polyacrylamide electrophoresis gels or capillary systems.

Phylogenetic analysis

1. Visually inspect color-printed electropherograms for mistakes made by the automated base-calling software, such as calling an incorrect number of bases to a series of peaks or high backgrounds obscuring correct base identification.
2. Import the sequences into a sequence-editing programme, e.g. Chromas (http://www.technelysium.com.au/chromas.html).

Delete poor-quality data and manually correct mistakes. Make a contiguous sequence by overlapping sequences of a single clone to form a full-length sequence. Check and correct mismatches. Prepare a contiguous sequence and export it to a new file.

3. Perform a FASTA, BLAST or comparable search against the Genbank, EMBL, DDBJ or RDP databases (www.ncbi.nlm.nih.gov/BLAST/, www.ebi.ac.uk/fasta33/, www.ddbj.nig.ac.jp http://rdp.cme.msu.edu/seqmatch/seqmatch _intro.jsp) to check the similarity of the sequences to previously published sequences, as well as for possible chimeric sequences.

4. Obtain aligned sequences from the RDP database (http://rdp.cme.msu.edu/hierarchy) which are closely related to the sequence(s) under study. Carefully align the sequences manually to the aligned sequences in a suitable sequence editing programme (such as GDE or ARB—see Chapter 15). Use secondary structure models for regions that are difficult to align and mask regions, which cannot be aligned unambiguously. Add also a sequence which is relatively closely related to, but obviously different from, the sequences under study as an outgroup to root phylogentic trees.

5. Derive phylogenetic trees from the aligned sequences using the parsimony, maximum likelihood and distance methods using suitable software packages (e.g PHYLIP; http://evolution.genetics.washington.edu/ph ylip/software.pars.html#PHYLIP). Bootstrap the data, typically 100–1000 replicates, for distance (use an evolutionary model, like the Jukes–Cantor model) and parsimony methods. Draw a neighbor-joining tree with bootstrap values. Compare trees obtained with different methods for consistency.

6. Visualize trees using a graphics viewing program (e.g. TREEVIEW; http://taxonomy.zoology.gla.ac.uk/rod/ treeview.html).

DNA fingerprinting of microbial communities

3

Andreas Felske and A. Mark Osborn

3.1 Introduction

The past two decades have seen a revolution in the application of molecular approaches to microbial ecology, offering the chance to study microbial communities and diversity via analysis of nucleic acids directly extracted from environmental samples (1). Moreover the application of a variety of DNA fingerprinting techniques, mostly adopted from other fields of molecular biology, has allowed investigation of both variation in microbial communities, in particular via study of ribosomal RNA (see Chapter 2) and variation between individual functional genes. The application of DNA fingerprinting has proved to be particularly valuable in the rapid generation of characteristic barcode-like patterns for entire microbial communities derived from large numbers of environmental samples. Fingerprinting is useful for readily monitoring changes in microbial communities in time and space with minimal effort, which is in stark contrast to the often astronomic diversity of microbial communities in natural habitats. The fingerprint is a drastically simplified representation of the microbial community, defined firstly by the specificity of the oligonucleotide primers used to generate the PCR amplicons of interest and secondly by the preferential amplification of DNA fragments during PCR. Separating complex PCR products into fingerprints yields single bands for predominant sequences but rare sequences may often remain undetected. However, in very complex environmental communities the few predominant species may well constitute only a minority while thousands of rare species contribute most of the biomass (and hence DNA) but no signal in the fingerprint. This loss of information should be considered as a drawback of any PCR-based investigation of environmental samples. By contrast, the application of fingerprinting avoids some of the other pitfalls of PCR (2) such as the formation of chimeric DNA amplicons during PCR (3; See also Chapter 2, Section 2.2.2). Such artificially created molecules when cloned can lead to problems in subsequent phylogenetic analysis of clone libraries (4).

Historically, the first DNA fingerprinting approach to be successfully applied to microbial ecology was denaturing gradient gel electrophoresis (DGGE) (5), and Muyzer and Ramsing (6) soon placed this technique at the center of their rRNA approach. Given that the vast majority of studies applying DNA fingerprinting to microbial communities are focused on ribosomal RNA (see Chapters 2 and 13), this chapter will mainly focus on ribosomal RNA and in particular on analysis of the 16S rRNA gene that encodes for 16S rRNA.

The growing popularity of DGGE paved the way for a number of other profiling approaches such as the related temperature gradient gel electrophoresis (TGGE) (7,8), the somewhat related single strand conformation polymorphism (SSCP) analysis (9,10,11) and two techniques: terminal-restriction fragment length polymorphism (T-RFLP) (12,13) and length heterogeneity PCR (LH-PCR) (14). Whilst DGGE and TGGE are based on the differential melting of GC-rich DNA stretches in the amplified DNA molecules, SSCP separates on the basis of different melting behavior of the secondary structures of single-stranded DNA. T-RFLP generates DNA fragment length variations via the presence of restriction sites and the LH-PCR takes advantage of the different length of DNA stretches in hyper-variable regions of the target gene and in particular for ribosomal RNA (15).

3.1.1 DGGE

This method was originally invented to detect point mutations in health-related studies (16). The separation power of DGGE (TGGE and SSCP alike) relies on the fact that single strand (ss) DNA migrates more slowly than double strand (ds) DNA during electrophoresis. This is due to an increased interaction of the branched structure of the single-stranded moiety with the environment (gel matrix) via the individual nucleotides hanging freely from the sugar phosphate backbone of the helix and then becoming entangled in the gel matrix. By contrast, dsDNA more than compensates for its 'double weight' by dense stacking of the nucleotides within the helix structure. Since the nucleotides are interacting with each other via hydrogen bonding, the dsDNA will pass much faster through a gel matrix than a ssDNA molecule. If we consider two almost identical PCR products (e.g. alleles), that differ by only one nucleotide within a low melting domain of the amplified sequence, these products will have different melting temperatures. A DGGE gel typically consists of a polyacrylamide gel across which an electric charge is passed. The gel is formed so that there is an increasing denaturing gradient formed by the addition of chemical denaturants (usually formamide and urea). When a dsDNA PCR product migrates through a DGGE gel, it is therefore subject to increasing denaturation. Depending on the composition of the sequence, in particular in terms of its %GC composition, the molecule will start to denature (melt) and the two strands separate. Thus migration of the molecule is retarded due to the growing interactions of the charged nucleotides with the charged gel matrix. However, the rate of migration will increase again once the two strands are completely separated. Therefore in DGGE, complete strand separation is prevented by the presence of a high melting domain in the DNA molecule, which is usually artificially created at one end of the molecule by the incorporation of a GC clamp (17). This is accomplished via the use of an oligonucleotide primer for PCR amplification that includes a 5' tail consisting of a sequence of ~40 guanine and cytosine residues. The GC clamp is usually positioned adjacent to the highest melting domain of the amplicon to efficiently force the progress of melting in only one direction on the molecule, thus avoiding the existence of different molecule formations that may cause either a smear or more than one band per sequence. Choosing the DGGE conditions that results in the denaturation of all of the amplified molecule except for the GC clamp will give the best results. For example, if we consider DGGE (or TGGE) analysis of a 440 bp amplicon that consists of a 400 bp amplified sequence with a 40 bp GC clamp attached, the molecule will not fall apart but will form a largely ssDNA fragment (800 nucleotides) with a 40 bp dsDNA bridge in the middle (*Figure 3.1*). Further migration of the molecule will be retarded and thus the molecule will effectively stop within the gel. Hence, for such analysis the length of time required for electrophoresis is not so critical as long as the gel is electrophoresed for sufficient time to result in denaturation of molecules. Although the theory behind DGGE is relatively simple, a significant amount of preparative work must be undertaken before this technique yields fingerprints that can successfully separate molecules that show sequence variation. Primers must be carefully chosen so that the variable region to be screened has discrete melting

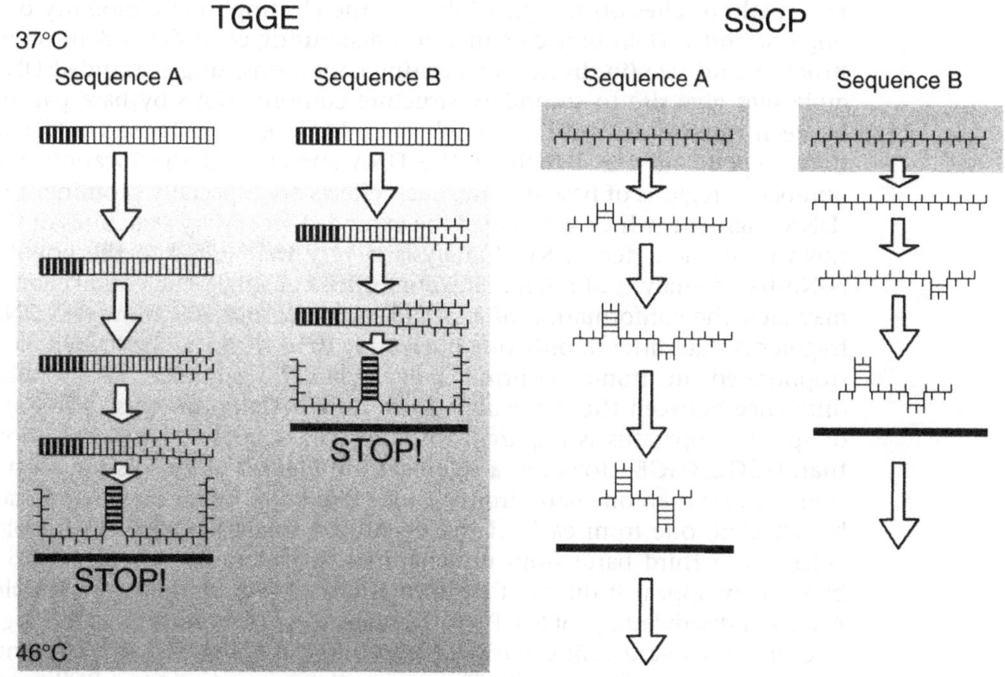

Figure 3.1

Molecular separation during TGGE (comparable to DGGE) and SSCP electrophoresis. Arrows of decreasing size indicate reduced migration speed. TGGE fragments are retarded along the temperature gradient and are finally stopped during electrophoresis. By contrast, in SSCP electrophoresis, ssDNA migration is affected by the conformation formed by secondary structures. The free nucleobases are shown as small hooks to symbolize their possible entanglement with the gel matrix (grid). The more free hooks available, the slower the migration of the fragment.

domains. Thus, full sequence data must be available to construct a denaturation or melt map of the molecule. The optimal gradient and gel running conditions have to be established and empirically fine-tuned to achieve the best resolution power.

3.1.2 TGGE

TGGE (7) is a variant of DGGE in which the increasing denaturing force applied across the gel is an increase in gel temperature towards the anode. A high concentration of chemical denaturing agents is included in the gel mix. The concentration of the chemical denaturation remains constant over the entire gel, in contrast to DGGE, with these denaturants only included to reduce the temperature that is required for full denaturation. This reduces the energy required and also stops the gel from drying out. The theory behind the separation of DNA molecules via TGGE is exactly the same as for DGGE, and its use in microbial ecology has been widespread since its initial applications in microbial ecology in 1997 (8,18).

3.1.3 SSCP analysis

This method relies on the principle that the electrophoretic mobility of a single-stranded DNA molecule in a non-denaturing gel is dependent on its structure and size (9). In non-denaturing conditions, single-stranded DNA molecules give rise to secondary structure conformations by base pairing between nucleotides within a single strand (*Figure 3.1*). These conformations depend on the length of the DNA stretch and the location and number of regions of base pairing. Such effects are especially prominent in rDNA fragments which will reflect the extended secondary structures of the rRNA molecules. Hence, SSCP analysis is very well suited to the popular rDNA-based analysis of microbial communities. A single nucleotide change may alter the conformation of a ssDNA molecule and will allow two DNA fragments that differ in only one nucleotide to be distinguished when electrophoresed in non-denaturing polyacrylamide gels due to mobility difference between the molecules. Since no GC clamp primers, gradients, or specific apparatus is required, SSCP appears at first to be more simple than DGGE/TGGE. However, a significant limitation of SSCP is the formation of more than one band from a single sequence. Often three bands can be detected, one from each of the denatured single-stranded DNA molecules and a third band from undenatured dsDNA molecules. Even more bands may appear if different conformations of one product are possible. A second disadvantage of SSCP for the analysis of communities is the high rate of DNA re-annealing during electrophoresis. This will reduce signal intensity for the ssDNA bands. The greater the concentration of PCR product that is loaded on a SSCP gel (i.e. which would be necessary for the analysis of a highly diverse community), the more the effects of renaturation will spoil the display of resolved products due to the formation of a large fraction of dsDNA molecules. However, following the first application of SSCP to environmental bacterial communities (10), a solution to these problems was developed by Schwieger and Tebbe (11). They amplified 400 bp fragments of the bacterial 16S rDNA using universal primers, for which one primer was phosphorylated at the 5'-end. The 5' phosphorylated strand of the PCR products can then be removed by λ-exonuclease treatment. Hence, the resulting signal will be produced from only one strand and any possible re-annealing or heteroduplex formation will be avoided. In this way, SSCP is much closer to the ideal 'one band = one sequence' display. The same authors also suggested using MDE polyacrylamide gel (FMC Bioproducts, Rockland, Maine, USA), a high-resolution gel matrix specifically formulated to separate DNA based upon conformational differences. Such sophisticated technical upgrades are advisable because SSCP electrophoresis has often shown a prominent tendency to produce smiling and smeary bands.

3.1.4 T-RFLP analysis

RFLP analysis of 16S rDNA, also known as amplified rDNA restriction analysis (ARDRA) (see Chapter 4), is well known for easy comparison of rDNAs from bacterial isolates or clone libraries. Briefly, PCR products are obtained by using universal 16S rDNA primers, and the product is digested with restriction enzymes with 4 bp recognition sites. The 16S rDNA PCR products from different bacteria, provided the same primers were used, would show only limited variation in length (see also LH-PCR below). However, the restriction sites may be found at very different positions within 16S rDNA sequences from different bacteria and hence cleave the PCR product into two or more fragments of different length. The discriminatory factor is the location of restriction sites within the 16S rDNA, which is sequence specific and therefore potentially taxon specific. Sequence databases may be searched for taxon-specific restriction sites and experiments can be readily customized by selecting the appropriate primers and restriction enzymes (19). While RFLP analysis for single isolates or clones is performed on low-resolution agarose gels, for community analysis, where potentially a large number of fragments would be expected, high-resolution devices are required, i.e. sequencing electrophoresis instruments. Such high-resolution equipment offers advantages for another concern, namely that it is desirable to have one band for one sequence, or one specific band for a certain taxon. Since PCR products amplified from different bacterial 16S rDNA will be split into two or more fragments, the RFLP fingerprint would be even more complex than that of the original PCR product. For instance, in a mixed community the presence of taxa represented by six restriction fragments could be overestimated compared to those that yield only two fragments. Within a complex community, the consideration of multiple bands per taxon would become even more confusing. A solution to this problem is to add a fluorescent dye to one of the primers that can be detected by the fluorescence-based DNA sequencer/genetic analyzer being used. The primer will be incorporated into the PCR product, but following restriction digestion only the terminal fragment containing the labeled primer will be detected by an automated DNA sequencing instrument. This approach was thus called terminal RFLP (T-RFLP) (12). This RFLP variant can be used with DNA from complex microbial communities and provides a valuable method to produce fingerprints of the general microbial community composition (13). A systematic evaluation of T-RFLP demonstrated the general reproducibility of this technique (20) but also identified some technical problems, not least the need to ensure complete digestion of the amplified products to avoid the generation of partially digested labeled fragments, or the formation of pseudo terminal restriction fragments (T-RFs) derived from single-stranded amplification products (21) that would lead to an overestimation of diversity. Other groups, in particular Dunbar *et al.* (22) have highlighted the need to analyze replicate T-RFLP profiles to provide the most accurate descriptions of microbial community structure.

T-RFLP does in principle offer the significant advantage that *in silico* predictions of T-RFs can be generated from DNA sequences. This has led to the development of programs such as TAP-T-RFLP (23) and PAT (24) which aim to allow predictive identification of microorganisms based on their T-RFs. However, such analyses are complicated for capillary electrophoresis, in particular by the fact that observed and expected T-RF sizes often differ (25) making definitive prediction problematic. One recent solution is to combine a microtitre-plate-based T-RFLP screening assay of clone libraries to identify clones that are representative of significant T-RFs in a total community profile (26).

3.1.5 LH-PCR (length heterogeneity PCR)

This fingerprinting approach takes advantage of naturally occurring sequence length variations. However, 16S rRNA genes from different species may typically only vary in length by a few base pairs. Hence, analysis of such variation requires the resolution power of a sequencing gel. The typical protocol involves PCR amplification of a small part of the target gene with a labeled primer and then electrophoresis of the labeled product on an automated fluorescence-detection-based sequencing device. An internal standard labeled with a different fluorescent dye is run together with the sample to allow determination of fragment length. This approach was first described by Suzuki *et al.* (14), with amplification of bacterial 16S rDNA sequences with universal primers between *E. coli* positions 8 and 355. This amplicon includes the highly variable regions V1 and V2 and yielded fragments of 312–355 bp in length, and identified up to 23 distinct length heterogeneity variants. The resolution limit is determined by the variability of the covered DNA stretch. Comparison of 16S rDNA sequences in the databases reveals that amplicon length in this region is not necessarily taxon specific, i.e. completely different bacterial taxa may share the same fragment length, with certain taxa having common amplicon lengths whilst other taxa yield amplicons that vary considerably in length. Interpretation of such data is improved if sequence information (i.e. from clones) is also available, not least so that primers can be chosen that will yield more discriminatory profiles. Inclusion of only one variable region may not provide sufficient resolution power but inclusion of a second hypervariable region may have the adverse effect of nullifying length heterogeneity in the first hypervariable region. For example, a certain taxon may include a specific insertion within the first variable region of two base pairs, but in the second variable region a corresponding deletion of two bases. Hence there will be no net change in amplicon length.

3.1.6 Comparison of the different methods

3.1.6.1 Reproducibility of electrophoresis

DGGE is perhaps the most commonly used environmental fingerprinting approach, yet differences in the preparation of the gradients can significantly affect profile reproducibility. Producing reproducible gradients can be difficult, and to prepare gradient gels that yield reproducible DGGE profiles requires a routine. In principle, the temperature gradient required for TGGE should be much easier to reproduce but even experienced TGGE users report on difficulties in maintaining the conditions necessary for high quality profile generation. When first starting DGGE/TGGE analysis there is also the considerable the time and effort required to optimize the gradient and electrophoresis conditions, which can vary considerably both between laboratories and different equipment. SSCP users on the other hand have the luxury of using conventional non-denaturing gels, and can also use commercially available pre-prepared gels. However, the use of temperature-controlled electrophoresis devices is still recommended for SSCP to produce data of sufficient quality for subsequent analysis, as temperature fluctuations can affect mobility of folded ssDNA molecules. Reliable comparison of different samples with D/TGGE and SSCP typically requires the electrophoresis of all relevant samples on the same gel to overcome the problem of gel-to-gel variability. In contrast, users of T-RFLP and LH-PCR can rely on the base-pair accuracy of the sequencing equipment and have the advantage of each sample containing its own internal size standard, to allow more accurate between-run comparison. Hence, comparison of T-RFLP or LG-PCR fingerprints from different experiments run at different times is facilitated.

3.1.6.2 Quality of signals and resolution

The fragments produced by T-RFLP or LH-PCR are limited to distinct lengths defined by variation in length and sequence of the amplicon. Therefore, a theoretical maximum number of different fragments can be determined based on the position of the primers. For T-RFLP, in particular, by a combination of different primers and/or enzymes the resolution power can be increased. However, with D/TGGE and SSCP, achieving high-resolution gels is influenced much more by skilled technical expertise, e.g. knowledge and understanding of the most suitable gradients and temperature ranges. Often such parameters must be determined empirically. Overall, it is difficult to say which of the various methods will provide the greatest resolution. For instance, Moeseneder *et al.* (27) reported higher resolution with T-RFLP but their comparison of T-RFLP plots was with relatively compressed and not very well separated DGGE fingerprints (44 distinct T-RFs vs. 28–36 DGGE bands). In practice, the resolution limit for the different methods appears to be quite similar. When sampling very complex communities, typically 10–20 dominant bands or T-RFs are seen, with approximately twice the number of faint T/DGGE bands or smaller T-RFs.

One particular problem with DGGE and TGGE is that some region of the gel may contain a smear of ssDNA that can obscure part of the fingerprint

(*Figure 3.2*). This is caused when some of the PCR products melt completely despite the presence of the GC clamp and hence form a ssDNA fraction. The art of gradient adjustment requires skill and experience to generate gels in which the ssDNA fraction and the DGGE fingerprint are separated. In contrast to D/TGGE, where the GC clamp maintains the integrity of the partially denatured amplicon, the conformation of the secondary structures formed from ssDNA during SSCP is dependent upon the primary sequence of the amplicon. Hence, SSCP analysis can yield multiple bands and smears for a single sequence. It is again necessary for users to empirically design

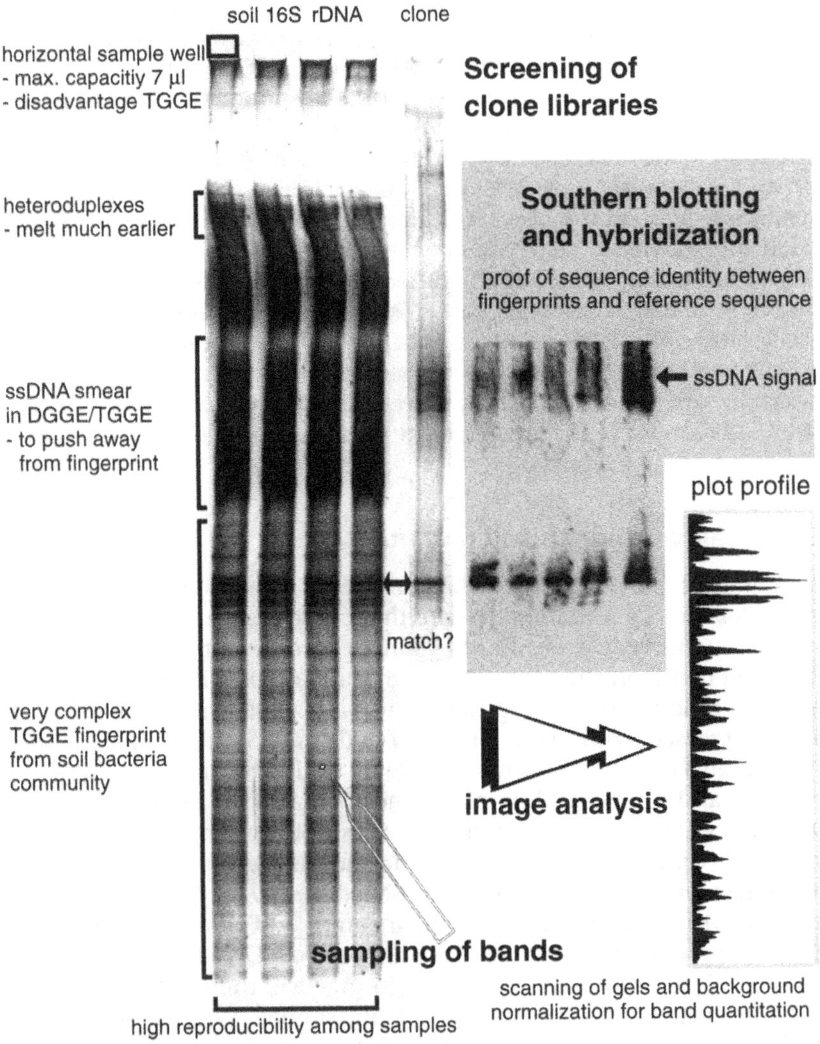

Figure 3.2

TGGE fingerprinting applications. The range of different TGGE applications is illustrated to show links to other methods. Common features of typical TGGE gels are also shown, e.g. reproducibility and also ssDNA smears.

and identify primer pairs that show good performance on SSCP gels. In contrast with T-RFLP and LH-PCR, separation and resolution efficiencies are determined by the automated sequencing instrument that is used. Whilst the principle of separation on the basis of sequence length does pose intrinsic problems, where peaks on T-RFLP and LH-PCR profiles are of similar sizes (i.e. in bp) there is the possibility that a very strong peak may overlap with adjacent weaker peaks, and hence obscure their signal.

3.1.6.3 Prediction of band positions

One of the major potential advantages of T-RFLP and LH-PCR is the potential to determine predicted sizes of T-RFs/amplicons for any known sequence by *in silico* amplification and restriction analysis of the respective database. In this way the best restriction enzymes or primers to investigate particular communities can be selected in advance. Whilst the melting behavior of DGGE/TGGE fragments can also be predicted this requires considerably more effort, and may fail if there are ambiguities or even 1 bp errors in the sequence. The possible secondary structure conformations generated during SSCP are readily calculated.

3.1.6.4 Access to DNA sequence information

Since DGGE/TGGE and SSCP separate on the basis of melting properties of the entire sequence, they may be used to directly separate individual amplicons prior to sequencing. In some but not all cases one D/TGGE/SSCP band will represent one sequence. Hence, single bands can be cut out, eluted and re-amplified for subsequent sequencing analysis. However, for complex community fingerprints one band may consist of two or more different sequences, necessitating a cloning step prior to sequencing. T-RFLP in contrasts separates fragments only on the basis of a small part of the sequence, defined by a restriction site. Hence, only for very simple bacterial communities does one T-RF (peak) necessarily correspond to one sequence type (28).

3.1.6.5 Limitation of DGGE/TGGE: the need for a GC clamp

The GC clamp on a primer (*Figure 3.3*) is an essential requirement for successful DGGE/TGGE analysis. Whilst GC-clamped primers consisting of an extended primer together with a GC-rich region at the 5' end of the primer can be readily purchased from commercial suppliers, the synthesis of such oligonucleotides (typically ~60 bp) is not trivial. Hence, primers should be purchased that are purified using the best available purification method; shortened GC clamps may spoil results by an extended ssDNA smear. Which particular GC clamp to choose is difficult to decide. For instance, Gerhard Muyzer, who first applied DGGE in microbial ecology, recognized limitations in the DGGE performance of his initially published GC clamp (5). However, TGGE users greatly appreciated the beautiful results obtained using the same GC-clamped primer (29). When designing a GC clamp, repetitive sequences should be avoided to ensure proper annealing of the GC clamp during the PCR. It is very likely that the GC clamp may

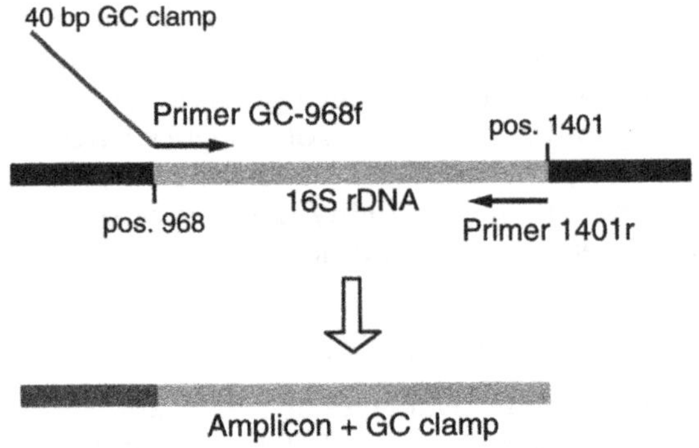

Figure 3.3

Application of GC clamps in DGGE/TGGE analysis. Tailed primers are used to attach the GC clamp to amplicons, to prevent formation of ssDNA during DGGE/TGGE analysis. The long GC tail will, however, reduce the efficiency of the PCR annealing step.

hamper primer annealing during PCR and reduce amplification efficiency. Hence, it is advisable to add GC clamps to standard primers that have already been shown to have high PCR efficiency.

3.1.6.6 DGGE or TGGE?

DGGE has clearly been more popular and therefore it has the advantage that it is also easier to find experienced DGGE users to consult with. There is also a broader range of manufacturers who supply relatively cheap gel apparatus. In contrast to the typical horizontal TGGE gel position, the traditional vertical electrophoresis used for DGGE allows large sample volumes to be more readily applied. However, the preparation of gradient gels for DGGE is more difficult, while TGGE gradients are built up automatically using water circulators or peltier elements. Another advantage of the physical TGGE gradient is the straightforward potential for miniaturization. Nowadays there are TGGE devices with gels of 20 × 20 × 0.5 mm size which may run three gels at once within 10 min (30).

3.2 Applications of microbial community fingerprinting

The use of environmental 16S rDNA cloning (4,31) revealed that bacterial communities in natural environments are much more diverse than was previously found with traditional culture methods. Moreover, the predominant organisms are mainly novel and have been commonly found to be from hitherto undefined higher taxa. Consequently the predominant fingerprint signals may well also represent such unknown bacteria. Therefore, fingerprinting is especially useful to monitor the uncultured majority within environmental bacterial communities. The ease of multiple sample analysis makes fingerprinting the method of choice for investigating the complex dynamics within microbial communities undergoing short-term or seasonal fluctuations or following perturbations. Temporal aspects of community fluctuations can be investigated by sampling the same sampling point several times. This can, however, be a problem with very small habitats which could in principle be completely consumed by sampling. Far less trivial is the study of spatial community variation. Environments of interest may constitute huge water bodies or extended landscapes providing enormous quantities of microbial life. However, in comparison, the samples used for DNA extraction may be extremely small. Conventional laboratory equipment for molecular biology is adapted to milliliter or microliter volumes, setting practical limits to nucleic acid extraction volumes, and hence the size of the sample to be treated. It is therefore important to assess which part or area of the sampling site is really represented by any given sample. For example, in terrestrial systems, should samples be taken at the millimeter, centimeter or meter scale or are microbial communities more or less constant for kilometers? Liquid environments do not pose such a significant problem: first, there is usually a constant homogenization of microorganisms; second, water samples are typically concentrated by filtration prior to DNA extraction. Preparation of representative mixed samples from various sampling points is also straightforward with liquid samples as they can be readily mixed. Solid substrate environments such as soils or sediments raise more serious difficulties; even the mixing of samples by sieving may not necessarily yield homogeneous samples. Moreover, molecular biologists are often forced to minimize the soil/sediment input for DNA extraction to reduce the impact of PCR-inhibiting soil compounds such as humic acids. Hence, there may be a huge discrepancy between the sample size and the landscape it is meant to represent. Thus ideally, much larger numbers of samples are required to cover the possible spatial variability. Fortunately, however, the situation is not that problematic, since the microbial soil communities (at least the dominant components) appear in some cases to be surprisingly constant. For example, a transect study covering a 1.5 km stretch of several grassland test fields using >600 soil samples showed that the composition of bacterial 16S rRNA molecules in a single homogeneous test field of several hundred square meters was comprehensively represented in a 1 g soil sample (29). The fingerprints from different samples of the same plot were almost perfectly identical. The similarity of the 16S rRNA fingerprints among the different test plots indicated that even long-distance spatial shifts of bacterial communities may not be dramatic, through a heterogeneous agricultural history. Hence, a substantially similar composition of bacterial communities in soils could even be expected at kilometer scales. This was borne out in a more recent study in which geographically separated agricultural soils (65 km apart) showed remarkably similar bacterial rRNA gene DNA DGGE profiles (~95% similarity). However, these similarities seemed to be defined by soil type, with bacterial communities on different soil types showing variation (32), as had previously been shown by Buckley and Schmidt (33). Nevertheless the observation that microbial communities may be common across significant distances has also been supported by other similar studies in other areas of the world, and makes it in principle feasible to tag an entire plot or landscape with one single individual DNA fingerprint. This is so reproducible that microbial community DNA profiles from soil may even serve forensic applications. In a preliminary study, T-RFLP fingerprints generated from small dirt samples recovered from properties such as shoes or cars were successfully compared to those from soil samples from a staged crime scene (34).

The following sections will outline specific strategies for the application of fingerprinting approaches.

3.2.1 Combination of cloning with fingerprinting

This approach represents a way to gain high-quality sequence data from sequences identified in fingerprints. When cloning and fingerprinting are applied in parallel from the beginning of an experiment they can overcome the limitations of each method (35). This is especially straightforward with DGGE/TGGE and SSCP. However, only relatively short sequences of up to ~500 bp are analyzed on DGGE/TGGE and SSCP gels, in contrast to the virtually complete (~1500 bp) 16S rDNA sequences (or of any other gene) by cloning. Additionally the sometimes difficult excision of fingerprint bands is not necessary. Comparison of amplicons from the cloned inserts on the same DGGE/TGGE/SSCP gel as the environmental fingerprint can be used to link clone and fingerprinting sequences. Thus clones that might represent dominant sequences in the community profile can be readily identified. However, it should be remembered that the clone library and the fingerprint represent any given microbial community in different ways. For example, if the 50 most abundant species make up only 30% of the microbial community and the remaining 70% is comprised of thousands of rare species, the fingerprint is likely to be composed of only the 50 abundant species while 70% of the clone library will be composed of the rare species (4). When investigating complex communities, potentially the majority of clones will be derived from the fraction of rare sequences and will fail to match with the fingerprint bands. Matching clones are more commonly found when the same environmental template DNA is used for both the generation of the fingerprint and the clone library. Even where co-migrating bands from a clone and in the community profile are identified, this is not necessarily evidence for sequence identity, given the possibility that a community fingerprint band may be comprised of different sequences. To address this, Southern blot hybridization with clone-specific probes can be used to confirm the presence of the cloned sequence within the fingerprint (see below).

3.2.2 Screening isolated strains next to fingerprints

This approach can similarly be used to identify whether isolated strains constitute important components of the microbial community. Technically, the same procedure is used as for clones, but for certain reasons it is not always as straightforward. As already discussed, there is a good chance that matching bands will indeed represent the same sequence band if they are based on the same template DNA. However, screening isolated strains may be based on a quite different 'DNA template'. This template is in effect the gene pool represented in the respective culture collection. It is well known that cultivation methods are highly selective, varying with the media and culture conditions used, and that they introduce a considerable bias in the representation of the bacterial community. In some cases, trying to match isolates to community profile bands in very complex microbial communities, e.g. from soil, may fail completely even with large culture collections (36). Conversely, identification of isolates representative of dominant components in simple microbial mixtures in artificial environments is relatively straightforward (37).

3.2.3 Southern blot hybridization and fingerprints

The identification of fingerprint bands either from clone libraries (see above) or following excision from gels (see below) should be confirmed by Southern blot hybridization. This is a standard molecular technique whereby the fingerprint is transferred from a gel to a membrane. The fingerprint DNA is fixed and a labeled DNA probe applied to it. This approach has been performed successfully to identify ammonia-oxidizing bacteria in DGGE fingerprints from soil (38). This common procedure will not be described here; however, the crucial consideration is the probe design (see also Chapter 8). Oligo probes will be designed on the basis of the sequence data and thus sequences of reliable quality must be used. A single sequencing error could change a useful signature nucleotide and render the probe useless. Cloning of PCR products may also introduce mutations that result from the cloning of an amplicon in which a base has been misincorporated by *Taq* polymerase (39). Alternatively excision and re-amplifying of bands from D/TGGE typically yields low quality sequences. Furthermore, in environmental DNA fingerprints there are often hitherto uncultured taxa represented with no close relatives in the 16S rDNA databases against which to compare, again raising concerns about the accuracy of the derived sequences and hence the resulting probe. There is however, a sequencing-independent way to generate probes, as described below.

3.2.3.1 PCR-generated polynucleotide probes for hyper-variable regions of the 16S rRNA

This method was developed for generating highly specific probes targeting the V6 region of the 16S rDNA without prior knowledge of DNA sequences (40). Within the 16S rDNA, the hypervariable V6 region, corresponding to positions 984–1047 of the *E. coli* 16S rDNA sequence, shows high variability combined with highly conserved adjacent regions. Hence, this V6 region can be amplified with universal bacterial primers at positions 971 and 1057. The ~90–100 bp product can be labeled and used directly as a probe. Other variable regions may also be adopted, but they require closely conserved flanking priming sites for amplification. Such conserved priming regions that flank truly effective variable probe regions constitute universal probes. Nevertheless such probes require the careful adjustment of hybridization stringency and may not differentiate between closely related species. Additionally following PCR synthesis of the probe the adjacent highly conserved regions must be removed to prevent hybridization of these conserved regions. This can be achieved using 5' exonuclease digestion, which is terminated by phosphorothioate bonds within modified primers. Such primers carry a phosphorothioate bond between bases at the 3' terminus, which will resist cleavage by T7 gene 6 exonuclease, a double-strand-specific 5'–3' exonuclease (41). Hence, the original primers can be selectively removed from the PCR product. The remaining complementary strand of the conserved priming region may now be removed by ssDNA-specific digestion with mung bean nuclease. Only the dsDNA-variable region of the PCR product will remain and can be used as probe. A standardized hybridization protocol readily provided high specificity with such truncated probes. Heuer *et al.* (40) were able to distinguish between bacteria which differ by only two bases within the probe target site (1.2% within the complete 16S rDNA). Using PCR products as probes has certain advantages but it must be considered that the plus and minus strands are in fact two probes. This may be important for quantification purposes where the V6 probe(s) can attach to each of the two strands of the denatured target molecule.

3.2.4 Multiple competitor RT-PCR/TGGE for rRNA

The quantitative representation of sequences within a fingerprint obeys the rules of exponential PCR amplification. Therefore, to quantify the original template concentrations via fingerprint bands, specialized quantitative approaches are required as adopted from quantitative PCR techniques. The only approach described to date that could quantify each of several different sequences in one fingerprint is a competitive RT-PCR with subsequent detection via TGGE (42). An *E. coli* rRNA standard of known concentration was added to the soil rRNA to allow a multiple competitor RT-PCR reaction. The signals of the single competitors (the sequences resulting in fingerprint bands) could be quantified by identifying the reaction in which one particular target signal (band) and the *E. coli* standard band showed the same intensity. After quantitative image analysis the resulting values can be used to relate the target concentration to the known *E. coli* standard rRNA. A suitable rRNA standard has to be from a bacterial strain (here: *E. coli*) that gives a TGGE signal somewhere in a band-less region of the environmental fingerprint and shows the same amplification efficiency as the environmental sequences. Within one competitive RT-PCR assay consisting of several parallel reactions with a series of different standard concentrations, numerous predominant bacterial rRNA sequences of complex bacterial communities could be quantified simultaneously. Since the absolute quantification of rRNA sequences is of questionable value (quantitative extraction of nucleic acids from environmental samples is problematic), a relative quantification is preferable. Bacteria which react to environmental changes in space or time by changing rRNA levels can be related to the total rRNA yield from all bacteria. This relative quantification with multiple competitor RT-PCR separated on high-resolution TGGE meets the demands of molecular microbial ecology to study numerous species. However, any possible PCR amplification bias has to be investigated (see also Chapter 2) and excluded in advance by various tests such as kinetic PCR, limiting dilution PCR and simulations using artificial rRNA mixtures (42). This again necessitates considerable effort to be expended on an otherwise simple approach, and the remaining uncertainties will still require careful interpretation of such data.

3.2.5 Environmental fingerprints with protein-coding genes

Screening of environmental samples for protein-coding genes promises to yield specific functional data (43). A major drawback for functional gene community profiling is the often limited number of reference sequences available, and therefore functional gene fingerprinting would often require extensive sequencing efforts to gain a suitable database for reliable primer design (see Chapter 12). An additional problem is the considerably higher degree of sequence variability that is found in functional genes than in 16S rDNA. In most cases, primer design will therefore be much more difficult and sometimes impossible. Depending on the target gene, one may also expect a variable size of PCR products, for example as shown for *mer* (mercury resistance) genes (44). This can raise question-marks concerning the authenticity of the PCR product as determined by the classical method of confirming the identity of the PCR product on the basis of fragment length during agarose electrophoresis. Hence, fingerprinting of unknown communities for certain genes should be accompanied by either a sequencing and/or hybridization (see Chapter 8) approach to confirm the authenticity of the bands. This approach may be more feasible by focusing on particular taxa to allow the selection of common priming sequences as Wawer and Muyzer (43) used to monitor the [NiFe] hydrogenase genes of *Desulfovibrio* species. Henckel *et al.* (45) screened for the genes for particulate methane monooxygenase (*pmoA*) and methanol dehydrogenase (*mxaF*) among methanotrophs. The identification, detection and enumeration of *Bifidobacterium* species via their transaldolase genes has also been described (46). A DGGE protocol for the ammonia monooxygenase gene (*amoA*) was developed to investigate the diversity of ammonia-oxidizing bacteria in different habitats (47). Similarly, T-RFLP has been used to analyze mercury resistance operons (*mer*) in soil (48) and nitrite reductase (*nirS*) diversity in marine sediments (49). Dahllöf *et al.* (50) found the highly conserved RNA polymerase β subunit (*rpoB*) fingerprinting to be a useful alternative to 16S rDNA fingerprinting. Despite the aforementioned difficulties, the growing availability of completely sequenced bacterial genomes and metagenome data (see Chapter 11) will increasingly provide a more suitable database for primer design and promote the future use of functional gene fingerprinting.

3.3 Analyzing microbial community fingerprints

The fingerprinting techniques discussed in this chapter translate the composition of complex microbial communities into abstract barcodes. Microbial communities and barcodes are however, quite different things, so the interpretation of fingerprinting data requires knowledge about the technical process, in particular the PCR. Since each fingerprint signal contains additional hidden sequence information, the further exploitation of fingerprints with other molecular techniques provides additional value. The following sections discuss additional analytical approaches, their merits and also potential pitfalls.

3.3.1 Cutting out fingerprint bands for sequencing

Once the mixture of environmental 16S rDNA amplicons has been resolved in a D/TGGE/SSCP fingerprint, individual single sequences can be directly accessed for identification. Single bands of interest can be excised from the gel, re-amplified and subsequently analyzed by sequencing. This approach appears to be a faster alternative to cloning, but certain problems have to be considered: firstly, the band must be excised accurately to avoid contamination by adjacent bands. This is particularly critical for extremely complex fingerprints (*Figure 3.2*). Moreover, not all of the PCR product will be separated during D/TGGE/SSCP into a defined fingerprint. For example, a few micrometers of the upper and lower surfaces of the acrylamide gel have a larger pore size and therefore lower separation power. Hence, all fingerprints will be 'sandwiched' by faint smears of differently separated PCR products. Secondly, the recovered amplicon will only represent a few hundred nucleotides of the target sequence, since only amplicons of <500 bp in size give good results on DGGE/TGGE and SSCP. This reduction in the corresponding length of sequence that can be generated prevents the more accurate phylogenetic analysis that could be undertaken with a full-length sequence. Finally, any one given band can be composed of two or more completely different sequences sharing identical migration properties (8, 51), resulting in unreadable sequences. In such cases cloning of the excised and re-amplified material would be required which would be subject to the same problems of base misincorporation discussed above. Hence, whilst in principle D/TGGE/SSCP offers the potential for direct identification of specific bands, the sequences generated from excised bands are typically short and often of low quality.

3.3.1.1 Using ribosomal RNA for community analysis

The ribosomal RNA sequence may be determined either from ribosomal RNA (rRNA) or from the *rrn* operons (rDNA) located in the genome. Although in principle containing the same sequence information, there are relevant qualitative and quantitative differences between rRNA and rDNA. Therefore, the separate analysis of both types of molecules from the same sample might yield different results (29). In contrast to rDNA, rRNA can be physically separated from other nucleic acids on the basis of their defined size, a useful feature during nucleic acid extraction. The rRNA content of a cell can vary considerably depending on the level of gene expression, i.e. protein synthesis, but also differences in *rrn* copy number. Hence, the study of rRNA focuses on actively growing cells (the active community), while rDNA will be present in all cells, even dead ones (the total community). The rRNA fraction will for most samples probably provide many more potential target sequences than genomic DNA. Since only 0.1–1% of the bacterial genome are *rrn* operons, any given amount of extracted rRNA will contain 100–1000-fold more targets than the same amount of DNA. However, this advantage may be lost again, because reverse transcription PCR (RT-PCR) is required to generate products for fingerprinting (and other) analyses. RT-PCR is also considerably more expensive and is neither as sensitive nor as robust as the PCR. One other consideration concerning

RT-PCR is that during cDNA synthesis each 16S rRNA molecule can only be used once as template, since the RNA is degraded during reverse transcription (52).

3.3.1.2 Influence of post-transcriptional RNA modification

Community analysis on rRNA can be affected by post-transcriptionally modified ribonucleotides in the ribosomes (53). Such altered nucleotides can interfere with RT and thus lower efficiency or even inhibit RT reactions. Therefore, the RT reaction performance should be carefully evaluated for every primer. For example, premature termination of RT often occurs due to the secondary structure of rRNA molecules at *E. coli* 16S rRNA-position 966/967 (54). Consequently, RT experiments using bacterial 16S rRNA should exclude this position and therefore only partial 16S rRNA sequences can be readily amplified and analyzed by RT-PCR.

3.3.1.3 Quantitative aspects of ribosomal RNA

It is known that the number of ribosomes per cell can vary considerably depending on the demands for protein synthesis. So in what way does the measurement of rRNA provide useful quantitative information? Ward *et al.* (55) suggested that the abundance of ribosomes in the environment should be a species-dependent function of the number of individual cells and their growth rates. Looking at a complex bacterial community, this should provide an estimate of the relative contribution of each species to the entire protein synthesis capacity of the community. This definition lacks any affiliation to bacterial cell numbers. The same amount of rRNA could be retrieved either from a species present in low cell number but high individual activity or another population at high cell number of low individual activity. Nevertheless, since protein synthesis depends on ribosomes and their rRNA, the large low-activity population might produce the same amount of proteins as the small high-activity population. However, it remains open to speculation to what extent individual components of a community obey this model, and to how variations between the different species present may bias such analyses. Carefully conducted fingerprint analysis on rRNA as compared to DNA should increasingly shift the focus more towards study of the active organisms. In some studies it has been observed that rRNA fingerprints are less complex than rDNA fingerprints (29). This fits the logical assumption that the number and diversity of active cells is (eventually much) lower than the number and diversity of present cells. However, as the PCR approach only amplifies the most dominant sequences in either the total or active communities, it can sometimes be possible for an active community profile to possess more bands than a total community profile.

3.3.2 The rRNA sequence heterogeneity

It has been established that one bacterial strain may contain several different 16S rDNA sequences. To date, a number of cases have been identified, for example in a comprehensive study of rDNA variation within a single

Clostridium perfringens strain (56). Therefore, two closely related rRNA sequences may originate from the same bacterial strain. Ueda *et al.* (57) concluded that certain helices of the 16S rRNA may be mutation tolerant, and that misincorporation during DNA replication and horizontal gene transfer could be the causes. This heterogeneity within a single strain has the potential to mislead interpretation of fingerprints. According to the most comprehensive multiple strain comparisons, the likelihood of detecting such heterogeneity in fingerprinting appears to be relatively limited: out of a collection of 115 different coryneform bacteria, only three *Curtobacterium* strains showed 16S rDNA fingerprint heterogeneity when analyzing PCR products spanning *E. coli* positions 968–1401 (58). However, this analysis did not include the regions of highest variability within the 16S rDNA. In the hypervariable regions the frequency of heterogeneity may be much higher, e.g. 33 of 475 *Streptomyces* strains showed within-strain 16S rDNA variability (57). DGGE/TGGE fingerprinting is in fact a very valuable tool for studying this phenomenon, for example as applied to determine rDNA heterogeneity in *Paenibacillus polymyxa* (59).

3.3.3 Quantitative analysis of fingerprint signals

Since the fingerprints produced by all of the different methods discussed here are based on PCR amplification of nucleic acid sequences, attempts to quantify fingerprint bands/peaks must consider the principles and drawbacks that are inherent to quantitative PCR (see also Chapter 6). The quantitative use of PCR is not straightforward since the DNA molecules are amplified exponentially during PCR. Calculating the amount of initial target molecules is only possible when there is reproducible amplification efficiency in the PCR. The amplification process is highly sensitive to differences in amplification efficiency, which may result in significant biases. The following sections discuss various aspects of this problem.

3.3.3.1 The C_0t effect

The relative amounts of PCR products that we observe within a fingerprint as different band intensities/peak sizes/areas may not reflect the relative proportions of template molecules. This is due to the PCR-inherent limitation of substrate depletion. In the early cycles of the PCR reaction there is a consistent exponential growth of product, but due to depletion of nucleotides and primers, products will no longer be generated in direct proportion to their initial starting concentrations. Considering 16S rDNA, it has been demonstrated that this effect causes a preferential amplification of less abundant sequences (60). This is due to annealing competition events, the so-called C_0t effect (61). During the annealing stage the primer binding sites on the template DNA may bind to the primers or with their complement on the complementary template DNA strand. In the early cycles of PCR this annealing competition will be won by the primers which are present in vast excess. However, as primers are depleted, complementary template strands are more likely to anneal, and therefore the template DNA re-hybridization process might now become a serious competitor to the primer and thus prevent DNA polymerization. This will firstly inhibit the amplification of the most abundant sequences, because their primer/amplicon ratio is less favorable than for the less abundant sequences.

3.3.3.2 Primer binding mismatches

Biases due to variable primer annealing efficiencies are difficult to predict and assess. In the first cycles of PCR, primers will anneal to the original template DNA. The success of this may be reduced by nucleotide mismatches between the primer and some of the target sites. Since the primer then becomes a part of the amplicon and therefore the primer sequence is introduced into the amplicon, the effects of reduced binding efficiencies due to mismatches is reduced with progressing cycle numbers. However, biases will by this stage have been introduced. To avoid such biases requires careful primer design, though this is again limited by the sequence databases being utilized. The use of degenerate primers may partially overcome such problems but it may cause problems with DGGE/TGGE and SSCP, for example multiple bands derived from one initial sequence. For users of universal primers for bacteria or Archaea, reducing the PCR annealing temperature to a minimum will reduce the impact of primer mismatches and additionally increase product yield. Under such conditions, it is important to determine that non-specific products are not amplified and therefore to check the size of the amplification products by agarose gel electrophoresis.

3.3.3.3 Multiple copies of *rrn* operons in bacterial cells

Within bacteria there is considerable variation in the number of *rrn* operons, ranging from 1 to 12 *rrn* operons per genome (62). This poses challenges for determining the relative proportions of uncultured bacterial cells using PCR-based 16S rDNA approaches. Fogel *et al.* (62) concluded that knowledge of *rrn* copy numbers for each genome would permit the enumeration of bacterial cells via 16S rDNA quantification. However, they did not consider the presence of replication forks in growing populations, i.e. the presence of more than one genome per cell (63). Replication of the bacterial genome starts at the *oriC* site (origin of replication). During chromosomal replication, every gene, beginning with those closest to the *oriC*, will be duplicated by DNA polymerase. Examining the genome maps of various bacteria, for example *Bacillus subtilis* (64), it is seen that *rrn* operons are often found to be located near to the *oriC*. Moreover, under favorable growth conditions the next generation of genome replication begins before the previous one is finished. Within one cell, the *oriC* site together with nearby genes may be present four, eight or even 16 times. In this way a highly active and fast growing cell may increase the *rrn* operon number by more than one order of magnitude, and increases of >100 *rrn* copies per cell could be feasible. Therefore, an absolute enumeration of 16S rDNA sequences in bacterial communities is of questionable value. Therefore when considering quantification of rDNA numbers, at least three variables should be considered: (i) the number of *rrn* operons per genome; (ii) the average number of genomes per cell; (iii) the relative genomic position of the *rrn* operons in relation to *oriC*. For most current environmental studies, variables cannot currently be defined.

3.3.4 Cluster analysis of DNA fingerprint signals

In recent years, computer-aided comparison of DNA fingerprints has become increasingly popular for quantifying the differences between fingerprint patterns. Scanning the gels, subtracting background, defining band positions and normalization of signals is generally tedious computational work. However, interpretation of such data is not as simple and straightforward as the output suggests. Differences in the relative strengths of PCR products and slight distortions within gels can make it difficult to correctly assign identical products as being 100% identical. The image analysis and clustering software may well yield deviations from 5 to 10% whereas the naked eye clearly confirms that theoretically identical signals are indeed identical. Cluster analysis appears sophisticated but data should be visually evaluated to identify anomalies that are obvious to the naked eye.

Moreover, the use of cluster analysis is not equally straightforward for the different fingerprinting methods. The major difference relates to signal uniqueness. For T-RFLP, in principle, cluster analysis should be readily applicable for where a particular fragment length implies a particular restriction site at the corresponding position. In this case the restriction site position has taxonomic value and can be correlated to sequence information, although the caveats discussed in Section 3.1.4 should be considered. In other words, if two T-RFs share the same peak size (bp) it is very likely that it represents a fragment generated by the presence of the same restriction site, if not necessarily the same taxon. By contrast, with methods based on melting behavior (DGGE, TGGE, SSCP) if two such fingerprints share a band at the same position, these may well represent completely different taxa, and as discussed may be comprised of different sequences. Whilst there is some tendency that closely related species run close to each other on DGGE/TGGE/SSCP gels there may well be large distances between their bands (36). Alternatively, there are many reports that quite different sequences form bands that are close to each other or even indistinguishable. Hence, cluster analysis could in some ways be considered to be a comparison of gel images but not necessarily providing such valuable information concerning the composition of the microbial community. However, such concerns are not so great if there is a close spatial and temporal relationship between the samples and if the resulting fingerprints are broadly similar. In such cases it can be reasonably assumed that bands at the same position will indeed often represent the same sequence.

3.3.5 Diversity and species richness

Diversity is a function of the species richness (number of present species) and the relative abundance of individuals per species. There is considerable interest in investigating environmental bacterial diversity, but for complex communities molecular fingerprinting can only show a fraction of the true species richness. The limited number of bands (usually up to 30) will not suffice to represent all the different species in any given environmental community. Therefore, the DNA fingerprinting approach does not display environmental bacterial diversity *per se*, but rather displays the molecular diversity of the represented predominant sequences (16S rDNA) or rRNA molecules. The emphasis on the 'predominant' sequences is due to the inherent limitation to genes that are accessible through successful cell lysis and amplification on the basis of sequences annealing to specific primers. Such surveys of bacterial diversity should consider these problems to be universal for molecular microbial ecology. Nevertheless, DNA fingerprinting should still be considered a useful tool for comparative determination of diversity and species richness in microbial communities (32), and in particular for communities that contain only a limited number of species (65).

3.3.6 The 'operational taxonomic unit'

When considering a 16S rDNA fingerprint band from an unknown bacterial community, it is not known which phylogenetic level this band represents, i.e. it may be specific for a genus, a species, or even a strain or alternatively provide no specific phylogenetic definition (55). Since it is impossible to define such relationships many researchers have used the term 'operational taxonomic unit' (OTU) to define each band, as an alternative to defining each band as a bacterial taxon. The use of arbitrarily defined OTUs is acceptable from a taxonomic point of view since the definition of bacterial species is itself somewhat arbitrary, a species being defined neither by its 16S rRNA sequence nor by any other single marker. The polyphasic taxonomic approach used today does not provide a regimented set of rules but rather takes a pragmatic and consensual approach to taxonomy via integration of all available data (66). Hence, the uncertainties in bacterial taxonomy provide the freedom to replace the assignment of bacterial taxa to bands by the more pragmatic definition of bands as OTUs.

3.4 Miscellaneous advice

(i) If you have never carried out fingerprinting analysis but are planning to do so, experience is still the best prerequisite for success. Budding fingerprinters are therefore encouraged to seek collaboration with experienced users and gain hands-on experience.

(ii) The resolution power of T-RFLP can be increased if the sequencing instrument is able to detect multiple fluorescent labels within the same lane. Forward and reverse primer can then both be labeled with different fluorescent markers, to yield two distinguishable signals per sequence (48).

(iii) Horizontal electrophoresis is more advanced than vertical devices (higher resolution, less smearing, more efficient miniaturization). However, when running TGGE, users may want to load as much sample as would be loaded in a vertical gel. Larger volumes can be applied to horizontal gel using application strips as are commonly used with isoelectric focusing (IEF) devices.

(iv) Silver staining is arguably the most sensitive and robust detection method for PAGE methods, and is especially suited for complex environmental fingerprints (i.e. from soil). SYBR Green and SYBR Gold are also increasingly used in fingerprinting studies. The image will not fade and the dried gel can be stored for years. Drying a silver-stained gel at 50°C will intensify the contrast and colors (for publication). However, where bands are to be excised and re-amplified it is advisable to dry the gel at room temperature.

(v) The combinations of denaturant/temperature gradients with acrylamide gradients can improve the resolution of DGGE/TGGE (67). Varying the acrylamide concentrations may also enhance the performance of the other fingerprinting methods.

(vi) Thinner gels provide higher resolution. For example, decreasing the width from 1 mm to 0.75 mm may significantly improve gel images. The drawback is that thinner gels are more sensitive to distortions, and more prone to tearing.

(vii) When designing primers for RT-PCR and subsequent DGGE/TGGE, the GC clamp should be added to the forward primer. The reverse primer will be involved in the more problematic RT stage and reaction efficiencies are likely to be reduced by the presence of a GC tail.

(viii) Some researchers use GC clamps on both of the primers used in the PCR with the suggestion that this can convert fuzzy signals to sharp bands (68).

References

1. Keller M, Zengler K (2004) Tapping into microbial diversity. *Nature Rev Microbiol* **2**: 141–150.
2. Witzingerode FV, Gobel UB, Stackebrandt E (1997) Determination of microbial diversity in environmental samples: pitfalls of PCR based rRNA analysis. *FEMS Microbiol Rev* **21**: 213–229.
3. Liesack W, Weyland H, Stackebrandt E (1991) Potential risks of gene amplification by PCR as determined by 16S rDNA analysis of a mixed-culture of strict barophilic bacteria. *Microb Ecol* **21**: 191–198.
4. Felske A, Weller R (2000) Cloning 16S rRNA genes and utilization to type bacterial communities. In: Akkermmans ADL, van Elsas JD, de Bruijn FJ (eds), *Molecular Microbial Ecology Manual,* Supplement 5, 3.3.3, pp. 1–16. Kluwer Academic Publishers, Dordrecht, The Netherlands.
5. Muyzer G, De Waal EC, Uitterlinden AG (1993) Profiling of complex microbial populations by denaturing gradient gel electrophoresis analysis of polymerase chain reaction-amplified genes coding for 16S rRNA. *Appl Environ Microbiol* **59**: 695–700.
6. Muyzer G, Ramsing NB (1995) Molecular methods to study the organization of microbial communities. *Wat Sci Tech* **32**: 1–9.
7. Rosenbaum V, Riesner D (1987) Temperature-gradient gel electrophoresis. Thermodynamic analysis of nucleic acids and proteins in purified form and in cellular extracts. *Biophys Chem* **26**: 235–246.
8. Felske A, Rheims H, Wolterink A, Stackebrandt E, Akkermans ADL (1997) Ribosome analysis reveals prominent activity of an uncultured member of the class Actinobacteria in grassland soils. *Microbiology* **143**: 2983–2989.
9. Orita M, Iwahana H, Kanazawa H, Hayashi K, Sekiya T (1989) Detection of polymorphisms of human DNA by gel electrophoresis as single-strand conformation polymorphisms. *Proc Natl Acad Sci USA* **86**: 2766–2770.

10. Lee D-H, Zo Y-G, Kim S-J (1996) Nonradioactive method to study genetic profiles of natural bacterial communities by PCR–single-strand conformation polymorphism. *Appl Environ Microbiol* **62**: 3112–3120.
11. Schwieger F, Tebbe CC (1998) A new approach to utilize PCR-single-strand-conformation polymorphism for 16S rRNA gene-based microbial community analysis. *Appl Environ Microbiol* **64**: 4870–4876.
12. Avaniss-Aghajani E, Jones K, Holtzman A, Aronson T, Glover N, Boian M, Froman S, Brunk CF (1996) Molecular technique for rapid identification of mycobacteria. *J Clin Microbiol* **34**: 98–102.
13. Liu WT, Marsh TL, Cheng H, Forney LJ (1997) Characterization of microbial diversity by determining terminal restriction fragment length polymorphisms of genes encoding 16S rRNA. *Appl Environ Microbiol* **63**: 4516–4522.
14. Suzuki M, Rappé MS, Giovannoni SJ (1998) Kinetic bias in estimates of coastal picoplankton community structure obtained by measurements of small-subunit rRNA gene PCR amplicon length heterogeneity. *Appl Environ Microbiol* **64**: 4522–4529.
15. Van de Peer Y, Chapelle S, De Wachter R (1996) A quantitative map of nucleotide substitution rates in bacterial rRNA. *Nucleic Acids Res* **24**: 3381–3391.
16. Fischer SG, Lerman LS (1979) Length-independent separation of DNA restriction fragments in two-dimensional gel electrophoresis. *Cell* **16**: 191–200.
17. Abrams ES, Murdaugh SE, Lerman LS (1990) Comprehensive detection of single base changes in human genomic DNA using denaturing gradient gel electrophoresis and a GC clamp. *Genomics* **7**: 463–475.
18. Heuer H, Kresk M, Baker P, Smalla K, Wellington EMH (1997) Analysis of actinomycete communities by specific amplification of genes encoding 16S rRNA and gel electrophoretic separation in denaturing gradients. *Appl Environ Microbiol* **63**: 3233–3241.
19. Moyer CL, Tiedje JM, Dobbs FC, Karl DM (1996) A computer-simulated restriction analysis of bacterial small-subunit rRNA genes: efficacy of selected tetrameric restriction enzymes for studies of microbial diversity in nature. *Appl Environ Microbiol* **62**: 2502–2507.
20. Osborn AM, Moore ERB, Timmis KN (2000) An evaluation of terminal-restriction fragment length polymorphism (T-RFLP) analysis for the study of microbial community structure and dynamics. *Environ Microbiol* **2**: 39–50.
21. Egert M, Friedrich MW (2003) Formation of pseudo-terminal restriction fragments, a PCR-related bias affecting terminal restriction fragment length polymorphism analysis of microbial community structure. *Appl Environ Microbiol* **69**: 2555–2562.
22. Dunbar J, Ticknor LO, Kuske CR (2000) Assessment of microbial diversity in four Southwestern United States soils by 16S rRNA gene terminal restriction fragment analysis. *Appl Environ Microbiol* **66**: 2943–2950.
23. Marsh TL, Saxman P, Cole J, Tiedje J (2000) Terminal restriction fragment length polymorphism analysis program: a web based research tool for microbial community analysis. *Appl Environ Microbiol* **66**: 3616–3620.
24. Kent AD, Smith DJ, Benson BJ, Triplett EW (2003) Web based phylogenetic assignment tool for analysis of terminal restriction fragment length polymorphism profiles of microbial communities. *Appl Environ Microbiol* **69**: 6768–6776.
25. Kaplan CW, Kitts CL (2003) Variation between observed and true terminal restriction fragment length is dependent on true TRF length and purine content. *J Microbiol Meth* **54**: 121–125.
26. Grant A, Ogilvie LA (2004) Name that microbe: rapid identification of taxa responsible for individual fragments in fingerprints of microbial community structure. *Mol Ecol Notes* **4**: 133–136.

27. Moeseneder MM, Arrieta JM, Muyzer G, Winter C, Herndl GJ (1999) Optimization of terminal-restriction fragment length polymorphism analysis for complex marine bacterioplankton communities and comparison with denaturing gradient gel electrophoresis. *Appl Environ Microbiol* **65**: 3518–3525.
28. Watts JE, Wu Q, Schreier SB, May HD, Sowers KR (2001) Comparative analysis of polychlorinated biphenyl-dechlorinating communities in enrichment cultures using three different molecular screening techniques. *Environ Microbiol* **3**: 710–719.
29. Felske A, Akkermans ADL (1998) Spatial homogeneity of the most abundant bacterial 16S rRNA molecules in grassland soils. *Microb Ecol* **36**: 31–36.
30. Biyani M, Nishigaki K (2001) Hundredfold productivity of genome analysis by introduction of microtemperature-gradient gel electrophoresis. *Electrophoresis* **22**: 23–28.
31. Pace NR, Stahl DA, Lane DJ, Olsen GJ (1986) The analysis of natural microbial communities by ribosomal RNA sequences. *Adv Microb Ecol* **9**: 1–55.
32. Girvan MS, Bullimore J, Pretty JN, Osborn AM, Ball AS (2003) Soil type is the primary determinant of the composition of the total and active bacterial communities in arable soils. *Appl Environ Microbiol* **69**: 1800–1809.
33. Buckley DH, Schmidt TM, (2001) The structure of microbial communities in soil and the lasting impact of cultivation. *Microb Ecol* **42**: 11–21.
34. Horswell J, Cordiner SJ, Maas EW, Martin TM, Sutherland KB, Speir TW, Nogales B, Osborn AM (2002) Forensic comparison of soils by bacterial community DNA profiling. *J Forensic Sci* **47**: 350–353.
35. Felske A, Wolterink A, van Lis R, Akkermans ADL (1998) Phylogeny of the main bacterial 16S rRNA sequences in Drentse A grassland soils (The Netherlands). *Appl Environ Microbiol* **64**: 871–879.
36. Felske A, Wolterink A, van Lis R, de Vos WM, Akkermans ADL (1999) Searching for predominant soil bacteria: 16S rDNA cloning versus strain cultivation. *FEMS Microbiol Ecol* **30**: 137–145.
37. Wagner-Döbler I, Lünsdorf H, Lübbehusen T, von Canstein HF, Li Y (2000) Structure and species composition of mercury-reducing biofilms. *Appl Environ Microbiol* **66**: 4559–4563.
38. Stephen JR, Kowalchuk GA, Bruns MA, McCaig AE, Phillips CJ, Embley TM, Prosser JI (1998) Analysis of β-subgroup proteobacterial ammonia oxidiser populations in soil by dematuring gradient gel electrophoresis and hierarchical phylogenetic probing. *Appl Environ Microbiol* **64**: 2958–2965
39. Speksnijder AGCL, Kowalchuk GA, De Jong S, Kline E, Stephen JR, Laanbroek HJ (2001) Microvariation artifacts introduced by PCR and cloning of closely related 16S rRNA gene sequences. *Appl Environ Microbiol* **67**: 469–472.
40. Heuer H, Hartung K, Wieland G, Kramer I, Smalla K (1999) Polynucleotide probes that target a hyper-variable region of 16S rRNA genes to identify bacterial isolates corresponding to bands of community fingerprints. *Appl Environ Microbiol* **65**: 1045–1049.
41. Kerr C, Sandowski PD (1972) Gene 6 exonuclease of bacteriophage T7 I Purification and properties of the enzyme. *J Biol Chem* **247**: 305–310.
42. Felske A, Akkermans ADL, de Vos WM (1998b) Quantification of 16S rRNAs in complex bacterial communities by multiple competitive reverse transcription-PCR in temperature gradient gel electrophoresis fingerprints. *Appl Environ Microbiol* **64**: 4581–4587.
43. Wawer C, Muyzer G (1995) Genetic diversity of *Desulfovibrio* spp in environmental samples analyzed by denaturing gradient gel electrophoresis of [NiFe] hydrogenase gene fragments. *Appl Environ Microbiol* **61**: 2203–2210.
44. Osborn AM, Bruce KD, Strike P, Ritchie DA (1993) Polymerase chain reaction-restriction fragment length polymorphism analysis shows divergence among

mer determinants from Gram negative soil bacteria indistinguishable by DNA-DNA hybridization. *Appl Environ Microbiol* **59**: 4024–4030.
45. Henckel T, Friedrich M, Conrad R (1999) Molecular analyses of the methane-oxidizing microbial community in rice field soil by targeting the genes of the 16S rRNA, particulate methane monooxygenase, and methanol dehydrogenase. *Appl Environ Microbiol* **65**: 1980–1990.
46. Requena T, Burton J, Matsuki T, Munro K, Simon MA, Tanaka R, Watanabe K, Tannock GW (2002) Identification, detection, and enumeration of human *Bifidobacterium* species by PCR targeting the transaldolase gene. *Appl Environ Microbiol* **68**: 2420–2427.
47. Nicolaisen MH, Ramsing NB (2002) Denaturing gradient gel electrophoresis (DGGE) approaches to study the diversity of ammonia-oxidizing bacteria. *J Microbiol Meth* **50**: 189–203.
48. Bruce KD (1997) Analysis of *mer* gene subclasses within bacterial communities in soils and sediments resolved by fluorescent-PCR-restriction fragment length polymorphism profiling. *Appl Envir Microbiol* **63**: 4914–4919.
49. Braker G, Ayala-del-Rio HL, Devol AH, Fesefeldt A, Tiedje JM (2001) Community structure of denitrifiers, bacteria, and archaea along redox gradients in Pacific Northwestern marine sediments by terminal restriction fragment length polymorphism analysis of amplified nitrite reductase (*nirS*) and 16S rRNA genes. *Appl Environ Microbiol* **67**: 1893–1901.
50. Dahllöf I, Baillie H, Kjelleberg S (2000) *rpoB*-based microbial community analysis avoids limitations inherent in 16S rRNA gene intraspecies heterogeneity. *Appl Environ Microbiol* **66**: 3376–3380.
51. Vallaeys T, Topp E, Muyzer G, Macheret V, Laguerre G, Soulas G (1997) Evaluation of denaturing gradient gel electrophoresis in the detection of 16S rDNA sequence variation in rhizobia and methanotrophs. *FEMS Microbiol Ecol* **24**: 279–285.
52. Gerard GF (1981) Mechanism of action of Moloney murine leukemia virus RNA-directed DNA polymerase associated RNase H (RNase H I). *Biochemistry* **20**: 256–265.
53. Noon KR, Bruenger E, McCloskey JA (1998) Post-transcriptional modifications in 16S and 23S rRNAs of the archaeal hyperthermophile *Sulfolobus solfataricus*. *J Bacteriol* **180**: 2883–2888.
54. Weller R, Weller JW, Ward DM (1991) 16S rRNA sequences of uncultivated hot spring cyanobacterial mat inhabitants retrieved as randomly primed complementary DNA. *Appl Environ Microbiol* **57**: 1146–1151.
55. Ward DM, Bateson MM, Weller R, Ruff-Roberts AL (1992) Ribosomal RNA analysis of microorganisms as they occur in nature. *Adv Microb Ecol* **12**: 219–286.
56. Shimizu T, Ohshima S, Ohtani K, Hoshino K, Honjo K, Hayashi H, Shimizu T (2001) Sequence heterogeneity of the ten rRNA operons in *Clostridium perfringens*. *Syst Appl Microbiol* **24**: 149–156.
57. Ueda K, Seki T, Kudo T, Yoshida T, Kataoka M (1999) Two distinct mechanisms cause heterogeneity of 16S rRNA. *J Bacteriol* **181**: 78–82.
58. Felske A, Vancanneyt M, Kersters K, Akkermans ADL (1999) Application of temperature gradient gel electrophoresis in taxonomy of coryneform bacteria. *Int J Syst Bacteriol* **49**: 113–121.
59. Nübel U, Engelen B, Felske A, Snaidr J, Wieshuber A, Amann RI, Ludwig W, Backhaus H (1996) Sequence heterogeneities of genes encoding 16S rRNAs in *Paenibacillus polymyxa* detected by temperature gradient gel electrophoresis. *J Bacteriol* **178**: 5636–5643.
60. Suzuki MT, Giovannoni SJ (1996) Bias caused by template annealing in the amplification of mixtures of 16S rRNA genes by PCR. *Appl Environ Microbiol* **62**: 625–630.

61. Mathieu-Daudé F, Welsh J, Vogt T, McClelland M (1996) DNA rehybridization during PCR: the 'C_0t effect' and its consequences. *Nucleic Acids Res* **24**: 2080–2086.
62. Fogel GB, Collins CR, Li J, Brunk CF (1999) Prokaryotic genome size and SSU rDNA copy number: estimation of microbial relative abundance from a mixed population. *Microb Ecol* **38**: 93–113.
63. Zyskind JW, Smith DW (1992) DNA replication, the bacterial cell cycle, and cell growth. *Cell* **69**: 5–8.
64. Kunst F, Ogasawara N, Maszer I *et al.* (1997) The complete genome sequence of the Gram-positive bacterium *Bacillus subtilis*. *Nature* **390**: 249–256.
65. Nübel U, Garcia-Pichel F, Kühl M, Muyzer G (1999) Quantifying microbial diversity: morphotypes, 16S rRNA genes, and carotenoids of oxygenic phototrophs in microbial mats. *Appl Environ Microbiol* **65**: 422–430.
66. Vandamme P, Pot B, Gillis M, De Vos P, Kersters K, Swings J (1996) Polyphasic taxonomy, a consensus approach to bacterial systematics. *Microbiol Rev* **60**: 407–438.
67. Petri R, Imhoff JF (2001) Genetic analysis of sea-ice bacterial communities of the Western Baltic Sea using an improved double gradient method. *Polar Biol* **24**: 252–257.
68. Gille C, Gille A, Booms P, Robinson PN, Nurnberg P (1998) Bipolar clamping improves the sensitivity of mutation detection by temperature gradient gel electrophoresis. *Electrophoresis* **19**: 1347–1350.

Protocol 3.1

The following protocol describes DNA fingerprint analysis of bacterial communities in environmental samples via T-RFLP analysis of amplified 16S rRNA genes.

MATERIALS

Reagents

Taq DNA polymerase and 10× buffer

Fluorescently labeled oligonucleotide primers:

FAM63F 5′-CAGGCCTAACACATGCAAGTC-3′

HEX518R 5′-CGTATTACCGCGGCTGCTCG-3′

[FAM and HEX are the phosphoramidite dyes 6-FAM and HEX (Applied Biosystems)] Note: fluorescently labeled primers should be stored as frozen aliquots to avoid repeated freeze–thaw cycles.

dNTPs (stock solution of 1 mM each of dATP, dCTP, dGTP and dTTP)

sterile milliQ water

0.2 ml, 0.5 ml and 1.5 ml Eppendorf tubes

Sterile Gilson tips

PCR product purification kit (Qiagen or similar)

*Alu*I and *Cfo*I restriction enzymes and 10× buffers

ROX labeled GS-500 internal size standard (Applied Biosystems)

Deionized formamide (molecular biology grade)

47 × 50 μm capillary, POP 4 polymer and electrophoresis buffer (Applied Biosystems)

Equipment

Gilson micropipettes

Vortex

PCR Thermal Cycler

Agarose gel electrophoresis apparatus and power supply

UV transilluminator

Microcentrifuge

Water bath (37°C)

Fluorescence-based DNA genetic analyzer, e.g. Applied Biosystems 310 with Genescan software.

Additional materials

Environmental DNA (see Chapter 1 for typical extraction procedures)

Agarose

TAE buffer (0.04 M Tris-acetate, 1 mM EDTA, pH 8.0)

DNA loading buffer (0.25% bromophenol blue, 0.25% xylene cyanol, 25% Ficoll type 400)

Molecular weight marker, e.g. 1 kb ladder (Invitrogen)

Ethidium bromide solution (0.5 μg ml^{-1})

Ice

METHODS

PCR amplification

1. Prepare a reagent master mix in a 1.5 ml Eppendorf tube to contain the following per each 50 μl per amplification reaction to be run:

   5 μl of 10 × *Taq* polymerase buffer

   5 μl of dNTPs (1 mM stock)

   2 μl of FAM63F (10 pmol μl^{-1} stock)

   2 μl of HEX518R (10 pmol μl^{-1} stock)

   0.5 IU Taq DNA polymerase (2.5 U μl^{-1})

   34.5 μl of sterile milliQ water

   Mix the master mix well using a vortex

   Note: Prepare enough of the master mix for *n* + 1 reactions where n is the number of PCR amplifications being set up.

2. Pipette 49 μl of the master mix into each of the 0.2 ml PCR tubes, and then add 1 μl of the environmental DNA to the reaction

tube. For a negative control add 1 μl of sterile milliQ water in place of DNA. Load into the PCR thermal cycler and run the reaction using the following conditions:

94°C for 2 min, followed by 30 cycles of: 94°C for 2 min, 55°C for 1 min, 72°C for 2 min, and a final elongation step of 72°C for 10 min.

Agarose gel electrophoresis

1. Prepare a 1% agarose gel in TAE buffer.
2. Mix 1.5 μl of DNA loading buffer with 5 μl of each PCR and load into the agarose gel. Load 5 μl DNA molecular weight marker into another lane. Run the electrophoresis at 80–100 V for ~45 min.
3. Stain the gel in the ethidium bromide solution (~15 min) and destain in water (~10 min), and then examine the gel on a transilluminator under UV light (302 nm). Ensure that no product has been amplified in the negative control reaction.

Purification of PCR products

1. For those reactions for which ~0.5 kb products have been generated, purify these products to remove dNTPs, primers and *Taq* polymerase using a PCR product purification kit (e.g. Qiagen).
2. Elute the purified DNA in a final volume of 50 μl of sterile milliQ water.
3. At this stage it is advised to run 5 μl of the purified PCR products on an agarose gel to check that the products have not been lost during purification on the column.

Restriction digest of PCR products

1. Prepare a restriction digestion master mix to contain the following per each 15 μl digestion:

   1.5 μl of 10 × restriction enzyme buffer

   2 μl of restriction enzyme; either *Alu*I or *Cfo*I

   6.5 μl of sterile milliQ water

   Mix well using a vortex.

Note: Prepare enough of the master mix for $n + 1$ reactions where n is the number of restriction digests being set up.

2. Pipette 10 µl of the master mix into each of the 1.5 ml Eppendorf tubes, and then add 5 µl of the purified PCR product.
3. Incubate at 37°C for 3 h. Digested products can then be frozen or analyzed immediately using a fluorescence based genetic analyzer.

Fluorescence-based electrophoretic analysis of T-RFLP products

1. Prepare a master mix for T-RFLP electrophoresis to contain the following for each digested PCR product:

   12 µl of deionized formamide

   0.5 µl of ROX labeled GS500 internal size standard.

   Mix well using a vortex.

Note: Prepare enough of the master mix for $n + 1$ reactions where n is the number of T-RFLPs being analyzed

2. Add 12.5 µl of the formamide/ROX standard mix to ~2 µl of the digested PCR product in a 0.5 ml Eppendorf tube. The volume of digested PCR product analyzed can be adjusted, e.g. between 1 and 5 µl, and will depend on the intensity of the purified PCR products; for weak products use more sample.
3. Heat the digests at 95°C for 5 min to denature the DNA and then transfer the tubes immediately to ice.
4. Transfer the samples into 0.5 ml Genetic analyzer sample tubes, and cap the tube with a septum (Applied Biosystems) and load into the ABI310.
5. Carry out electrophoresis of the samples using an injection time of 3–5 s at 15 kV and electrophoresis for 24–30 min at 15 kV in a 47 cm capillary containing POP4 (Applied Biosystems).
6. Analyze the data using Genescan to visualize electropherograms and tabular data.

Suggestions for successful data analysis

1. It is recommended that T-RFLP profiles for any given sample are conducted at least in duplicate. This then allows profiles to be compared to identify terminal restriction fragments (T-RFs; peaks) that are reproducibly generated from any given sample.
2. The first stage of the analysis involves the identification of individual T-RFs within a profile. A cut-off can be used to exclude T-RFs that have a peak height or area below a certain threshold (e.g. exclude all peaks with a peak height of <50 units). Where duplicate or triplicate profiles are run, T-RFs that do not have a peak height greater than the threshold in any one of the profiles can be excluded. (However, this will mean that some data are being discarded.) It is strongly recommended that users also learn how the different analysis parameters in the Genescan program will affect the output that they generate. Once the user is satisfied that they have found the best analysis parameters for their data they can export the Genescan tabular data into Excel, and data tables can then be edited manually with reference to the electropherograms to produce a table that will consist of two variables for each T-RF, namely T-RF size (in bp), and peak area (fluorescence units).
3. These numerical data can then be analyzed using a variety of macroecological methods, e.g. to generate diversity indices, evenness measurements, or to produce dendrograms based either on presence or absence of T-RFs, or additionally by taking into account the relative abundance of each T-RF as a component of the total profile. For methods that are based on both presence/absence and relative abundance data the following calculation is used.

$$\text{Relative abundance of a T-RF} = \frac{\text{peak area of the individual T-RF}}{\text{sum of the total peak areas of all T-RFs}}$$

These relative abundance values can be readily calculated in Excel and then exported into other analysis programs.

4. It is recommended that multiple primer/enzyme combinations are used in T-RFLP analysis, i.e. that both primers in the original amplification reaction are labeled with a different dye, and that PCR products are then digested (separately) using at least two different restriction enzymes (typically 4 bp cutters). The use of multiple enzymes will provide greater discrimination between different phylogenetic groups.

5. T-RFLP profiles can also be generated for cloned sequences in a clone library to allow presumptive identification of individual T-RFs in a community profile, via sequencing of clones that have T-RFs that correspond (i.e. in size in bp) to T-RFs found in the total bacterial community profile. Note: T-RF sizes for clones should be determined experimentally, in particular when using capillary-based systems such as the ABI310, as experimentally determined T-RF sizes are nearly always smaller than the T-RF sizes that would be predicted from *in silico* analysis of DNA sequences.

   Further comments on the reproducibility of T-RFLP and the use of replicates are discussed in the literature (20,22).

Molecular typing of environmental isolates

4

Jan L. W. Rademaker, Henk J. M. Aarts and Pablo Vinuesa

4.1 Introduction

The characterization of microorganisms is of essential importance for microbial ecology studies. Classification, identification and differentiation of microbes have traditionally relied on tests based on phenotypic characteristics. However, DNA- and more specifically polymerase chain reaction (PCR)-based fingerprinting methods have emerged as the most rapid, reliable and simple alternatives to characterize and differentiate microorganisms.

In typing, the objective is to reveal diversity within taxa. Taxa or taxonomic units are groups arranged on the basis of similarities or defined relationships that distinguish them from other organisms in a process called classification. Nomenclature refers to the assignment of taxonomic names to the taxa. Identification refers to assignment of unknown isolates to a distinct taxonomic unit and naming it accordingly. Classification, nomenclature and identification make up taxonomy with the 'species' as the central taxonomic unit. Complementary to taxonomy, typing enables differentiation of isolates within species or subspecies, i.e. at the strain level.

A variety of phenotypic and genotypic methods are employed for microbial typing, identification and classification. Each method has its advantages and disadvantages, in terms of ease of application, reproducibility, a requirement for sophisticated equipment, mode of action and level of phylogenetic and taxonomic resolution (1–6) (*Table 4.1*). DNA-based typing may involve specific or aspecific PCR amplification, restriction enzyme digestion and always fragment length analysis. PCR-based typing methods enable scanning of part of, or the entire, microbial genome structure. The typing methods yield banding or fingerprint profiles that are generally amenable to computer-assisted analysis and comparative typing or database-mediated identification of bacteria (*Figures 4.1–4.4*) (1,7,8).

In this Chapter, PCR amplification-based microbial typing methods for cultured organisms will be discussed such as repetitive sequence-based (rep)-PCR genomic fingerprinting, arbitrarily primed PCR (AP-PCR) fingerprinting, random amplified polymorphic DNA (RAPD), multi-locus sequence typing (MLST) and amplified fragment length polymorphism (AFLP) analysis and PCR-restriction fragment length polymorphism (PCR-RFLP). The Chapter concludes with protocols for analysis of the whole genome and ribosomal operon using the latter two techniques respectively.

4.1.1 Role of PCR amplification-based typing for microbial ecology

PCR-based typing methods have a widespread and important role in studies of environmental, agricultural, medical and industrial microbial ecology. These complex microbial environments comprise bulk soil, the phyllosphere, endosphere and rhizosphere of plants, salt and fresh water systems, human and animal bodies, including the skin, oral, gastrointestinal, and urogenital tracts, feces, feed and food, and many other man-made environments such as those found in biotechnology, bioremediation and sewage treatment facilities. PCR-based microbial typing methods have been applied to study the diversity in all these settings, including different kinds of isolates such as saprophytes, endophytes, commensals, pathogens, symbionts and organisms involved in biotechnological production.

Historically many food products are preserved by fermentation, or entail in some way an incubation period involving microorganisms. Microbes play an important role in determining flavor, texture and appearance of products such as cheese, yoghurt, sour cream, beer, wine, olives, soy sauce and meat alternatives, such as tofu and tempe. In food manufacturing, starter cultures for these products can constitute single or multiple defined strains, or undefined mixtures of organisms. PCR-based typing is applied to improve the quality of these fermentation products. Moreover, PCR-mediated typing methods are applied in the food industry to link spoilage or pathogenic microorganisms in foods to contamination sources such as the environment, equipment, processing or ingredients (2). From a microbial ecological perspective, the production chain can be seen as an enrichment culture for microbes that survive the processing conditions. Pristine and disturbed soils, fresh or saltwater aquatic ecosystems ranging from temperate to extreme environments have also been studied using molecular typing methods. PCR-based typing can be used to track and trace the contaminant microorganisms as well as, for instance, to monitor genetically modified organisms or to detect agents of bioterrorism. Microbial mats, biofilms causing fouling of pipes and waterlines, and other mixed microbial populations are of increasing interest. PCR-based typing is utilized to study the effect of pre- or pro-biotics as well as antimicrobial agents on health and the intestinal flora composition of farm animals and humans. Studies can focus on the extensive diversity present by total community fingerprinting (see Chapter 3) or using PCR-based typing methods directed to individual cultured isolates (this Chapter). Thus, PCR-based methods have proven to be of considerable value in the molecular characterization and typing of individual strains and complex communities of microbes.

4.2 Microbial typing methods

This section includes a brief discussion of the methods, the time required for various experimental protocols, key parameters, and the benefits and disadvantages of each method. Most typing methods can be applied to organisms obtained in pure culture. In this Section, AFLP, rep-PCR genomic fingerprinting, RAPD, DNA amplification fingerprinting (DAF) and AP-PCR fingerprinting, and PCR–RFLP including protocols for analysis of the ribosomal operon and MLST will be discussed.

Initially the most commonly used fingerprinting approaches were based on digestion of genomic DNA with rare cutting site endonucleases, followed by pulsed field gel electrophoresis (PFGE; 9,10 and references therein), and other RFLP-based methods, employing DNA hybridization with selected probes. Currently the best standardized whole genome DNA fingerprinting technique is PFGE (see http://www.cdc.gov/pulsenet/). This procedure requires the enzymatic digestion of DNA with rare cutting restriction endonucleases followed by agarose gel electrophoresis. In general the method has a high taxonomic resolution, is highly reproducible and it has been applied successfully to a large array of prokarya and eukarya (10–11 and references therein). However, PFGE requires high-quality DNA, obtained by a cumbersome DNA preparation method, specific electrophoresis conditions and equipment for separation of large DNA fragments.

The discovery of PCR has revolutionized DNA fingerprinting methodologies. Selected fragments of any DNA molecule (i.e. from a microbial genome) can be amplified and subjected to subsequent analyses. Combined analyses of highly conserved domains and hypervariable DNA segments are applied to reveal differences between closely related strains. PCR fingerprinting methods have become valuable tools for microbial ecologists and taxonomists because of their discriminatory power, speed, ease of application and feasibility to process large numbers of isolates. The most discriminatory PCR-based fingerprinting methods presently used are based on whole genome analysis.

4.2.1 AFLP analysis

A relatively new and promising genomic fingerprint technique is AFLP. This PCR-based method, initially developed for plant breeding purposes, is also widely used for the typing of bacteria, fungi and yeast (12; see also 5,13–17 and references therein). AFLP combines universal applicability, discriminative power and a high level of reproducibility (15). The AFLP procedure (see the protocol in this Chapter) involves restriction enzyme digestion of a small amount of purified genomic DNA (~50 ng) with one, two and sometimes more enzymes. Usually, one enzyme with an average cutting frequency (e.g. the 6 bp cutter *Eco*RI) and a second with a higher cutting frequency (e.g. the 4 bp cutters *Mse*I or *Taq*I) is applied. Subsequently, restriction site-specific adapter molecules are ligated to the generated restriction fragments. The conditions used will allow simultaneous restriction enzyme digestion and ligation while preventing the formation of unintended restriction fragment concatemers. Moreover, the adapters are designed in such a way that the adapter-restriction fragments will not be cut because the restriction sites are not restored after ligation. AFLP involves the subsequent amplification of a subset of these adapter-restriction fragments. The amplification is accomplished by using primers that are complementary to the adapter sequence and may additionally contain 0, 1 or 2 nucleotides at their 3' ends. Statistically, the extension of 1 selective nucleotide will select for amplification of 1 out of 4 ligated fragments. The generation of complex AFLP fingerprint profiles is facilitated by high-resolution separation of the amplification products on denaturing polyacrylamide gels or capillary-based systems that are identical to those used for DNA sequence analysis. AFLP analysis has several advantages compared to other fingerprinting techniques (18). The technique combines the benefits of PCR with the robustness and reliability of RFLP analysis. AFLP results can be obtained within 8–12 h starting from isolates available in pure culture, using a capillary or slab-gel-based automatic sequencer, respectively. AFLP genomic fingerprint analysis can be applied for DNA of any origin and complexity such as DNA viruses, intracellular bacteria, Gram-positive, Gram-negative bacteria and eukarya (see *Table 4.1*). No sequence knowledge of the organism under investigation is required. AFLP fingerprints exhibit a high complexity, i.e. display a high number of bands, typically 40–200 fragments. The number of fragments generated can be easily regulated by the selection of the restriction enzymes or nucleotide extensions at the primers, for example 0 or 1 for bacteria and 1 to 2 for fungal typing. The AFLP technique is applicable at various taxonomic levels such as strain, subspecies and species, but cannot discriminate between different genera. Like most other PCR typing methods, contamination of host or environmental DNA can interfere with the fingerprint results (19–21). A disadvantage of AFLP is that sophisticated equipment and software is necessary to analyze the complex profiles. The application of fluorescent labels circumvents the use of radioactive labels or silver staining profiles (22) and the use of automatic DNA sequencers significantly reduces the effort and time necessary to analyze the AFLP fingerprint profiles but on the other hand increases reagent costs. Successful application of AFLP requires the use of a standardized amount of highly purified DNA. In general, AFLP typing results show a high correlation at the subspecies/strain level with conventional taxonomic studies. Moreover, AFLP-generated genomic fingerprints yield results that are in close agreement with DNA–DNA homology studies, as determined by cluster analyses (indirect), and by a direct comparison of primary similarity values (23). This clearly indicates that AFLP analysis can be used with other core typing techniques in a polyphasic approach to taxonomy and microbial ecology.

Agarose gel-based assays have also been described, and use restriction enzymes with lower cutting frequencies (e.g. 6 or 8 bp cutters) that result in fewer fragments and, therefore, less complex fingerprints (19,24). This application is less costly than using a DNA sequencer and can be used for a variety of applications.

Several AFLP fingerprint typing applications are described below including studies of *Salmonella*, *Listeria* and *Yersinia*. Among the food-borne pathogens, *Salmonella* is one of the main causes of human enteric diseases. For *Salmonella*, 2000 different serotypes are recognized (25) and AFLP has been evaluated as an alternative to serotyping. Different *Salmonella* strains belonging to 63 serotypes were analyzed by AFLP using radioactively labeled primers and manual processing (26). In our experience, AFLP typing is less demanding and has a higher discriminatory power as compared to serotyping. Each serotype could be characterized by a unique banding profile and in some cases even strains belonging to one serotype could be distinguished. *Eco*RI and *Mse*I were used as restriction enzymes in combination with the E11 [*Eco*RI primer (E00) with extension AA] and M00 (*Mse*I

Table 4.1 Typing and target characteristics, electrophoresis platforms and references of several typing techniques

	Characterization of:	Target used for characterization	Electrophoresis platform	Methods and reviews/application references
AFLP	Whole genome or plasmid	Restriction site and additional nucleotide	Polyacrylamide[a] or agarose	5,12–17/5, 12–17,19–21,23,24, 26,27,30–38,40,41, 45,46,48,128, 144,174,184,188
rep-PCR Genomic fingerprinting	Whole genome	Repetitive element oligonucleotide sequence	Agarose or polyacrylamide[a]	1–3,7,8,51,53,57, 63/23,51,52,59-62, 65-68,70,72,73,75, 78,79,81,106,122, 135, 189,190
ARDRA	Ribosomal gene	Oligonucleotide sequence and restriction site	(MetaPhor) Agarose	116/121–124, 127–129, 135
ITS-PCR-FLP	Inter-transfer ribosomal gene spacer sequence	Oligonucleotide sequence and fragment length	Agarose or polyacrylamide[a]	-/132
IGS-PCR-FLP	Inter-ribosomal gene spacer sequence	Oligonucleotide sequence and fragment length	Agarose or polyacrylamide[a]	119,146,150/126, 130
IGS-PCR-RFLP	Inter-ribosomal gene spacer sequence	Oligonucleotide sequence and restriction site	Agarose	119,146,150/31, 118,122,124
T-RFLP	(Ribosomal) gene	Oligonucleotide sequence and restriction site	Polyacrylamide[a]	159,160/167–173
SSCP	(Ribosomal) gene	Specific oligonucleotide and total fragment sequence	Polyacrylamide[a]	162/161,163,164, 166
T/DGGE	(Ribosomal) gene	Specific oligonucleotide and total fragment sequence	Polyacrylamide	165/-
RAPD/AP-PCR/DAF	Whole genome	Repetitive random oligonucleotide sequence	Agarose	97–99 (RAPD), 100–102 (AP-PCR), 81,103,104 (DAF)/33,37,41, 85,98,101,102,105
PFGE	Whole genome	Low frequency restriction site	Agarose	9–11/105
MLST	Several genes	DNA sequence	DNA-sequencer	177–179/180–185
Plasmid profiling	Plasmid	Plasmid sizes	Agarose	191

[a]Fluorescent labels and capillary electrophoresis can be applied.

primer without extension) primer. Alternatively the enzyme combination *Bgl*II–*Mfe*I was used for AFLP analysis of *Salmonella enterica* subsp. *enterica* isolates (27).

Listeria monocytogenes is a ubiquitous microorganism (28) and known as an opportunistic food-borne pathogen that causes listeriosis. As *Listeria monocytogenes* has the ability to grow at refrigerated temperatures (29) it poses a serious health risk especially in view of the present-day consumer preference for ready-to-eat products. Only a few serotypes (1/2a, 1/2b and 4b) are responsible for the majority of the listeriosis cases. Serotyping does not allow further differentiation of potentially pathogenic strains. In order to improve epidemiological studies we have evaluated the applicability of AFLP. In one study 106 different strains belonging to 14 serotypes were characterized by AFLP (30) using automatic laser fluorescence analysis (ALFA). The results clearly showed that AFLP had a higher discriminatory power compared to serotyping as each *Listeria monocytogenes* strain analyzed was characterized by a unique profile. For these analyses *Eco*RI and *Mse*I were used in combination with primer E01 (E00 + A) and primer M02 (M00 + C). The profiles were resolved using an ALF-Express sequencer and a 6% denaturing polyacrylamide gel. The obtained digital images of the profiles

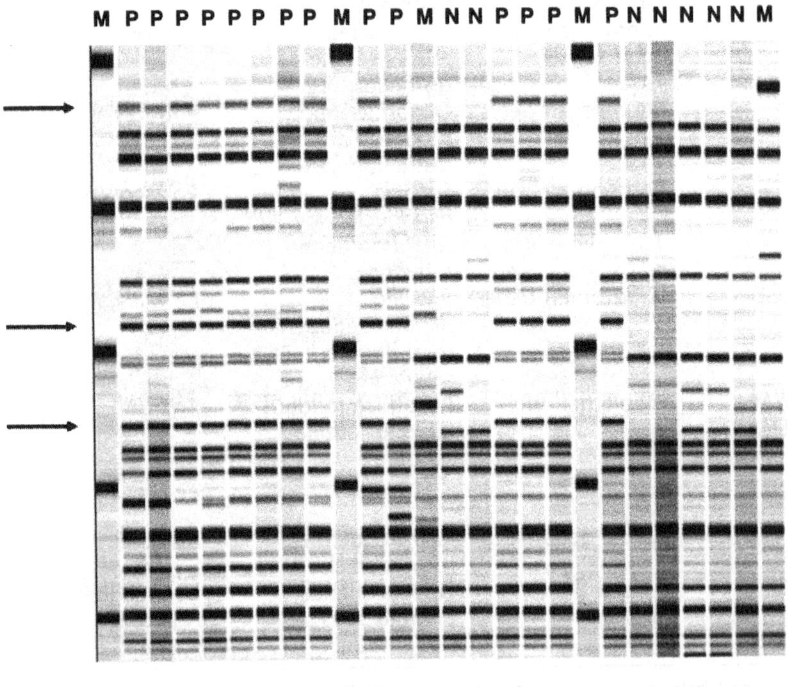

Figure 4.1

AFLP fingerprints of *Yersinia enterocolitica* strains run on an automatic DNA sequencer (Alfexpress). Genomic DNA was digested by *Eco*RI and *Mse*I. After adapter ligation, a subset of fragments was amplified by a selective PCR using the Cy5™-labeled E01 primer and the non-labeled M02 primers. Pathogen-correlated bands are indicated by an arrow. M: marker; P: pathogenic; NP: non-pathogenic.

were analyzed by the software package ImageMaster® 1D Elite (Amersham Pharmacia Biotech).

Yersinia enterocolitica is a small rod-shaped Gram-negative bacterium causing yersiniosis, infecting humans mostly through consumption of contaminated meat. Pathogenic *Yersinia enterocolitica* strains and non-pathogenic strains have been investigated by AFLP to identify pathogen-specific fragments. The application of the restriction enzymes *Eco*RI and *Mse*I, primer M02 (M00 + C) and the Cy5™ labeled E01 (E00 + A) primer resulted in highly characteristic fingerprints (*Figure 4.1*). The AFLP fingerprint profiles of the pathogenic strains were close to identical and several fragments could, in comparison to the profiles of the non-pathogenic strains, be identified as pathogen specific. These fragments may be exploited to develop pathogen-specific PCR tests. In a similar fashion, AFLP proved useful for typing and epidemiological studies of *Chlamydia* (19–21) and allowed the identification of genomic markers of ruminant *Chlamydia psittaci* strains (19).

Additional applications of AFLP are found in the taxonomy and identification of *Bradyrhizobium* strains nodulating legumes (31), and plant pathogens such as *Erwinia* (32) and *Xanthomonas* (23) as well as lactobacilli (33). Moreover, AFLP analysis was used for the systematics of *Burkholderia cepacia* (34), *Aeromonas* (14,35) and *Acinetobacter* (36–39). In the analysis of outbreaks of (food-borne) diseases, AFLP has been applied to study streptococci (40), enterococci (41) *Campylobacter* (42,43) and Legionnaires' disease (24,44). Furthermore AFLP analysis has been used to study genetic variation between strains of *Bacillus cereus, B. thuringiensis* and *B. anthracis* (45–47), *Clostridium perfringens* (48) and fungi (17,49,50).

4.2.2 rep-PCR fingerprinting

Repetitive sequence-based PCR (rep-PCR) genomic fingerprinting is a less complex alternative to AFLP typing for whole genome typing. rep-PCR uses DNA primers complementary to naturally occurring, repetitive DNA sequences, dispersed throughout most bacterial genomes. The rep-PCR amplicons are resolved in a gel matrix, resulting in complex and highly specific genomic fingerprints (*Figure 4.2*) (51,52). Three families of repetitive sequences have been identified. These include the 35–40 bp repetitive extragenic palindromic (REP) sequence, the 124–127 bp enterobacterial repetitive intergenic consensus (ERIC) sequence (51), and the 154 bp BOX element of *Streptococcus pneumoniae* (51). The repetitive elements may be present in both orientations and are located in distinct, intergenic positions around the genome. Oligonucleotide primers have been designed to prime PCR-mediated DNA synthesis outward from the inverted repeats in REP and ERIC (51) and from the box A subunit of BOX (51). The use of these primer(s) and PCR leads to the selective amplification of distinct genomic regions located between REP, ERIC or BOX elements. The amplified fragments can be resolved in a gel matrix, yielding profiles referred to collectively as rep-PCR genomic fingerprints (*Figure 4.2*). The specific protocols are denominated as REP-PCR, ERIC-PCR and BOX-PCR genomic fingerprinting, respectively (1,3,4,7,8,51,53). The rep-PCR genomic fingerprints generated from bacterial isolates permit differentiation to the species,

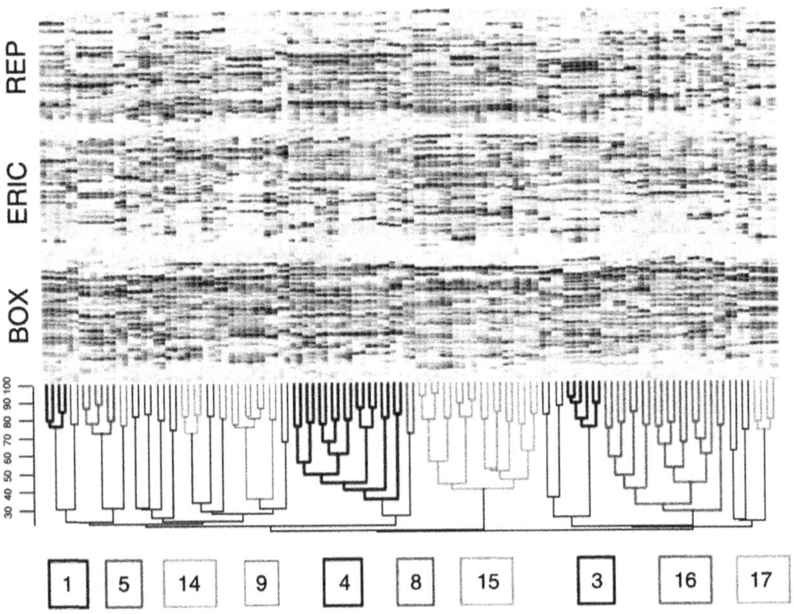

Figure 4.2

Rep-PCR genomic fingerprinting as applied for the classification and identification of *Xanthomonas*. Clusters of 59 *Xanthomonas* strains correspond to DNA–DNA homology groups, as indicated in squares, or species using combined BOX-, ERIC- and REP-PCR genomic fingerprints.

subspecies, and strain level. Moreover, the method is widely applicable to a wide variety of microorganisms, yielding complex fingerprint profiles in a variety of Gram-negative and Gram-positive bacteria (see *Table 4.1* and references therein). No previous knowledge of the genomic structure or nature of indigenous repeated sequences is necessary. Eukaryal applications have also been described (54–57). The need to identify suitable arbitrary primers (by trial and error) for large strain collections, for RAPD, AP-PCR and DAF protocols (see below) is not normally necessary for rep-PCR-based genomic fingerprinting. Another distinct advantage is that rep-PCR genomic fingerprinting can yield specific fingerprints in samples where host DNA (human, animal or plant) is present. Of additional convenience is that several methods of template preparation can be used for rep-PCR genomic fingerprinting. The method of choice can depend on the nature of the microorganisms being analyzed, the ease with which they lyse, the number of isolates to be analyzed, the level of resolution desired, and the time available. rep-PCR fingerprints have been obtained from purified DNA, cells in liquid culture, bacterial isolates from agar plates, directly from organisms in cerebral spinal fluid, and from extracts of plant lesions and nodules (1,3,4,51,58-69). In whole cell applications, relatively few cells, for example an amount barely visible on a disposable 1 μl inoculation loop, can yield enough DNA for a rep-PCR reaction (1,3,4,53). In fact, using too many cells can have an adverse effect on rep-PCR profiles resulting in the generation of a background smear.

rep-PCR genomic fingerprinting is simple, rapid and robust. The method is performed using ~50 ng of target organism DNA per 25 μl PCR reaction; the amplification requires 5–7 h (1,3,4,53). Electrophoresis on agarose gels can be performed in 8 h, but 18–19 h on 25 cm long gels is preferred for better resolution of the complex fingerprints (1,3,4,53). Alternatively, fluorescent-enhanced rep-PCR (FERP) fingerprints can be resolved on polyacrylamide gels using an ABI sequencer in ~2 h (3,4,68,70–73). Rep-PCR fingerprinting, including computer-assisted or visual pattern analysis (1,7,8), can be performed in ~2 days; however, with a few modifications it is possible to complete a rep-PCR assay within 24 h. Recent developments that apply Agilent's 'lab-on-a-chip' technology (http://www.agilent.com) for rep-PCR fingerprinting may increase speed and standardization of the analyses (74) but also increase costs per analysis. This system has now been developed as a commercial application by Spectral Genomics as the Diversilab™ system (http://www.bacbarcodes.com/).

In many cases sufficient characterization and differentiation of isolates is obtained using a single primer set. Robust and, commonly, highly complex fingerprints are obtained using the BOX primer. The REP primer set usually generates fragment profiles of lower complexity. The ERIC primer set frequently yields highly complex profiles, but is more sensitive to PCR conditions, such as the presence of contaminants in the DNA preparations. A small pilot experiment can be carried out to identify the optimum primer set for a given application. When isolates are expected to be highly similar or if a more robust assessment of genetic relatedness is to be generated, analysis of fingerprints generated with two or more primer sets is preferred.

Various typing studies have included a combination of primer sets. For example, BOX and ERIC primers were used to evaluate bacterial isolates from a hospital cafeteria-associated outbreak of gastroenteritis due to *Salmonella* (75). BOX-PCR fingerprinting has been used alone and in combination with ERIC-PCR fingerprinting to assess the diversity of fluorescent pseudomonads in soil as well as from other environmental and clinical settings (76,77). BOX-, ERIC- and REP-PCR genomic fingerprints were used in a study of bacteria obtained from pasteurized vegetable purées (78) and tomato plants (79) and enabled *X. populi* pathovars *populi* and *sacalicis* to be distinguished from each other and from other *Xanthomonas* species (80) as well as more extensive classification of the genus (23). ERIC and REP primers were successfully applied to determine the origin and diversity of mesophilic lactobacilli in cheese (81). BOX and REP primers were used to differentiate *E. coli* isolates from human and animal origin (59) and to allow identification of *E. coli* O157:H7 isolates of food-borne and environmental origin (82). BOX-PCR genomic fingerprint profiles were sufficient to determine the sources of fecal pollution (59), enabled differentiation of benzoate and nonbenzoate degrading *Rhodopseudomonas palustris* strains (83), and could characterize strains of the newly defined species *Halomonas muralis* from mural paintings in a chapel (84). ERIC-PCR genomic fingerprints were effective in differentiating virulent from avirulent *Bacillus anthracis* strains (85) and *Aeromonas* strains from diverse sources (86,87) as well as sulfate-reducing bacteria from marine sediments

(88). *Pseudomonas pseudoalcaligenes* isolates, able to utilize phenol as a sole carbon source, were found to belong predominantly to a non-clonal cluster of REP-PCR genomic fingerprint profiles (89).

Rep-PCR fingerprinting, including the more expensive FERP version, has proven to be a microbial typing method of high discriminatory power and reproducibility, and is a valuable tool for studying microbial ecology. Results of rep-PCR fingerprints were shown to be in close agreement with DNA–DNA homology studies either by cluster analyses (indirect), or by a direct comparison of primary similarity values (23).

Rep-PCR typing has also been used at low annealing temperatures to include the typing of fungi (55–57,90), but at a cost of reduced specificity. Under these conditions, rep-PCR will be susceptible to the same problems described below for arbitrary primers (91) such as poor reproducibility. Nevertheless, using lower temperatures can broaden the application of rep-PCR genomic fingerprinting.

4.2.3 Mini- and microsatellite fingerprinting

Single primers directed against mini- and microsatellite sequences have been used in a similar fashion using PCR to generate characteristic fingerprint profiles of a wide array of prokaryotic and eukaryotic organisms. Minisatellites also referred to as variable number of tandem repeats (VNTRs) include the core sequence of the wild type phage M13 (5'-GAGGGTG GCGGTTCT-3') (92), whilst microsatellites consist of simple or short sequence (di, tri, tetra etc.) repeats (SSRs), also referred to as simple tandem repeats (STRs), and include primers $(GACA)_4$ (54,93), $(CA)_8$, $(CT)_8$, $(CAC)_5$, $(GTG)_5$ (51) and $(GATA)_4$ (94). A large number of fungal species and strains, including *Penicillium*, *Trichoderma*, *Leptosphaeria*, *Saccharomyces*, *Candida* and *Cryptococcus* have been differentiated using these oligonucleotides in PCR. A high correlation of the PCR fingerprinting results with serotypes was found for *Cryptococcus neoformans* (95). Short-sequence DNA repeats of prokaryotes have been reviewed by Van Belkum *et al.* (96). The fingerprinting methods applying the primers detailed above can be regarded as an intermediate between rep-PCR genomic fingerprinting and the random or arbitrary primed PCR fingerprinting described in the following section. This intermediacy relates to the complexity of fingerprint profiles generated, their reproducibility, the time required to generate the profiles and their overall performance.

4.2.4 Fingerprinting methods using random primers (RAPD, AP-PCR, DAF)

Short random or arbitrary primers are used with PCR to generate genomic fingerprint profiles in methods referred to as random amplified polymorphic DNA (RAPD) (97–99), arbitrarily primed PCR (AP-PCR) (100–102) or DNA amplification fingerprinting (DAF) (10,103,104). The different names describe variations on a theme of PCR-based fingerprinting, which use a small nonspecific single primer at low annealing temperatures (i.e. ≤35°C). The primers anneal to multiple regions of the genome simultaneously and amplification occurs when the 3′ ends of the annealed primers face one another on opposite strands of DNA flanking regions up to several kilobases apart. Essentially, the techniques scan genomes at a low stringency for small inverted repeats with intervening DNA sequences of variable length subsequently amplified. Due to the high speed (within 24 h) and low technical demand of the method, RAPD, AP-PCR and DAF-based fingerprinting have been widely applied for species and strain differentiation. A disadvantage of the method is that the reaction conditions must be optimized rigorously for each specific application, with a high degree of standardization required, in particular when large collections of isolates are to be analyzed. Moreover the methods are highly sensitive to temperatures, reaction times, and source of polymerase, $MgCl_2$ concentration, quality and quantity of template DNA, and sequence, size and concentration of primers. A large set of primers and a number of reaction conditions should be evaluated to set up a random or arbitrary primed PCR fingerprinting experiment. Often several sets of primers need to be applied in the final analysis in order to obtain the most discriminating patterns between species or strains. The amount of DNA template required is 20–50 ng, similar to that required for rep-PCR and AFLP genomic fingerprint analysis. Primers can be purchased from various manufacturers (e.g. the OPA series from Operon Technologies; http://www.operon.com) or designed randomly. RAPD, AP-PCR and DAF allow rapid detection of polymorphic DNA markers using small agarose gels with the profiles comprising a limited number of bands, typically five fragments or fewer. Diversity is usually evident using one or more primers, but reliable similarity data usually require the use of multiple primers. Up to 10 or more primers may be required to obtain a set of markers with reliable discriminatory power. For example, six primers were needed in one RAPD fingerprinting study to match the typing performance of single AFLP genomic fingerprints (37).

The random and arbitrary PCR fingerprinting methods are particularly suited for rapid and easy comparative typing of relatively small sets of isolates, in particular because the fingerprints obtained in the same PCR amplification are by definition less sensitive to reproducibility factors. DNA samples, however, need to be of a good quality. The random and arbitrary PCR fingerprinting methods are applicable to a wide variety of pro- and eukarya, yielding complex fingerprint profiles in a variety of Gram-negative and Gram-positive bacteria (33,37,41,85,101,105), fungi and other eukarya (98,102).

Computer-assisted cluster analysis of four combined RAPD profiles has been shown to have a remarkable congruence with cluster analysis of two AFLP profiles and to some extent correlation with origin, pathogenicity and bacteriogenicity in a characterization of 78 *Enterococcus feacium* isolates (41). RAPD-PCR fingerprinting using at least three different primers followed by computer-assisted cluster analysis of the combined patterns was successful in differentiating *Lactobacillus acidophilus* and related species, as was similarly found using AFLP analysis (33). As with ERIC-PCR fingerprinting, RAPD analysis was found to be effective in differentiating virulent from avirulent *Bacillus anthracis* isolates (85). Moreover, RAPD analysis has also been successfully applied for typing of homofermentative lactobacilli, and compared with analysis using REP-PCR genomic fingerprinting and PFGE approaches (105). RAPD analysis has been applied to a large variety of other bacteria such as *Borellia, Actinobaccillus, Pseudomonas, Proteus* and *Staphylococcus* and many others such as various food- and plant-associated bacteria (see references in 2,102–104,106). When higher levels of discrimination and the assessment of relationships are required, AFLP and rep-PCR genomic fingerprinting methods are more suitable, particularly for extensive surveillance studies that generate databases of profiles (1,7,8).

4.2.5 Ribosomal RNA gene fingerprinting methods

The PCR-based whole genome fingerprinting methods discussed above (AFLP, rep-PCR, RAPD and AP-PCR) yield highly discriminatory, subspecific and frequently strain-specific genomic fingerprints (i.e. provide a high taxonomic resolution). To compare isolates above species level, various typing methods based on amplified ribosomal DNA (restriction) fragment length polymorphism (RFLP) analysis have been extensively used in molecular microbial ecology and systematic studies. For several reasons, ribosomal RNA genes (rRNA genes or rDNA) have contributed largely to our current understanding and perception of microbial ecology, diversity and evolution (107–110) (see also Chapter 2):

(i) orthologs of rRNA genes are found in every living organism, and encode rRNA molecules that are components of ribosomes which carry out polypeptide synthesis;
(ii) rRNAs display highly conserved secondary structure motifs (stems and loops) (111) that allow for the alignment of rRNA sequences from distantly related taxa (112);
(iii) domains within the ribosomal genes and/or operon evolve at different rates, allowing phylogenetic analyses to be performed at various levels of taxonomic resolution, including the deepest branches (109,110).

Ribosome-related data services can be obtained from the Ribosomal Database Project (RDP; http://rdp.cme.msu.edu/) (113), including online data analysis, rRNA-derived phylogenetic trees, and aligned and annotated rRNA sequences (e.g. using the ARB program; http://www.arb-home.de/) (see also Chapter 15). The use of ribosomal operons as target for a variety of PCR-based typing methods is discussed here, while phylogenetic sequence analyses are discussed in Chapter 2.

As detailed below, different regions of the ribosomal operons have been targeted for PCR amplicon length or RFLP analyses of microbial isolates. The general structure of the rDNA operon (*rrn*) of prokaryotes is shown in *Figure 4.3*. The *rrf* , *rrs* and *rrl* rRNA genes encode for the structural rRNA molecules required for ribosome assembly and function (5S, 16S and 23S rRNAs, respectively) (114). However, various structures have been reported, including split *rrn* operons with different configurations such as linked *rrs-rrl* and separated *rrf* genes, or separated *rrs* and linked *rrl–rrf* genes (see 154 for a review). For bacteria, the 16S, Small Sub Unit (SSU or *rrs*), 23S, Large Sub Unit (LSU or *rrl*), ribosomal gene sequences, inter 16S–23S rRNA or *rrs–rrl* spacers, and inter 23S–5S rRNA or *rrl–rrf* spacers, can be targeted for analysis. The targets are PCR-amplified followed by (restriction) fragment length polymorphism analysis.

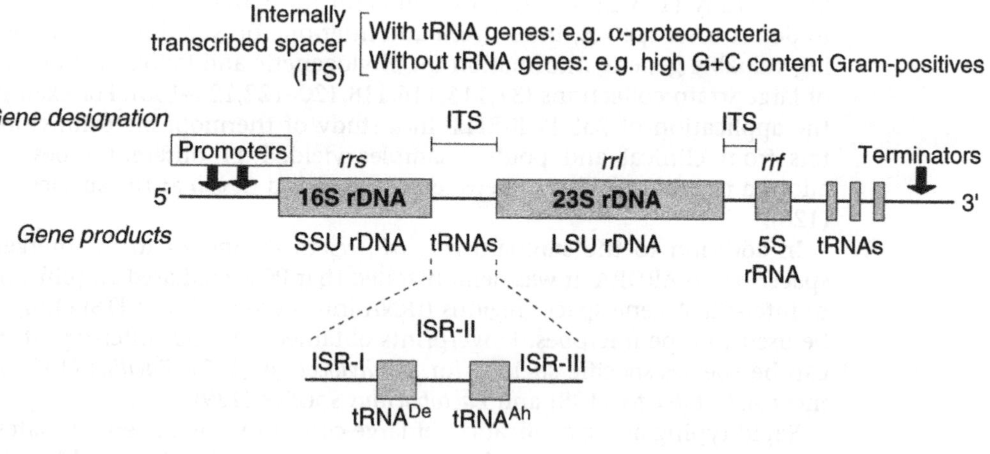

Figure 4.3

General structure of the prokaryotic ribosomal rRNA operon (*rrn*). Note that the 16S-23S IGS region may or may not contain tandem $tRNA^{Ile}$ and $tRNA^{Ala}$ genes. Several groups of prokaryotes do not contain such genes linked to the *rrn* operon, and others contain only one of the two tRNA genes ($tRNA^{Ile}$ or $tRNA^{Ala}$) or alternatively the single $tRNA^{Glu}$ gene. Several species have been reported to contain split *rrn* operons, in which *rrs*, *rrl* and *rrf* are not all linked to each other (for a review see 154).

In amplified ribosomal DNA restriction enzyme analysis (ARDRA) (115,116) specific (or moderately degenerate) PCR primers targeting highly conserved domains in the sequences of *rrs* and *rrl* genes are used to amplify partial or nearly full-length 16S, 23S or intergenic spacer rDNA fragments (117–120). The high specificity and low degeneracy of these primers makes the amplification of the selected targets possible from different sources, including liquid or plate cultures, without a previous DNA purification step (121). The PCR products are generally single bands of a well-defined size (but see discussion below), making the analysis of the specificity and efficiency of the amplification reaction a straightforward and rapid undertaking, using standard agarose gel electrophoresis. Subsequently one or two frequently cutting, tetrameric (i.e. having a 4 bp recognition site) restriction enzymes are used to digest the rDNA amplicons. There is no need to purify the PCR products before their restriction with endonucleases. Electrophoresis of the resulting restriction fragments is often performed using MetaPhor agarose for enhanced resolution of the DNA fragments (122), but preparation of these gels is more complex than for standard agarose gels. However, if large collections of diverse isolates are to be typed, the use of MetaPhor agarose is recommended. In contrast to the earlier-mentioned whole genome fingerprinting methods that yield complex fingerprint profiles, only a limited number of fragments are obtained per RFLP assay. In general terms the PCR-RFLP typing procedure is straightforward and highly reproducible, and can be performed for many isolates in <2 days. Important parameters in the analysis are the selection of restriction enzymes and primers, preferably guided by computer-simulated digestion of the complete rDNA sequences of organisms related to the ones under study (123,124). ARDRA of different *rrn* markers has been reported to detect inter-species, inter-strain and inter-operon variability, thus yielding valuable genetic information for phylogenetic and taxonomic analyses of large strain collections (31,115,116,118,120–122,124–135). For example, the application of 23S PCR-RFLP in a study of thermophilic campylobacters from clinical and poultry samples yielded 11 different types, and allowed the differentiation between *C. jejuni* and *C. coli* at the species level (128).

In addition to the standard *rrs*-, *rrl*-, *rrs* to *rrl* and *rrl* to *rrf* intergenic spacer-based ARDRA, it was demonstrated that PCR-mediated amplification of inter-tRNA gene spacer regions (tRNA-inter-repeat-PCR or ITS) (136) can be used to type microbes. Fingerprints obtained by tRNA-inter-repeat-PCR can be species-specific such as for *Staphylococcus* (132), *Bacillus* (137) and most *Acinetobacter* (138) and *Lactobacillus* species (139).

Rapid typing and classification of large collections of bacterial isolates at the genus to species levels of taxonomic resolution has been achieved by PCR-RFLP analysis of *rrn* operon markers coupled with computer-assisted pattern analysis (116). This is illustrated by several studies on diversity and classification of rhizobia and other bacteria of ecological, industrial and clinical interest (121,122,124,127). The endosymbiotic nodule bacteria from leguminous plants comprise a diverse and polyphyletic assemblage of α- and β-Proteobacteria currently classified in 12 genera grouped in nine families (140,141). The 16S rDNA and 23S rDNA sequences or RFLP data enable grouping of strains at the genus to species levels of taxonomic refinement, but frequently do not provide enough taxonomic resolution to study infraspecific diversity. This applies particularly to genera displaying low *rrs* sequence divergence (142), such as the genus *Bradyrhizobium* (143,144,145). The 16S–23S internally transcribed spacer region is more variable in sequence, thus providing significantly greater taxonomic resolution than is found for the *rrs* and *rrl* genes, as shown in several recent studies dealing with diversity, taxonomy and phylogeny of different bacterial groups, and discussed in recent reviews (10,31,76,77,119,122,124,130,141,144, 146–151). We have shown that the combined analysis of the *rrs*, IGS and *rrl* PCR-RFLPs using just three enzymes for each region provides taxonomic resolution to at least two or three times as many groups as is found for full-length *rrs* sequence analysis of bradyrhizobia. The overall topology of the RFLP-based dendrograms is in good agreement with those derived from IGS nucleotide sequence phylogenetic reconstruction (*Figure 4.4*) (145). However, incongruent tree topologies derived from sequence or RFLP data of different *rrn* markers is an indication of potential horizontal gene transfer and recombination events (152). This can also affect the entire *rrn* operon, as elegantly demonstrated by Yap *et al.* (153).

All DNA sequence analyses of inter-ribosomal gene spacer sequence IGS regions of rhizobia to date contain tandem $tRNA^{Ile}$ and $tRNA^{Ala}$ genes (see *Figure 4.3*). In other bacterial groups, such as most high G + C bacteria (Actinobacteria), these genes may be absent in the operon. The tRNA genes display some sequence variability in fast-growing rhizobia (*Rhizobium* spp. and *Sinorhizobium* spp.), but are highly conserved in slow-growing *Bradyrhizobium* strains (151). In both groups, IGS sequences contain highly variable intergenic spacer regions (ISRs) located upstream, between, and

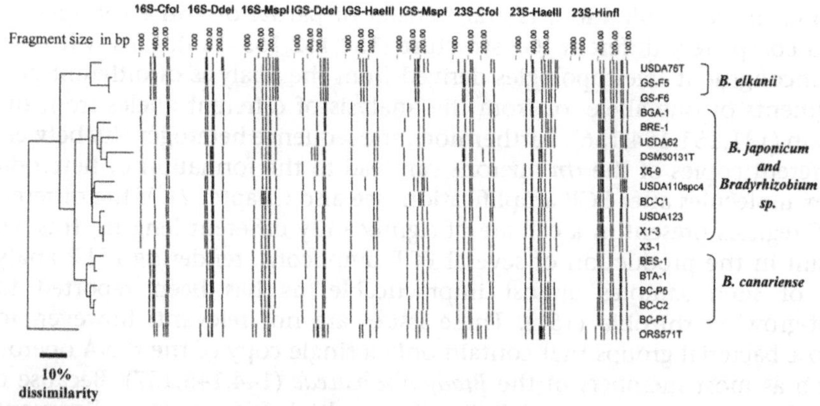

Figure 4.4

Dice-neighbor joining analysis of the combined 16S, ITS and 23S rDNA PCR-RFLPs of a selected set of *Bradyrhizobium* spp. strains. The Canarian *Bradyrhizobium* cluster (112) is recognizable with 100% bootstrap support by IGS sequence analysis, but not by full-length *rrs* sequences (unpublished). Strains BGA-1 and USDA123 are indistinguishable by their 16S rDNA sequence, but are clearly resolved by the RFLP analysis, demonstrating its higher resolution power. The Venezuelan soybean isolates GS-F5 and GS-F6 are phylogenetically related to *B. elkanii*, whereas the Chinese soybean isolates X1-3, X3-1 and X6-9 were classified as *B. japonicum* strains, illustrating its value in the classification of novel bacterial isolates. *Azorhizobium caulinodans* strain ORS571 forms an outgroup to the *Bradyrhizobium* spp. strains, as expected from phylogenies reconstructed from 16S rDNA and other gene sequences.

downstream of the $tRNA^{Ile}$ and $tRNA^{Ala}$ genes, respectively (see *Figure 4.3*). This variability constitutes nucleotide substitutions at particular sites, as well as insertions and deletions of varying lengths. As a consequence, IGS amplification of rhizobia often results in conspicuous intra-genus and even intra-species length polymorphisms, as also reported for other bacteria (154). For example, the IGS amplification products generated from a world-wide collection of 75 *Bradyrhizobium* strains ranged in size between 900 and 1200 bp, with some isolates displaying a much shorter IGS amplification product (124). This variation can be exploited for the development of strain-specific PCR primers, targeted to the strain-specific singleton sites, which can then be used to track particular strains in the environment. The use of rDNA IGS sequences and RFLP analyses is increasingly applied in high-resolution typing and phylogenetic analyses of bacterial isolates, and holds great promise for studies on molecular bacterial population genetics (118,119,130,141, 145–150,154 and references therein).

Several properties of *rrn* operons should be considered when *rrn* markers are chosen for PCR-RFLP typing or phylogenetic inferences. First it is important to notice that many bacterial genera contain multiple copies of the *rrn* operon (e.g. 10 in *Bacillus subtilis*, nine in *Staphylococcus aureus*, seven in *E. coli* and *Salmonella*, six in *Streptomyces nodosus* and *Enterococcus faecalis*, three in *Sinorhizobium meliloti* and *Mesorhizobium loti* (155). The rDNA alleles on each copy can be divergent, due to nucleotide substitutions, localized intragenic recombination, lateral transfer of partial or entire *rrn* operons, or a completely different IGS structure (151,152,153–155). This will result in incongruent tree topologies derived from the analysis of different gene segments of one allele, or from the analysis of different alleles from one strain (131,151,154,156). Furthermore, the sequence heterogeneity between different copies of the *rrn* operons can lead to the formation of heteroduplex molecules after PCR amplification (see also Chapter 2). If the different IGS regions present in a cell are of significantly different lengths, this will result in the production of several PCR amplicons, rendering RFLP analysis of such samples almost impracticable, as has been reported for fast-growing rhizobia (151). These issues are not relevant, however, for those bacterial groups that contain only a single copy of the rRNA operon, such as most members of the *Bradyrhizobiaceae* (144,145,157). Because of highly variable non-coding ISR domains, multiple IGS sequence alignments may contain regions of low confidence with respect to the reliability of the alignment, even when strains from the same genus or species are compared. Therefore these regions are generally excluded for phylogenetic analyses, but are useful for diversity studies and the formulation of strain- or group-specific PCR primers for ecological studies (151). Indel regions, however, contain significant phylogenetic information, which can be taken into account either by coding gapped sites as a fifth character state (145), or by the construction of an indel matrix using different coding schemes (157). It should also be noted that very strong among-site rate variation exists in *rrs*, *rrl* and IGS sequences containing tRNA genes. Therefore a phylogenetic reconstruction based on these sequences using model-based methods (i.e. distance, maximum likelihood and Bayesian) should correct for this variation by choosing substitution models that include the gamma and/or proportion of invariant sites parameters (131, 145,158).

Terminal restriction fragment length polymorphism analysis (T-RFLP) (159,160), separation of PCR-amplified genes by single-strand conformation polymorphism (SSCP) (161–164) and temperature or denaturing gradient gel electrophoresis (T/DGGE) (see 165 and references therein) can detect subtle sequence variation in specific genetic loci. The results of these analyses are typically limited to the mobility of a single band for each isolate. This limitation has been exploited by typing several isolates simultaneously in the same gel lane such as applied in the genetic profiling of whole microbial communities (see Chapter 3). However, these methods have been shown to be useful in typing single isolates. T-RFLP has been used for identifying clinically important bacterial strains (159) whilst PCR-SSCP analysis has been applied to characterize the genomovars of the *Burkholderia cepacia* complex (166). In our laboratory T-RFLP enabled differentiation of *Alyciclobacillus* strains that are known to spoil acidic food and ingredients and that may produce off-flavor, for example in fruit juices (unpublished data). In addition, population structure and dynamics of silage, cheese and intestinal microflora could be determined using T-RFLP analysis (167,168 and unpublished data). PCR (T-)RFLP analysis of ribosomal sequences for genetic typing has several benefits such as the availability of large sequence databases, such as the Ribosomal Database Project (RDP) (113), and well-characterized phylogenetic primers. Several alternative genes have been used in T-RFLP analysis. Polymorphisms present on a variety of DNA segments including the *mer* operon (169), the *amoA* gene (170), the *nirS* gene (171) and *mcrA* genes (172,173) have been used for typing. Additional polymorphisms may be present in housekeeping genes, *recA* (174), flagellin genes such as *fla*, antimicrobial resistance genes or pathogenicity and symbiotic islands (34,127,175,176). Also mobile genetic elements, such as transposons, integrons, insertion sequences (IS), bacteriophages and plasmids may be used to obtain or may contribute to the complexity of the restriction patterns (see references in 147). Increasingly popular for characterization and strain differentiation is the analysis of short DNA sequences of multiple loci in a procedure referred to as multi-locus sequence typing (MLST) (177–179). Short DNA sequences of ~450 bp, of several loci, often about seven, have been used for MLST typing (180) of *Campylobacter jejuni*, *Streptococcus pneumonia* (see references in 147), *S. pyrogenes*, *Staphylococcus aureus*, *Haemophilus influenzae*, *Salmonella* (181,182) *Enterococcus feacium* (183), *Neisseria meningitidis* (184) and *Candida albicans* (185). MLST typing is highly reproducible, versatile, quick and easy when high-throughput sequencing facilities can be afforded. It is amenable to computer-assisted analysis, database construction and data can be easily exchanged between laboratories (181) and is therefore increasingly being adopted for bacterial typing studies.

4.3 Computer-assisted analysis of PCR-generated fingerprint profiles

When a large number of diverse DNA fingerprint profiles of high complexity need to be compared, computer-assisted analysis becomes imperative and standardization becomes critical. Moreover, the quality of the data is essential to obtain reliable comparisons, especially when the profiles are obtained from multiple gels over a long time and/or if large databases are to be generated. Standardized conditions should include: sample preparation and processing; DNA isolation methods; and PCR, restriction digestion and ligation reaction conditions. The use of standardized electrophoresis conditions, size markers and image capturing is also essential. For computer-assisted analysis of DNA typing profiles, many hardware and software combinations are available (7,8). Comparison and cluster analysis of a collection of profiles can be carried out using band-based as well as curve-based approaches (1,7,8). In general, a well-defined fingerprint of low complexity, such as PFGE and PCR-RFLP, can be defined as an array of peaks or bands. Nevertheless it is tedious, laborious and subjective, especially for highly complex profiles such as AFLP and rep-PCR genomic fingerprints. Alternatively, a curve-based analysis, such as the product-moment correlation (186) can be used. This coefficient permits a direct comparison of the densitometric curves and is better suited for identification of DNA fingerprinting profiles than band-matching algorithms (1,7,8,187). The product moment correlation values obtained among rep-PCR and AFLP fingerprint profiles have shown high correlation with DNA-DNA homology values supporting the high relevance of the product moment (23,147). Subsequent cluster analysis requires a simplification of the original data by reducing a set of DNA fingerprint profiles to a proximity matrix. The choice of clustering method is not always obvious, and different similarity coefficients and more than one clustering method can be applied to compare the resulting groupings in order to chose the representation of the data that is most appropriate. In summary, cluster analysis is not a statistical approach to test a hypothesis. However, it is a valuable representation method instrumental for the interpretation of complex data sets.

4.4 Concluding remarks

In this Chapter several important features of PCR-based typing techniques have been reviewed, including reproducibility, discriminatory ability, and relative ease of use, versatility, and amenability to comparative and library typing. Moreover, differences in technical expertise required, time to completion, and instrumentation required have been compared. The typing methods are capable of characterizing the whole genome, or subcomponents of the genome, such as single gene or intergenic DNA sequences, based on primer sequence and, for example, restriction sites. Some of these parameters and methods as well as application references are summarized in *Table 4.1*.

In general, many factors including financial budget, the scientific question being addressed, and the organisms being studied will all influence which techniques are applied. A research study investigating diversity within a eukaryotic species may require a different technique than a study of diversity within a prokaryotic species. AFLP or rep-PCR genomic fingerprint analyses are often useful for the determination of subspecies relationships. DNA-DNA homology values have shown a high correlation with these typing methods, suggesting that these genomic fingerprinting methods truly reveal genomic relationships between organisms. RAPD, AP-PCR and DAF fingerprinting are rapid and relatively inexpensive but also less reproducible with primers and experimental conditions needing to be optimized for each new application. The random or arbitrary primed PCR fingerprint profiles are typically useful for obtaining diversity information. The combination of several (i.e. four to 10) RAPD or AP-PCR profiles may yield reliable relationship information. The reliability of the typing results can usually be ensured using a number of independent methods and can be confirmed by incorporating a suitable collection of well-characterized reference strains. AFLP genomic fingerprinting is technically more demanding, cannot be performed on whole cells, and will yield fragments of DNA contaminants, such as the host of the microbes being studied. PCR-mediated genomic fingerprinting and computer-assisted analysis facilitates the study of the taxonomic and phylogenetic relationships between microbes. The continuing advancements in typing and analysis methods will lead to further insights into the distribution and characteristics of microorganisms.

More recent developments include the increasing application of MLST and the advent of high-density DNA arrays and DNA chips, which, coupled with additional benefits of the -omics era (see Chapter 10), will permit more extensive comparative analysis and characterization of microbial genomes. These methods enable combinations of simultaneous detection, identification and typing of organisms in combination with the characterization of large spectra of functional determinants.

Acknowledgments

We gratefully acknowledge Meike te Giffel for critical reading of this manuscript and Arjen Wagendorp for excellent assistance in preparing this manuscript.

References

1. Rademaker JLW, de Bruijn FJ (1997) Characterization and classification of microbes by rep-PCR genomic fingerprinting and computer assisted pattern analysis. In: Caetano-Anollés G, Gresshoff PM (eds) *DNA Markers: Protocols, Applications and Overviews*, pp. 151–171. J. Wiley & Sons, New York.
2. Rademaker JLW, Savelkoul PHM (2004) PCR amplification based microbial typing. In: Pershing DH, Tenover FC, Versalovic J, Tang YW, Unger ER, Relman DA, White TJ (eds) *Diagnostic Molecular Microbiology, Diagnostic Principles and Practice*, pp. 197–221. ASM Press, Washington.
3. Rademaker JLW, Louws FJ, de Bruijn FJ (1998) Characterization of the diversity of ecologically important microbes by rep-PCR genomic fingerprinting. In: Akkermans ADL, van Elsas JD, de Bruijn FJ (eds) *Molecular Microbial Ecology Manual*, Supplement 3, Chap. 343, pp. 1–26. Kluwer Academic Publishers, Dordrecht.
4. Rademaker JLW, Louws FJ, Versalovic J, de Bruijn FJ (2004) Characterization of the ecological important microbes by rep-PCR genomic fingerprinting. In: Kowalchuk GA, de Bruijn FJ, Head IM, Akkermans AD, van Elsas JD (eds) *Molecular Microbial Ecology Manual*, 2nd edn, Chap. 532, pp. 1–33. Kluwer Academic Publishers, Dordrecht.
5. Savelkoul PHM, Aarts H, Duim B, Dijkshoorn L, de Haas J, Otsen M, Rademaker J, Schouls L, Lenstra JA (1999) Amplified fragment length polymorphism (AFLP™): the state of an art. *J Clin Microbiol* 7, 3083–3091.
6. Vandamme P, Pot B, Gillis M, de Vos P, Kersters K, Swings J (1996) Polyphasic taxonomy, a consensus approach to bacterial systematics. *Microbiol Rev* **60**: 407–438.
7. Rademaker JLW, Louws F J, Rossbach U, Vinuesa P, de Bruijn FJ (1999) Computer-assisted pattern analysis of molecular fingerprints and database construction In: Akkermans ADL, van Elsas JD, de Bruijn FJ (eds) *Molecular Microbial Ecology Manual*, Supplement 4, Chap. 713, pp. 1–33. Kluwer Academic Publishers, Dordrecht.
8. Rademaker JLW, de Bruijn FJ (2004) Computer-assisted pattern analysis of electrophoretic fingerprints and database construction. In: Kowalchuk GA, de Bruijn FJ, Head IM, Akkermans AD, van Elsas JD (eds) *Molecular Microbial Ecology Manual*, 2nd edn, Chap. 175, pp. 1–50 Kluwer Academic Publishers, Dordrecht.
9. Schwartz DC, Cantor CR (1984) Separation of yeast chromosome-sized DNAs by pulsed field gradient gel electrophoresis. *Cell* 37: 67–75.
10. Goering RV (1998) The molecular epidemiology of nosocomial infection: an overview of principles, application, and interpretation. In: Specter S, Bendinelli M, Friedman H (eds) *Rapid Detection of Infectious Agents*, pp. 131–147. Plenum Publishing Co., New York.
11. Goering RV (2004) Pulsed-field gel electrophoresis. In: Pershing DH, Tenover FC, Versalovic J, Tang YW, Unger ER, Relman DA, White TJ (eds) *Diagnostic Molecular Microbiology, Diagnostic Principles and Practice*, pp. 185–196. ASM Press, Washington, DC, USA.
12. Vos P, Hogers R, Bleeker M, Reijans M, Lee van de T, Hornes M, Frijtens A, Pot J, Peleman J, Kuiper M, Zabeau M (1995) AFLP: a new technique for DNA fingerprinting. *Nucleic Acid Res* **23**: 4407–4414.

13. Blears MJ, De Grandis SA, Lee H, Trevors JT (1998) Amplified fragment length polymorphism (AFLP): review of the procedure and its applications. *J Industr Microbiol Biotechnol* **21**: 99–114.
14. Huys G, Coopman R, Janssen P, Kersters K (1996) High-resolution genotypic analysis of the genus *Aeromonas* by AFLP fingerprinting. *Int J Syst Bacteriol* **4**: 572–580.
15. Janssen P, Coopman R, Huys G, Swings J, Bleeker M, Vos P, Zabeau M, Kersters K, (1996) Evaluation of the DNA fingerprinting method AFLP as a new tool in bacterial taxonomy. *Microbiology* **142**: 1881–1893.
16. Lin JJ, Kuo J, Ma J (1996) A PCR-based DNA fingerprinting technique- AFLP for molecular typing of bacteria. *Nucleic Acids Res* **24**: 3649–3650.
17. Majer D, Mitchen R, Lewis BG, Vos P, Oliver RP (1996) The use of AFLP fingerprinting for the detection of genetic variation in fungi. *Mycol Res* **100**: 1107–1111.
18. Van Belkum A, Struelens M, De Visser A, Verbrugh H, Tibayrenc M (2001) Role of genomic typing in taxonomy, evolutionary genetics and microbial epidemiology. *Clin Microbiol Rev* **14**: 547–560.
19. Boumedine KS, Rodolakis A (1998) AFLP allows the identification of genomic markers of ruminant *Chlamydia psittaci* strains useful for typing and epidemiological studies. *Res Microbiol* **149**: 735–744.
20. Meijer A, Morré SA, Van den Brule AJC, Savelkoul PHM, Ossewaarde JM (1999) Genomic relatedness of *Chlamydia* isolates determined by amplified fragment length polymorphism analysis. *J Bacteriol* **181**: 4469–4475.
21. Morré SA, Ossewaarde JM, Savelkoul PHM, Stoof J, Meijer CJLM, Van den Brule AJC (2000) Analysis of heterogeneity in *Chlamydia trachomatis* clinical isolates of serovars D, E, and F by amplified fragment length polymorphism. *J Clin Microbiol* **38**: 3463–3466.
22. Chalhoub BA, Thibault S, Laucou V, Rameau C, Hofte H, Cousin R (1997) Silver staining and recovery of AFLP amplification products on large denaturing polyacrylamide gels. *Biotechniques* **22**: 216–218.
23. Rademaker JLW, Hoste B, Louws FJ, Kersters K, Swings J, Vauterin L, Vauterin P, de Bruijn FJ (2000) Comparison of AFLP and rep-PCR genomic fingerprinting with DNA–DNA homology studies: *Xanthomonas* as a model system. *Int J Syst Evol Microbiol* **50**: 665–677.
24. Jonas D, Meyer HG, Matthes P, Hartung D, Jahn B, Daschner FD, Jansen B (2000) Comparative evaluation of three different genotyping methods for investigation of nosocomial outbreaks of Legionnaires' disease in hospitals. *J Clin Microbiol* **38**: 2284–2291.
25. World Health Organization, Centre for Reference and Research on *Salmonella* (1980) Antigenic formulae of the *Salmonella*. WHO International *Salmonella* Center, Institute Pasteur, Paris.
26. Aarts HJM, Lith van LAJT, Keijer J (1998) High-resolution genotyping of *Salmonella* strains by AFLP-fingerprinting. *Lett Appl Microbiol* **26**: 131–135.
27. Lindstedt BA, Heir E, Vardund T, Kapperud G (2000) A variation of the amplified-fragment length polymorphism (AFLP) technique using three restriction endonucleases, and assessment of the enzyme combination *Bgl*II-*Mfe*I for AFLP analysis of *Salmonella enterica* subsp *enterica* isolates. *FEMS Microbiol Lett* **189**: 19–24.
28. Weis J, Seelinger HPR (1975) Incidence of *Listeria monocytogenes* in nature. *Appl Microbiol* **30**: 29–32.
29. Seelinger HPR, Jones D (1986) *Listeria* In: Sneath PHA, Mair NS, Sharpe ME, Holt JG (eds) *Bergey's Manual of Systematic Bacteriology*, Vol. 2, pp. 1235–1245. Williams and Wilkins, Baltimore, MA.
30. Aarts HJM, Hakemulder LE, Hoef van AMA (1999) Genomic typing of *Listeria*

monocytogenes strains by automated laser fluorescence analysis of AFLP fingerprint patterns. *Int J Food Microbiol* **49**: 95–102.
31. Doignon-Bourcier F, Willems A, Coopman R, Laguerre G, Gillis M, de Lajudie P (2002) Genotypic characterization of *Bradyrhizobium* strains nodulating small Senegalese legumes by 16S-23S rRNA intergenic gene spacers and amplified fragment length polymorphism fingerprint analyses. *Appl Environ Microbiol* **66**: 3987–3997.
32. Avrova AO, Hyman LJ, Toth RL, Toth IK (2002) Application of amplified fragment length polymorphism fingerprinting for taxonomy and identification of the soft rot bacteria *Erwinia carotovora* and *Erwinia chrysanthemi*. *Appl Environ Microbiol* **68**: 1499–1508.
33. Gancheva A, Pot B, Vanhonacker K, Hoste B, Kersters KA (1999) Polyphasic approach towards the identification of strains belonging to *Lactobacillus acidophilus* and related species. *Syst Appl Microbiol* **22**: 573–585.
34. Coenye T, LiPuma JJ, Henry D, Hoste B, Vandemeulebroecke K, Gillis M, Speert DP, Vandamme P (2001) *Burkholderia cepacia* genomovar VI, a new member of the *Burkholderia cepacia* complex isolated from cystic fibrosis patients. *Int J Syst Evol Microbiol* **51**: 271–279.
35. Huys G, Altwegg M, Hanninen ML, Vancanneyt M, Vauterin L, Coopman R, Torck U, Luthy Hottenstein J, Janssen P, Kersters K (1996) Genotypic and chemotaxonomic description of two subgroups in the species *Aeromonas eucrenophila* and their affiliation to *A. encheleia* and *Aeromonas* DNA hybridization group 11. *Syst Appl Microbiol* **19**: 616–623.
36. Janssen P, Maquelin K, Coopman R, Tjernberg I, Bouvet P, Kersters K, Dijkshoorn L (1997) Discrimination of *Acinetobacter* genomic species by AFLP fingerprinting. *Int J Syst Bacteriol* **47**: 1187–1187.
37. Koeleman JGM, Stoof J, Biesmans DJ, Savelkoul PHM, Vandenbroucke-Grauls CMJE (1998) Comparison of ARDRA, RAPD and AFLP fingerprinting for identification of Acinetobacter genomic species and typing of *Acinetobacter baumannii*. *J Clin Microbiol* **36**: 2522–2529.
38. Koeleman JGM, Stoof J, van der Bijl MW, Vandenbroucke-Grauls CMJE, Savelkoul PHM (2001) Identification of epidemic strains of *Acinetobacter baumannii* by integrase gene PCR. *J Clin Microbiol* **39**: 8–13.
39. Spence RP, van der Reijden TJK, Disjkshoorn L, Towner KJ (2004) Comparison of *Acinetobacter baumannii* isolates from United Kingdom hospitals with predominant northern European genotypes by amplified fragment length polymorphism analysis. *J Clin Microbiol* **42**: 832–834.
40. Desai M, Tanna A, Wall R, Efstratiou A, George R, Stanley J (1998) Fluorescent amplified-fragment length polymorphism analysis of an outbreak of group A streptococcal invasive disease. *J Clin Microbiol* **36**: 3133–3137.
41. Vancanneyt M, Lombardi A, Andrighetto C *et al.* (2002) Intraspecies genomic groups in *Enterococcus faecium* and their correlation with origin and pathogenicity. *Appl Environ Microbiol* **68**: 1381–1391.
42. Duim B, Wassenaar TM, Rigter A, Wagenaar JA (1999) High resolution genotyping of *Camplylobacter* strains isolated from poultry and humans with AFLP fingerprinting. *Appl Environm Microbiol* **65**: 2369–2375.
43. Siemer BL, Harrington CS, Nielsen EM, Borck B, Nielsen NL, Engberg J, On SLW (2004) Genetic relatedness among *Campylobacter jejuni* serotyped isolates of diverse origin as determined by numerical analysis of amplified fragment length polymorphism (AFLP) profiles. *J Appl Microbiol* **96**: 795–802.
44. Fry NK, Bangsborg JM, Bergmans A *et al.* (2002) Designation of the European Working Group on *Legionella* infection (EWGLI) amplified fragment length polymorphism types of *Legionella pneumophila* serogroup 1 and results of intercentre proficiency testing using a standard protocol. *Eur J Clin Microbiol Inf Dis* **21**: 722–728.

45. Ticknor LO, Kolsto AB, Hill KK, Keim P, Laker MT, Tonks M, Jackson PJ (2001) Fluorescent amplified fragment length polymorphism analysis of Norwegian *Bacillus cereus* and *Bacillus thuringiensis* soil isolates. *Appl Environ Microbiol* **67**: 4863–4873.
46. Van der Zwet WC, Savelkoul PHM, Stoof J, Vandenbroucke-Grauls CMJE (2000) An outbreak of *Bacillus cereus* infections in a neonatal intensive care unit traced to balloons used in manual ventilation. *J Clin Microbiol* **38**: 4131–4136.
47. Hill KK, Ticknor LO, Okinaka RT *et al.* (2004) Fluorescent amplified fragment length polymorphism analysis of *Bacillus anthracis, Bacillus cereus* and *Bacillus thuringiensis* isolates. *Appl Environ Microbiol* **70**: 1068–1080.
48. McLauchlin J, Ripabelli G, Brett MM, Threlfall EJ (2000) Amplified fragment length polymorphism (AFLP) analysis of *Clostridium perfringens* for epidemiological typing. *Int J Food Microbiol* **56**: 21–28.
49. Abd-Elsalam KA, Schnieder F, Khalil MS, Asran-Amal A, Verreet JA (2003) Use of AFLP fingerprinting to analyze genetic variation within and between populations of *Fusarium* spp derived from Egyptian cotton cultivars. *J Plant Pathol* **85**, 99–103.
50. Martinez SP, Snowdon R, Pons-Kuhnemann J (2004) Variability of Cuban and international populations of *Alternaria solani* from different hosts and localities: AFLP genetic analysis. *Eur J Plant Pathol* **110**: 399–409.
51. Versalovic J, Koeuth T, Lupski JR (1991) Distribution of repetitive DNA sequences in eubacteria and application to fingerprinting of bacterial genomes. *Nucleic Acids Res* **19**: 6823–6831.
52. Lupski JR, Weinstock GM (1992) Short, interspersed repetitive DNA sequences in prokaryotic genome. *J Bacteriol* **174**: 4525–4529.
53. Rademaker JLW (2002) Rep-PCR genomic fingerprint typing. In: van Belkum A, Duim B, Hays J (eds) *Experimental Approaches for Assessing Genetic Diversity among Microbial Pathogens*, pp. 49–70. Dutch Working party on Epidemiological Typing (WET) and the Dutch Society of Microbiology (NVVM), Ponsen en Looijen, Wageningen.
54. Ali S, Muller CR, Epplen JT (1986) DNA fingerprinting by oligonucleotide probes specific for simple repeats. *Hum Genet* **74**: 239–243.
55. Leenders A, van Belkum A, Janssen S, de Marie S, Kluytmans J, Wielenga J, Lowenberg B, Verbrugh H (1996) Molecular epidemiology of apparent outbreak of invasive *Aspergillosis* in a hematology ward. *J Clin Microbiol* **34**: 345–351.
56. Van Belkum A, Quint WGV, de Pauw BE, Melchers WJG, Meis JF (1993) Typing of *Aspergillus fumigatus* isolates by interrepeat polymerase chain reaction. *J Clin Microbiol* **31**: 2502–2505.
57. Versalovic J, Lupski JR (1996) Distinguishing bacterial and fungal pathogens by repetitive sequence-based PCR. *Lab Med Int* **8**: 12–15.
58. Collins MD, East AK (1998) Phylogeny and taxonomy of the food-borne pathogen *Clostridium botulinum* and its neurotoxins. *J Appl Microbiol* **84**: 5–17.
59. Dombek PE, Johnson LK, Zimmerley ST, Sadowsky MJ (2000) Use of repetitive DNA sequences and the PCR to differentiate *Escherichia coli* isolates from human and animal sources. *Appl Environ Microbiol* **66**: 2572–2577.
60. Judd AK, Schneider M, Sadowsky MJ, de Bruijn FJ (1993) Use of repetitive sequences and the polymerase chain reaction technique to classify genetically related *Bradyrhizobium japonicum* serocluster 123 strains. *Appl Environ Microbiol* **59**: 1702–1708.
61. Louws FJ, Fulbright DW, Stephens CT, de Bruijn FJ (1994) Specific genomic fingerprints of phytopathogenic *Xanthomonas* and *Pseudomonas* pathovars and strains generated with repetitive sequences and PCR. *Appl Environ Microbiol* **60**: 2286–2295.
62. Louws FJ, Fulbright DW, Stephens CT, de Bruijn FJ (1995) Differentiation of

genomic structure by rep-PCR fingerprinting to rapidly classify *Xanthomonas campestris* pv *vesicatoria*. *Phytopathology* **85**: 528–536.

63. Louws FJ, Schneider M, de Bruijn FJ (1997) Assessing genetic diversity of microbes using repetitive-sequence-based PCR (rep-PCR). In: Toranzos G (ed.) *Nucleic Acid Amplification Methods for the Analysis of Environmental Samples*, pp. 63–94. Technomic Publishing, Lancaster.
64. Nick G, Lindstrom K (1994) Use of repetitive sequences and the polymerase chain reaction to fingerprint the genomic DNA of *Rhizobium galegae* strains and to identify the DNA obtained by sonicating the liquid cultures and root nodules. *Syst Appl Microbiol* **17**: 265–273.
65. Poh CL, Ramachandran V, Tapsal JW (1996) Genetic diversity of *Neisseria gonorrhoeae* IB-2 and IB-6 isolates revealed by whole-cell repetitive element sequence-based PCR. *J Clin Microbiol* **34**: 292–295.
66. Sadowsky MJ, Kinkel LL, Bowers JH, Schottel JL (1996) Use of repetitive intergenic DNA sequences to classify pathogenic and disease-suppressive *Streptomyces* strains. *Appl Environ Microbiol* **62**: 3489–3493.
67. Snelling AM, Gerner-Smidt P, Hawkey PM, Heritage J, Parnell P, Porter C, Bodenham AR, Inglis T (1996) Validation of Use of whole-cell repetitive extragenic palindromic sequence-based PCR (REP-PCR) for typing strains belonging to the *Acinetobacter calcoaceticus–Acinetobacter baumannii* complex and application of the method to the investigation of a hospital. *J Clin Microbiol* **34**: 1193–1202.
68. Versalovic J, de Bruijn FJ, Lupski JR (1998) Repetitive sequence-based PCR (rep-PCR) DNA fingerprinting of bacterial genomes. In: de Bruijn FJ, Lupski JR, Weinstock GM (eds) *Bacterial Genomes: Physical Structure and Analysis*, pp. 437–456. Chapman & Hall, New York.
69. Woods CR, Versalovic J, Koeuth T, Lupski JR (1993) Whole-cell repetitive element sequence-based polymerase chain reaction allows rapid assessment of clonal relationships of bacterial isolates. *J Clin Microbiol* **31**: 1927–1931.
70. De Bruijn FJ, Schneider M, Rossbach U, Louws FJ (1997) Automated fluorescent and conventional rep-PCR genomic fingerprinting and multiplex PCR to classify bacteria and track genes. In: Martins MT, Zanolisato MI, Tiedje JM, Nortron Hagler LC, Dobereiner J, Sanchez PS (eds) *Proceedings of the 7th International Symposium on Microbial Ecology*, pp. 197–205. Brazilian Society for Microbiology, Santoz.
71. Del Vecchio VG, Petroziello JM, Gress MJ, McCleskey FK, Melcher GP, Crouch HK, Lupski R (1995) Molecular genotyping of methicillin-resistant *Staphylococcus aureus* via fluorophore-enhanced repetitive-sequence PCR. *J Clin Microbiol* **33**: 2141–2144.
72. Versalovic J, Lupski JR (1995) DNA Fingerprinting of *Neisseria* strains by rep-PCR. *Meth Mol Cell Biol* **5**: 96–104.
73. Versalovic J, Kapur V, Koeuth T, Mazurek GH, Whittam TS, Musser JM, Lupski JR (1995) DNA fingerprinting of pathogenic bacteria by fluorophore-enhanced repetitive sequence-based polymerase chain reaction. *Arch Pathol Lab Med* **119**: 23–29.
74. Heirro N, Gonzalez A, Maz A, Guillamon JM (2004) New PCR-based methods for yeast identification. *J Appl Microbiol* **97**: 792–801.
75. Johnson JR, Clabots C, Azar M, Boxrud DJ, Besser JM, Thurn JR (2001) Molecular analysis of a hospital cafeteria-associated salmonellosis outbreak using modified repetitive element PCR fingerprinting. *J Clin Microbiol* **39**: 3452–3460.
76. Cho JC, Tiedje JM (2000) Biogeography and degree of endemicity of fluorescent *Pseudomonas* strains in soil. *Appl Environ Microbiol* **66**: 5448–5456.
77. Dawson SL, Fry JC, Dancer BN (2002) A comparative evaluation of five typing techniques for determining the diversity of fluorescent pseudomonads. *J Microbiol Methods* **50**: 9–22.

78. Guinebretiere MH, Berge O, Normand P, Morris C, Carlin F, Nguyen-The C (2001) Identification of bacteria in pasteurized zucchini purees stored at different temperatures and comparison with those found in other pasteurized vegetable purees. *Appl Environ Microbiol* **67**: 4520–4530.
79. Bouzar H, Jones JB, Stall RE, Louws FJ, Schneider M, Rademaker JLW, de Bruijn FJ, Jackson LE (1999) Multiphasic taxonomy of bacteria causing spot disease on tomato and pepper plants in the Caribbean and Central America reveals distinct predominance of *Xanthomonas axonopodis vesicatoria* strains. *Phytopathology* **89**: 328–335.
80. McDonald JG, Wong E (2001) Use of a monoclonal antibody and genomic fingerprinting by repetitive-sequence-based polymerase chain reaction to identify *Xanthomonas populi* pathovars. *Can J Plant Pathol* **23**: 47–51.
81. Berthier F, Beuvier E, Dasen A, Grappin R (2001) Origin and diversity of mesophilic lactobacilli in Comté cheese, as revealed by PCR with repetitive and species-specific primers. *Int Dairy J* **11**: 293–305.
82. Hahm BK, Maldonado Y, Schreiber E, Bhunia AK, Nakatsu CH (2003) Subtyping of foodborne and environmental isolates of *Escherichia coli* by multiplex-PCR, rep-PCR, PFGE, ribotyping and AFLP. *J Microbiol Meth* **53**, 387–399.
83. Oda Y, Wanders W, Huisman LA, Meijer WG, Gottschal JC, Forney LJ (2002) Genotypic and phenotypic diversity within species of purple nonsulfur bacteria isolated from aquatic sediments. *Appl Environ Microbiol* **68**: 3467–3477.
84. Heyrman J, Balcaen A, De Vos P, Swings J (2002) *Halomonas muralis* sp nov, a new species isolated from microbial biofilms colonising the walls and murals of the Saint-Catherine chapel (castle Herberstein, Austria). *Int J Syst Evol Microbiol* **52**: 2049–2054.
85. Shangkuan YH, Chang YH, Yang JF, Lin HC, Shaio MF (2001) Molecular characterization of *Bacillus anthracis* using multiplex PCR, ERIC-PCR and RAPD. *Lett Appl Microbiol* **32**: 139–145.
86. Sechi LA, Deriu A, Falchi MP, Fadda G, Zanetti S (2002) Distribution of virulence genes in *Aeromonas spp* isolated from Sardinian waters and from patients with diarrhoea. *J Appl Microbiol* **92**: 221–227.
87. Szczuka E, Kaznowski A (2004) Typing of clinical and environmental *Aeromonas* sp strains by random amplified polymorphic DNA, PCR, repetitive extragenic palindromic PCR, and enterobacterial repetitive intergenic consensus sequence PCR. *J Clin Microbiol* **42**: 220–228.
88. Wieringa EB, Overmann J, Cypionka H (2000) Detection of abundant sulphate-reducing bacteria in marine oxic sediment layers by a combined cultivation and molecular approach. *Environ Microbiol* **2**: 417–427.
89. Whiteley AS, Wiles S, Lilley AK, Philp J, Bailey MJ (2001) Ecological and physiological analyses of Pseudomonad species within a phenol remediation system. *J Microbiol Methods* **44**: 79–88.
90. Arora DK, Hirsch PR, Kerry BR (1996) PCR-based molecular discrimination of *Verticillium chlamydosporium* isolates. *Mycological Res* **100**: 801–809.
91. Tyler KD, Wang G, Tyler SD, Johnson WM (1997) Factors affecting reliability and reproducibility of amplification-based DNA fingerprinting of representative bacterial pathogens. *J Clin Microbiol* **35**: 339–346.
92. Vassart G, Georges M, Monsieur R, Brocas H, Lequarre AS, Christophe D (1987) A sequence in M13 phage detects hypervariable minisatellites in human and animal DNA. *Science* **235**: 683–684.
93. Epplen JT (1988) On simple repeated GATCA sequences in animal genomes: a critical reappraisal. *J Hered* **79**: 409–417.
94. Vogel JM, Scolnik PA (1997) Direct amplifications from microsatellites: detection of simple sequence repeat-based polymorphisms without cloning. In: Caetano-Anollés G, Gresshoff PM (eds) *DNA Markers: Protocols, Applications and Overviews*, pp. 133–150. J, Wiley & Sons, New York.

95. Meyer W, Mitchell TG (1995) Polymerase chain reaction fingerprinting in fungi using single primers specific to minisatellites and simple repetitive DNA sequences: strain variation in *Cryptococcus neoformans*. *Electrophoresis* **16**: 1648–1656.
96. Van Belkum A, Scherer S, van Alphen L, Verbrugh H (1998) Short-sequence DNA repeats in prokaryotic genomes. *Microbiol Mol Biol Rev* **62**: 275–293.
97. Welsh JW, McClelland M (1991) Genomic fingerprinting using arbitrary primed PCR and a matrix of pairwise combinations of primers. *Nucleic Acids Res* **19**: 5275–5279.
98. Rafasaki JA (1997) Randomly amplified polymorphic DNA (RAPD) analysis. In: Caetano-Anollés G, Gresshoff PM (eds) *DNA markers: Protocols, Applications and Overviews*, pp. 75–83. J. Wiley & Sons, New York.
99. Williams JGK, Kubelik Livak AR, Rafalski JA, Tingey SV (1990) DNA polymorphisms amplified by arbitrary primers are useful as genetic markers. *Nucleic Acids Res* **18**: 6531–6535.
100. Welsh JW, McClelland M (1990) Fingerprinting genomes using PCR with arbitrary primers. *Nucleic Acids Res* **18**: 7213–7218.
101. Welsh JW, Pretzman CP, Postic DP, Saint Girons I, Baranton G, McClelland M (1992) Genomic fingerprinting by arbitrarily primed polymerase chain reaction resolves *Borrelia burgdorferi* into three distinct phyletic groups. *Int J Syst Bacteriol* **42**: 370–377.
102. Vogt T, Mathieu-Daude F, Kullmann F, Welsh J, McClelland M (1997) Fingerprinting of DNA and RNA using arbitrarily primed PCR. In: Caetano-Anollés G, and Gresshoff PM (eds) *DNA markers: Protocols, Applications and Overviews*, pp. 55–74. J. Wiley & Sons, New York.
103. Caetano-Anollés G, Gresshoff PM (eds) (1997) *DNA Markers: Protocols, Applications and Overviews*, pp. 1–364. J. Wiley & Sons, New York.
104. Micheli MR, Bova R (1997) *Fingerprinting Methods Based on Arbitrarily Primed PCR*, pp. 1–441. Springer-Verlag, New York.
105. Bouton Y, Guyot P, Beuvier E, Tailliez P, Grappin R (2002) Use of PCR-based methods and PFGE for typing and monitoring homofermentative lactobacilli during Comté cheese ripening. *Int J Food Microbiol* **76**: 27–38.
106. Louws FJ, Rademaker JLW, de Bruijn FJ (1999) The three Ds of PCR-based genomic analysis of phytobacteria: diversity, detection, and disease diagnosis. *Annu Rev Phytopath* **37**: 85–125.
107. Amann R, Ludwig W (2000) Ribosomal RNA-targeted nucleic acid probes for studies in microbial ecology. *FEMS Microbiol Rev* **5**: 555–565.
108. Pace NR (1997) A molecular view of microbial diversity and the biosphere. *Science* **276**: 734–740.
109. Woese CR (1987) Bacterial evolution. *Microbiol Rev* **51**: 221–271.
110. Woese CR, Kandler O, Wheelis ML (1990) Towards a natural system of organisms: proposal for the domains Archaea, Bacteria, and Eucarya. *Proc Natl Acad Sci USA* **87**: 4576–4579.
111. Brosius J, Dull TJ, Sleeter DD, Noller HF (1981) Gene organization and primary structure of a ribosomal RNA operon from *Escherichia coli*. *J Mol Biol* **148**: 107–127.
112. Cannone JJ, Subramanian S, Schnare MN *et al.* (2002) The Comparative RNA Web (CRW) Site: an online database of comparative sequence and structure information for ribosomal, intron, and other RNAs. *BMC Bioinformatics* **3**: 2.
113. Maidak BL, Cole JR, Lilburn TG *et al.* (2000) The RDP (Ribosomal Database Project) continues. *Nucleic Acids Res* **28**: 173–174.
114. Lafontaine DL, Tollervey D (2001) The function and synthesis of ribosomes. *Nat Rev Mol Cell Biol* **2**: 514–520.
115. Vaneechoutte M, De Beenhouwer H, Claeys G, Verschraegen G, De Rouck A,

Elainchouni A, Portaels F (1993) Identification of *Mycobacterium* species by using amplified ribosomal DNA restriction analysis. *J Clin Microbiol* **31**: 2061–2065.

116. Heyndrickx M, Vauterin L, Vandamme P, Kersters K, De Vos P (1996) Applicability of combined amplified ribosomal DNA restriction analysis (ARDRA) patterns in bacterial phylogeny and taxonomy. *J Microbiol Methods* **26**: 247–259.
117. Weisburg WG, Barns SM, Pelletie DA, Lane DJ (1991) 16S ribosomal DNA amplification for phylogenetic study. *J Bacteriol* **173**: 697–703.
118. Navarro E, Simonet P, Normand P, Bardin R (1992) Characterization of natural populations of *Nitrobacter* spp using PCR/RFLP analysis of the ribosomal intergenic spacer. *Arch Microbiol* **157**: 107–115.
119. Gürtler V, Stanisich VA (1996) New approaches to typing and identification of bacteria using the 16S-23S rDNA spacer region. *Microbiology* **142**: 3–16.
120. Tilsala-Timisjarvi A, Alatossava T (2001) Characterization of the 16S-23S and 23S-5S rRNA intergenic spacer regions of dairy propionibacteria and their identification with species-specific primers by PCR. *Int J Food Microbiol* **68**: 45–52.
121. Laguerre G, Allard MR, Revoy F, Amarger N (1994) Rapid identification of rhizobia by restriction fragment length polymorphism analysis of PCR-amplified 16S rRNA genes. *Appl Environ Microbiol* **60**: 56–63.
122. Vinuesa P, Rademaker JLW, de Bruijn FJ, Werner D (1998) Genotypic characterization of *Bradyrhizobium* strains nodulating endemic woody legumes of the Canary Islands by 16S rDNA- and 16S-23S rDNA-intergenic spacer-PCR/RFLP analysis, rep-PCR genomic fingerprinting and partial 16S rDNA sequencing. *Appl Environ Microbiol* **64**: 2096–2104.
123. Moyer CL, Tiedje JM, Dobbs FC, Karl DM (1996) A computer-simulated restriction fragment length polymorphism analysis of bacterial small-sub-unit rRNA genes: efficacy of selected tetrameric restriction enzymes for studies of microbial diversity in nature. *Appl Environ Microbiol* **62**: 2501–2507.
124. Vinuesa P, Rademaker JLW, de Bruijn FJ, Werner D (1999) Characterization of *Bradyrhizobium* spp strains by RFLP analysis of amplified 16S rDNA and rDNA intergenic spacer regions. In: Martínez E, Hernández G (eds) *Highlights on Nitrogen Fixation*, pp. 275–279. Plenum Publishing Corporation, New York.
125. Grundmann GL, Neyra M, Normand P (2000) High-resolution phylogenetic analysis of NO($_2^-$)-oxidizing *Nitrobacter* species using the *rrs-rrl* IGS sequence and *rrl* genes. *Int J Syst Evol Microbiol* **50**: 1893–1898.
126. Jensen MA, Webster JA, Straus N (1993) Rapid identification of bacteria on the basis of polymerase chain reaction-amplified ribosomal DNA spacer polymorphisms. *Appl Environ Microbiol* **59**: 945–952.
127. Laguerre G, Mavingui P, Allard MR, Charnay MP, Louvrier P, Mazurier SI, Rigottier-Gois L, Amarger N (1996) Typing of rhizobia by PCR and PCR-restriction fragment length polymorphism analysis of chromosomal and symbiotic gene regions: application to *Rhizobium leguminosarum* and its different biovars. *Appl Environ Microbiol* **62**: 2029–2036.
128. Moreno Y, Ferrus MA, Vanoostende A, Hernandez M, Montes RM, Hernandez J (2002) Comparison of 23S polymerase chain reaction-restriction fragment length polymorphism and amplified fragment length polymorphism techniques as typing systems for thermophilic campylobacters. *FEMS Microbiol Lett* **211**: 97–103.
129. Moyer CL, Dobbs FC, Karl DM (1994) Estimation of diversity and community structure through restriction fragment length polymorphism distribution analysis of bacterial 16S rRNA genes from a microbial mat at an active, hydrothermal vent system. Loihi Seamount, Hawaii. *Appl Environ Microbiol* **60**: 871–879.

130. Scheinert P, Krausse R, Ullman U, Soller R, Krupp G (1996) Molecular differentiation of bacteria by PCR amplification of the 16S-23S rRNA spacer. *J Microbiol Meth* **26**: 103–117.
131. Terefework Z, Nick G, Suomalainen S, Paulin L, Lindstrom K (1998) Phylogeny of *Rhizobium galegae* with respect to other rhizobia and agrobacteria. *Int J Syst Bacteriol* **48**: 349–356.
132. Welsh J, McClelland M (1992) PCR-amplified length polymorphisms in tRNA intergenic spacers for categorizing staphylococci. *Mol Microbiol* **6**, 1673–1680.
133. Aquilanti L, Mannazzu I, Papa R, Cavalca L, Clementi F (2004) Amplified ribosomal DNA restriction analysis for the characterization of Azotobacteraceae: a contribution to the study of these free living nitrogen-fixing bacteria. *J Microbiol Meth* **57**: 197–206.
134. Kurabachew M, Enger O, Sandaa RA, Lemma E, Bjorvatn B (2003) Amplified ribosomal DNA restriction analysis in the differentiation of related species of mycobacteria. *J Microbiol Meth* **55**: 83–90.
135. Vinuesa P, Silva C, Werner D, Martínez-Romero E (2004) Population genetics and phylogenetic inference in bacterial molecular systematics: the roles of migration and recombination in *Bradyrhizobium* species cohesion and delineation. *Mol Phylogenet Evol*, **34**: 29–54.
136. Welsh JW, McClelland M (1991) Genomic fingerprints produced by PCR with consensus tRNA gene primers. *Nucleic Acids Res* **19**: 861–866.
137. Cherif A, Borin S, Rizzi A, Ouzari H, Bourdabous A, Daffonchio D (2003) *Bacillus anthracis* diverges from related clades of the *Bacillus cereus* group in 16S-23S ribosomal DNA intergenic transcribed spacers containing tRNA genes. *Appl Environ Microbiol* **69**: 33–40.
138. Wiedman-Al-Ahmad M, Tichy HV, Schon G (1994) Characterization of *Acinetobacter* type strains and isolates obtained from wastewater treatment plants by PCR fingerprinting. *Appl Environ Microbiol* **60**: 4066–4071.
139. Baele M, Vaneechoutte M, Verhelst R, Vancanneyt M, Devriese LA, Haesebrouck F (2002) Identification of *Lactobacillus* species using tDNA-PCR. *J Microbiol Methods* **50**: 263–271.
140. Sawada H, Kuykendall LD, Young JM (2003) Changing concepts in the systematics of bacterial nitrogen-fixing legume symbionts. *J Gen Appl Microbiol* **49**: 155–179.
141. Vinuesa P, Silva C (2004) Species delineation and biogeography of symbiotic bacteria associated with cultivated and wild legumes. In: Werner D (ed.), ***Biological Resources and Migration***, pp. 143–155. Springer-Verlag, Berlin.
142. Fox GE, Wisotzkey JD, Jurtshuk Jr P (1992) How close is close: 16S rRNA sequence identity may not be sufficient to guarantee species identity. *Int J Syst Bacteriol* **42**: 166–170.
143. Van Berkum P, Fuhrmann JJ (2000) Evolutionary relationships among the soybean bradyrhizobia reconstructed from 16S rRNA gene and internally transcribed spacer region sequence divergence. *Int J Syst Evol Microbiol* **50**: 2165–2172.
144. Willems A, Coopman R, Gillis M (2001) Comparison of sequence analysis of 16S-23S rDNA spacer regions, AFLP analysis and DNA–DNA hybridizations in *Bradyrhizobium. Int J Syst Evol Microbiol* **51**: 623–632.
145. Vinuesa P, León-Barrios M, Silva C, Willems A, Jarabo-Lorenzo A, Pérez-Galdona R, Werner D, Martínez Romero E (2004) *Bradyrhizobium canariense* sp nov, an acid-tolerant endosymbiont that nodulates endemic genistoid legumes (Papilionoideae: Genisteae) growing in the Canary Islands, along with *B. japonicum* bv *genistearum*, *Bradyrhizobium* genospecies α and *Bradyrhizobium* genospecies β. *Int J Syst Evol Microbiol*, **55**: 569–575.
146. Garcia-Martinez J, Silvia A, Acinas G, Anton AI, Rodriguez-Valera F (1999) Use

of the 16S-23S ribosomal genes spacer region in studies of prokaryotic diversity. *J Microbiol Meth* **36**: 55–64.
147. Gürtler V, Mayal BC (2001) Genomic approaches to typing, taxonomy and evolution of bacterial isolates. *Int J Sys Evol Microbiol* **51**: 3–16.
148. Kostman JR, Alden MB, Mair M, Edlind TD, LiPuma JJ, Stull TL (1995) A universal approach to bacterial molecular epidemiology of by polymerase chain reaction ribotyping. *J Inf Dis* **171**: 204–208.
149. Kostman JR, Edlind TD, LiPuma JJ, Stull TL (1992) Molecular epidemiology of *Pseudomonas cepacia* determined by polymerase chain reaction ribotyping. *J Clin Microbiol* **30**: 2084–2087.
150. Nagpal ML, Fox KF, Fox A (1998) Utility of 16S-23S rRNA spacer region methodology: how similar are interspace regions within a genome and between strains for closely related organisms? *J Microbiol Meth* **33**: 211–219.
151. Tan Z, Hurek T, Vinuesa P, Muller P, Ladha JK, Reinhold-Hurek B (2001) Specific detection of *Bradyrhizobium* and *Rhizobium* strains colonizing rice (*Oryza sativa*) roots by 16S-23S ribosomal DNA intergenic spacer-targeted PCR. *Appl Environ Microbiol* **67**: 3655–3664.
152. Gürtler V, Rao Y, Pearson SR, Bates SM, Mayall BC (1999) DNA sequence heterogeneity in the three copies of the long 16S-23S rDNA spacer of *Enterococcus faecalis* isolates. *Microbiology* **145**: 1785–1796.
153. Yap WH, Zhang Z, Wang Y (1999) Distinct types of rRNA operons exist in the genome of the actinomycete *Thermonospora chromogena* and evidence for horizontal transfer of an entire rRNA operon. *J Bacteriol* **181**: 5201–5209.
154. Boyer SL, Flechtner VR, Johansen JR (2001) Is the 16S-23S rRNA internal transcribed spacer region a good tool for use in molecular systematics and population genetics? A case study in cyanobacteria. *Mol Biol Evol* **18**: 1057–1069.
155. Klappenbach JA, Saxman PR, Cole JR, Schmidt TM (2001) rrndb: the ribosomal RNA operon copy number database. *Nucleic Acids Res* **29**: 181–184.
156. Haukka K, Lindström K, Young JPW (1996) Diversity of partial 16S rRNA sequences among and within strains of African rhizobia isolated from *Acacia* and *Prosopis*. *Syst Appl Microbiol* **19**: 352–359.
157. Simons MP, Ochoterena H (2000) Gaps as characters in sequence-based phylogenetic analyses. *Syst Biol* **49**: 369–381.
158. Yang Z (1996) Among-site rate variation and its impact on phylogenetic analyses. *Trends Ecol Evol* **11**: 367–372.
159. Avaniss-Aghajani E, Jones K, Chapman D, Brunk C (1994) A molecular technique for identification of bacteria using small subunit ribosomal RNA sequences. *Biotechniques* **17**: 144–149.
160. Liu WT, Marsh TL, Cheng H, Forney LJ (1997) Characterization of microbial diversity by determining terminal restriction fragment length polymorphisms of genes encoding 16S rRNA. *Appl Environ Microbiol* **63**: 4516–4522.
161. Orita M, Iwahana H, Kanazawa H, Hayashi K, Sekyia T (1989) Detection of polymorphisms of human DNA by gel electrophoresis as single-strand conformation polymorphisms. *Proc Nat Acad Sci USA* **86**: 2766–2770.
162. Hayashi K (1991) PCR-SSCP: a simple and sensitive method for detection of mutations in the genomic DNA. *PCR Methods Appl* **1**: 34–38.
163. Widjojoatmodjo MN, Fluit A, Verhoef J (1994) Rapid identification of bacteria by PCR-single stranded conformation polymorphism. *J Clin Microbiol* **32**: 3002–3007.
164. Widjojoatmodjo MN, Fluit AC, Verhoef J (1995) Molecular identification of bacteria by fluorescence-based PCR-single-strand conformation polymorphism analysis of the 16S rRNA gene. *J Clin Microbiol* **33**: 2601–2606.
165. Muyzer G (1999) DGGE/TGGE, a method for identifying genes from natural ecosystems. *Curr Opin Microbiol* **2**: 317–322.

166. Moore JE, Millar BC, Jiru X, McCappin J, Crowe M, Elborn JS (2001) Rapid characterization of the genomovars of the *Burkholderia cepacia* complex by PCR-single-stranded conformational polymorphism (PCR-SSCP) analysis. *J Hosp Infect* **48**: 129–134.
167. Rademaker JLW, Driehuis F, te Giffel MC, van Wikselaar PG, Becker PM (2002) Measurement of microbial population dynamics of silages by T-RFLP DNA fingerprints. In: *Proceedings of the XIIIth International Silage Conference*, September 11–13, 2002, The Scottish Agricultural College (SAC), Auchincruive, Scotland, pp. 376–377.
168. Rademaker JLW, Peinhopf M, Rijnen L, Bockelmann W, Noordman WH (2005) The surface microflora dynamics of bacterial smear-ripened Tilsit cheese determined by T-RFLP DNA population fingerprint analysis. *Int Dairy J*, **15**: 785–794.
169. Bruce KD (1997) Analysis of *mer* gene subclasses within bacterial communities in soils and sediments resolved by fluorescent-PCR-restriction fragment length polymorphism profiling. *Appl Environ Microbiol* **63**: 4914–4919.
170. Horz HP, Rotthauwe JH, Lukow T, Liesack W (2000) Identification of major subgroups of ammonia-oxidizing bacteria in environmental samples by T-RFLP analysis of *amoA* PCR products. *J Microbiol Methods* **39**: 197–204.
171. Braker G, Ayala-Del-Rio HL, Devol AH, Fesefeldt A, Tiedje JM (2001) Community structure of denitrifiers, bacteria, and archaea along redox gradients in pacific northwest marine sediments by terminal restriction fragment length polymorphism analysis of amplified nitrite reductase (*nirS*) and 16S rRNA genes. *Appl Environ Microbiol* **67**: 1893–1901.
172. Lueders T, Chin KJ, Conrad R, Friedrich M (2001) Molecular analyses of methyl-coenzyme M reductase alpha-subunit (*mcrA*) genes in rice field soil and enrichment cultures reveal the methanogenic phenotype of a novel archaeal lineage. *Environ Microbiol* **3**: 194–204.
173. Leuders T, Friedrich MW (2003) Evaluation of PCR amplification bias by terminal restriction fragment length polymorphism analysis of small-subunit rRNA and *mcrA* genes by using defined template mixtures of methanogenic pure cultures and soil DNA extracts. *Appl Environ Microbiol* **69**: 320–326.
174. Vandamme P, Mahenthiralingam E, Holmes BM, Coenye T, Hoste B, De Vos P, Henry D, Speert DP (2000) Identification and population structure of *Burkholderia stabilis* sp. nov. (formerly *Burkholderia cepacia* Genomovar IV). *J Clin Microbiol* **38**: 1042–1047.
175. Miele A, Bandera M, Goldstein BP (1995) Use of primers selective for vancomycin resistance genes to determine van genotype in enterococci and to study gene organization in VanA isolates. *Antimicrob Agents Chemother* **39**: 1772–1778.
176. Salisbury SM, Sabatini LM, Spiegel CA (1997) Identification of methicillin-resistant staphylococci by multiplex polymerase chain reaction assay. *Am J Clin Pathol* **107**: 368–373.
177. Maiden MCJ, Bygraves JA, Feil E *et al.* (1998) Multilocus sequence typing: a portable approach to the identification of clones within populations of pathogenic microorganisms. *Proc Natl Acad Sci USA* **95**: 3140–3145.
178. Urwin R, Maiden MC (2003) Multi-locus sequence typing: a tool for global epidemiology. *Trends Microbiol* **11**: 479–487.
179. Cooper JE, Feil EJ (2004) Multilocus sequence typing—what is resolved? *Trends Microbiol* **12**: 373–377.
180. Smith JM, Feil EJ, Smith NH (2000) Population structure and evolutionary dynamics of pathogenic bacteria. *BioEssays* **22**: 1115–1122.
181. Chan MS, Maiden MC, Spratt BG (2001) Database-driven multi locus sequence typing (MLST) of bacterial pathogens. *Bioinformatics* **17**: 1077–1083.
182. Kotetishvili M, Stine OC, Kreger A, Morris JG Jr, Sulakvelidze A (2002)

Multilocus sequence typing for characterization of clinical and environmental salmonella strains. *J Clin Microbiol* **40**: 1626–1635.

183. Homan WL, Tribe D, Poznanski S, Li M, Hogg G, Spalburg E, Van Embden JD, Willems RJ (2002) Multilocus sequence typing scheme for *Enterococcus faecium*. *J Clin Microbiol* **40**: 1963–1971.
184. Hookey JV, Arnold CA (2001) Comparison of multilocus sequence typing and fluorescent fragment-length polymorphism analysis genotyping of clone complex and other strains of *Neisseria meningitidis*. *J Med Microbiol* **50**: 991–995.
185. Bougnoux ME, Morand S, d'Enfert C (2002) Usefulness of multilocus sequence typing for characterization of clinical isolates of *Candida albicans*. *J Clin Microbiol* **40**: 1290–1297.
186. Pearson K (1926) On the coefficient of racial likeness. *Biometrika* **18**: 105–117.
187. Häne BG, Jäger K, Drexler H (1993) The Pearson product-moment correlation coefficient is better suited for identification of DNA fingerprinting profiles than band matching algorithms. *Electrophoresis* **14**: 967–972.
188. Duim B, Wassenaar TM, Rigter A, Wagenaar JA (1999) High resolution genotyping of *Camplylobacter* strains isolated from poultry and humans with AFLP fingerprinting. *Appl Environm Microbiol* **65**: 2369–2375.
189. De Bruijn FJ (1992) Use of repetitive (repetitive extragenic palindromic and enterobacterial repetitive intergenic consensus) sequences and the polymerase chain reaction to fingerprint the genomes of *Rhizobium meliloti* isolates and other soil bacteria. *Appl Environ Microbiol* **58**: 2180–2187.
190. De Bruijn FJ, Rademaker JLW, Schneider M, Rossbach U, Louws FJ (1996) Rep-PCR genomic fingerprinting of plant-associated bacteria and computer-assisted phylogenetic analyses. In: Stacey G, Mullin B, Gresshoff PM (eds) *Proceedings of the 8th International Congress of Molecular Plant-Microbe Interactions*, pp. 497–502. APS Press, St Paul, MN.
191. Eckhardt T (1978) A rapid method for the identification of plasmid deoxyribonucleic acid in bacteria. *Plasmid* **1**: 584–588.
192. Aarts HJM, Keijer J (1999) Genomic fingerprinting of micro-organisms by automatic laser fluorescence analysis (AFLP™). In: Akkermans ADL, van Elsas JD, de Bruijn FJ (eds) *Microbial Ecology Manual*, Chap. 349, pp. 1-10. Kluwer Academic Publishers, Dordrecht.
193. Boom R, Sol CJA, Salimans MMM, Jansen CL, Wertheim-van Dillen PME, Noordaa van der J (1990) Rapid and simple method for purification of nucleic acids. *J Clin Microbiol* **28**: 495–503.
194. Dice LR (1945) Measures of the amount of ecological association between species. *J Ecol* **26**: 297–302.
195. Sokal RR, Michener CD (1958) A statistical method for evaluating systematic relationships. *Univ Kansas Sci Bull* **38**: 1409–1438.
196. Saitou N, Nei M (1987) The neighbor-joining method: a new method for reconstructing phylogenetic trees. *Mol Biol Evol* **4**: 406–425.

Protocol 4.1: AFLP analysis

AFLP analysis was modified from (192).

MATERIALS

Reagents

DNA isolation reagents: Commercial DNA isolation kit suitable for bacteria such as Wizard genomic purification (Promega) or see (1,193)

Digestion and ligation reagents:

5 × RL⁺ buffer [50 mM Tris-HCl (pH 7.5), 50 mM MgAc, 250 mM KAc, 25 mM DTT and 250 ng μl^{-1} BSA)

Restriction enzymes such as *Eco*RI, *Mse*I, *Hin*dIII, *Taq*I (store at –20°C)

T4 DNA ligase (1 unit μl^{-1}: Invitrogen), store at –20°C

Adapters (see *Table 4.2*): IC adapter 5 pmol μl^{-1}, FC adapter 50 pmol μl^{-1} (where IC = infrequent cutter, FC = frequent cutter)

0.1 M ATP

Sterile double-distilled or milliQ water (sH_2O)

PCR amplification reagents: PCR primers; Cy5™-labeled IC primer (see Table 4.2), FC primer

dNTP mix (dATP, dCTP, dGTP, dUTP; 5 mM each), store at –20°C

Table 4.2 Nucleotide sequence of *Eco*RI, *Mse*I, *Hin*dIII and *Taq*I AFLP adapters and primers, modified from (192)

	Adapter	Primer-core sequence[a]
*Eco*RI (IC)	5′-CTCGTAGACTGCGTACC-3′ 3′-CTGACGCATGGTTAA-5′	5′-GACTGCGTACCAATTCX-3′
*Mse*I (FC)	5′-GACGATGAGTCCTGAG-3′ 3′-TACTCAGGACTCAT-5′	5′-GATGAGTCCTGAGTAAX-3′
*Hin*dIII (IC)	5′-CTCGTAGACTGCGTACC-3′ 3′-CTGACGCATGGTCGA-5′	5′-GACTGCGTACCAGCTTX-3′
*Taq*I (FC)	5′-GACGATGAGTCCTGAC-3′ 3′-TACTCAGGACTGGC-5′	5′-CGATGAGTCCTGACCGAX-3′

[a]X = 3′-extension of 0, 1 or 2 nucleotides.

Taq DNA polymerase, store at –20°C

PCR buffer, provided by the supplier of *Taq* DNA polymerase

Polyacrylamide gel-electrophoresis reagents:

6% ReadyMix denaturing polyacrylamide gel (Amersham Pharmacia Biotech, Roosendaal, The Netherlands)

0.6 × TBE (0.06 M Tris-base, 0.05 M Boric Acid, 0.6 mM EDTA)

Loading buffer (100% dionized formamide and 5 mg ml^{-1} Dextran Blue 2000)

TEMED (NNN'N'-tetramethylethylenediamine)

10% (w/v) APS (ammonium persulphate, freshly prepared with distilled water)

Equipment

Thermal cycler

ALF (Automatic Laser Fluorescence)-express sequencer (Amersham Pharmacia Biotech, Roosendaal, The Netherlands)

Software

Appropriate software to analyze and convert .alx files

METHODS

Isolation of bacterial DNA

To obtain unambiguous AFLP™ patterns, the strains under investigation should be pure. DNA is generally isolated from overnight cultures and several commercial DNA isolation kits are available. An often-used alternative is a protocol described by Boom *et al.* (193). This protocol involves the lysis of the bacterial cells by guanidine isothiocyanate and the binding of the released DNA to silica particles (or glass-milk).

AFLP™ using two restriction enzymes

Restriction and ligation:

1. Digest 100 ng of bacterial DNA in a total reaction volume of 20 µl containing:

   4 units of an infrequent cutting (IC) restriction enzyme

4 units of a frequent cutting (FC) restriction enzyme

4 µl 5 × RL⁺ buffer

2. Incubate 37°C for 1 h
3. Add ligation mix:

Ligation mix: total volume 5 µl, including:

0.5 µl IC-adapter* (5 pmol μl^{-1})

0.5 µl FC-adapter* (50 pmol μl^{-1})

0.05 µl 0.1 M ATP

1.0 µl 5 × RL^+ buffer

0.5 µl T4 DNA ligase

2.5 µl sH_2O

*Adapters can be prepared by mixing the partially complementary top and bottom oligonucleotides in equimolar amounts and heating the mixture to 94°C and slowly cooling at ~1°C per minute to room temperature. Adapters can be prepared in advance for a number of reactions.

4. Incubate overnight at 4°C, followed by 70°C for 5 min, or alternatively 37°C for 3 h followed by 70°C for 5 min.
5. Dilute the restriction/ligation mix 10-fold with sH_2O.

Selective amplification of restriction-adapter fragments:

1. Prepare Mix A, this mix contains per reaction:

5 ng Cy5™-labeled IC primer

50 ng FC-primer

0.8 µl 5 mM each of dNTPs

made up to a total volume of 5 µl with sH_2O.

2. Prepare Mix B by combining, per reaction:

2 µl 10 × standard PCR buffer

0.4 units *Taq* polymerase

made up to a total volume of 10 µl with sH_2O.

3. Assemble amplification reaction mixture by mixing

5 µl diluted restriction/ligation mix

5 µl mix A

10 µl mix B

4. PCR amplification is carried out using a PTC200 Thermal cycler (MJ Research, Watertown, USA) or a comparable apparatus by using the following protocol. Incubate for 10 cycles of 94°C for 60 s, 65°C for 60 s (including lowering the annealing temperature by –1°C per cycle) and 72°C for 90 s, followed by 22 cycles of 94°C for 30 s, 56°C for 30 s and 72°C for 60 s.

Preparation of polyacrylamide gel and electrophoresis:

1. Prepare a 6% ReadyMix denaturing polyacrylamide gel according to the recommendations of the supplier.
2. Add an equal amount (20 μl) of loading buffer to the reaction mixtures.
3. Denature the DNA by incubating the samples for 95°C for 3 min followed by immediate transfer to ice.
4. Load 4 μl of each sample onto the gel.
5. Perform the electrophoresis for 5.5 h at 60 mA, 1500 V, 25 W, 55°C using 0.6 × TBE.

Data processing:

1. Convert .alx files to TIFF files and import the file into an appropriate software package for further analysis and comparison.

Protocol 4.2: PCR-RFLP analysis

MATERIALS

Reagents

PCR reagents:

PCR buffer (10 ×; commercial source; note, some companies provide 10 × PCR buffer containing 15 mM $MgCl_2$)

25 mM $MgCl_2$

DMSO (dimethyl sulfoxide; keep at room temperature protected from light; caution, it is toxic)

dNTP mix (dATP, dCTP, dGTP and dTTP, 10 mM each), store at –20°C

PCR primers: prepare 10 μM working solutions and store at –20°C

PCR primers for nearly full-length 16S rDNA amplification (127):

fD1: 5′-ccgaattcgtcgacaacAGAGTTTGATCCT-GGCTCAG-3′ and

rD1: 5′-cccgggatccaagcttAAGGAGGTGATCC-AGCC-3′

The lower case letters correspond to sequences that tag PCR products with terminal restriction sites for cloning purposes. These regions can be omitted from the oligonucleotides if not required.

PCR primers for the amplification of partial 23S rDNA regions (131):

P3: 5′-CCGTGCGGGAAAGGTCAAAAGTACC-3′ and

P4: 5′-CCCGCTTAGATGCTTTCAGC-3′

Primers for the 16S-23S rDNA intergenic spacer region amplification (118):

FGPS1490: 5′-TGCGGCTGGATCACCTCCTT-3′ and

FGPL132′: 5′-CCGGGTTTCCCCATTCGG-3′

Taq DNA polymerase

Sterile double-distilled or milliQ water (sH_2O)

Mineral oil or paraffin

Agarose electrophoresis reagents:

Standard agarose, and MetaPhor or other high resolution agarose (for high resolution of RFLP patterns; Sigma-Aldrich)

TBE electrophoresis buffer (1 litre of 5 × stock solution contains 54 g of Tris base, 27.5 g of boric acid, and 20 ml of 0.5 M EDTA; pH 8.0).

100 bp molecular weight ladder (Gibco BRL)

10 × DNA Loading buffer [0.25% Bromophenol Blue, 25% Ficoll (type 400) in H_2O]

Restriction analysis:

10 × restriction buffer (commercial source)

Tetrameric or pentameric restriction enzymes such as *Cfo*I, *Dde*I, *Sau*3AI, *Msp*I, *Hin*fI, *Hae*III, (store at –20°C)

Equipment

Thermal cycler

Horizontal (submarine) agarose electrophoresis apparatus

Electrophoresis power supply

METHODS

Polymerase chain reaction (PCR)

1. Prepare a PCR mastermix by adding the following reagents for each 50 µl reaction, and scale up according to your sample number:

sH_2O (double distilled or MilliQ water)	37.25 µl
10 × PCR buffer (with 15 mM $MgCl_2$)	5 µl
DMSO*	2.5 µl
dNTPs mix (10 mM each)	1 µl
sense primer (10 pmol μl^{-1})	1.5 µl
antisense primer (10 pmol μl^{-1})	1.5 µl
Taq DNA polymerase (10 IU μl^{-1})	0.25 µl

*DMSO is toxic. This reagent can be replaced by sH_2O, but may aid for efficient amplification of G + C-rich DNA templates.

Dispense 49 µl into each reaction vial and add 1 µl of template DNA (10–50 ng μl^{-1})

When a thermal cycler without heated lid is used, overlay the reaction mixtures with 2 drops of mineral oil or paraffin.

2a. Incubate 95°C for 3 min, followed by 30 cycles of 94°C for 45 s, 55°C for 45 s and 72°C for 2 min, with a final extension for 72°C for 5 min for 16S and 23S rDNA amplifications, using primer pairs fD1/rD1 and P3/P4, respectively.

2b. Incubate 95°C for 3 min, followed by 30 cycles of 94°C for 45 s, 55°C for 40 s and 72°C for 1 min, with a final extension of 72°C for 5 min for IGS rDNA amplifications, using primers FGPS1490 and FGPL123′.

Endonuclease digestion of PCR products and electrophoresis

1. The successful PCR amplifications of the different *rrn* markers are routinely checked by loading 5 µl of each reaction onto 1% TAE agarose gels.
2. Typically 8–10 µl of the PCR products are digested with 8–10 IU of a restriction enzyme in 20–30 µl reaction mixtures, for ≥37°C for 2 h. Previous PCR product purification is not required. We routinely use *Cfo*I, *Dde*I, *Sau*3AI and *Msp*I to digest 16S rDNA amplicons, *Dde*I, *Hae*III and *Msp*I, for the IGS region, and *Hae*III, *Hin*fI and *Msp*I for the 23S rDNA amplification products.
3. The resulting restriction fragments are resolved by electrophoresis using ~10–12 cm-long 2% MetaPhor agarose gels in TBE at 8 V/cm. A 100 bp ladder is applied at both sides and in the central lane of each gel, for accurate normalization of profiles on multiple gels. The gels are stained after electrophoresis using ethidium bromide and documented as described in the next section.

Computer-assisted RFLP pattern analysis

Cluster analysis of the RFLP patterns is performed with specialized commercial software such as the GelCompar or BioNumerics software packages (see www.applied-maths.com). The basic steps to perform the analysis are as follows [for detailed descriptions see (1,7,53)].

1. Ethidium bromide stained gel images are digitized with a charge-coupled device (CCD) video camera and stored to disk as TIFF files. Alternatively, Polaroid photographs of the gels can be scanned with a flatbed or laser scanner.
2. The TIFF files are imported into the software package for track definition and labeling, and database construction.
3. The raw gel image files are then normalized using the molecular size markers included in each gel with respect to the reference positions previously defined for each database.
4. Subsequently, the cluster analysis can be performed. The analysis of fingerprint patterns requires a mathematical simplification of the original data via the generation of a proximity matrix based on (dis)similarity criteria. Such proximity matrices can be established by a wide array of (dis)similarity coefficients. For the analysis of RFLP patterns, a band-matching algorithm is used, defining manually (or automatically) the band positions and choosing the Dice (194) coefficient of similarity for the calculation of pairwise similarity matrices. Clustering of the similarity matrices is performed using the unweighted pair group method using arithmetic averages (UPGMA) (195) or neighbor-joining algorithms (196).

RT-PCR and mRNA expression analysis of functional genes

5

Balbina Nogales

5.1 Introduction

Microbial communities are responsible for processes that are essential for the functioning of ecosystems. The activity of a microbial community as a whole is regulated by environmental factors and depends on the activities of their individual members and populations and interactions amongst them. Microbial communities are thus complex, metabolically flexible and highly adaptable to changing environmental conditions, with their function finely regulated at the molecular level. The complexity of communities makes the analysis of microbial processes in the environment an important challenge for microbial ecologists. In addition to classical ecophysiological approaches that rely on the detection of products resulting from a particular process and the measurement of transformation rates (1), we now possess sophisticated and powerful molecular biology techniques to analyze microbial function in the environment (2,3), such as the analysis of gene expression via detection of mRNA after reverse transcription–polymerase chain reaction (RT-PCR). Since prokaryotic gene expression is a finely regulated process (4,5), detection of transcripts for a given gene constitutes significant evidence of the occurrence of a given biological process within the environment. Recent years have seen a significant increase in the number of studies reporting the analysis of microbial gene expression by RT-PCR in environmental systems. The majority of these studies can be subdivided into three main groups with respect to the type of genetic systems being analyzed:

(i) analysis of gene expression in pathogenic bacteria, e.g. the human pathogens *Staphylococcus aureus* and *Helicobacter pylori* (6,7);
(ii) detection and analysis of expression of genes involved in relevant biogeochemical processes such as methanotrophy, nitrogen fixation, nitrification, denitrification and carbon fixation (8–11);
(iii) investigation of the expression of genes involved in the biodegradation of environmental pollutants, such as aromatic hydrocarbons (12–14).

5.2 Advantages and limitations of the RT-PCR analysis of bacterial functional genes

Several environmentally relevant questions can be approached by using RT-PCR-based techniques as shown in *Figure 5.1*. RT-PCR can be used to detect transcription of a gene of interest and to determine the diversity of the transcripts being expressed in the environment (10,14). The individual microorganisms in which specific gene expression is occurring can be identified by *in situ* RT-PCR techniques, and these organisms can be enumerated and their spatial distribution determined (15). Thirdly, the effect of environmental parameters in gene expression can be explored by quantitative RT-PCR and by using techniques for global gene expression analysis such as RNA fingerprinting by arbitrarily primed PCR (RAP-PCR) (16) or differential display (DD) (17), which allow for the analysis of modulation of gene expression in response to changing environmental conditions. Moreover, RT-PCR methods are not intrusive (no incubation of samples, nor addition of substrates required), although there is potential for the combination of RT-PCR of mRNA genes with stable isotope probing (see Chapter 7). RT-PCR-based analysis of mRNA is also culture independent, sensitive, specific, rapid, reproducible and can be adapted to high-throughput systems when the analysis of many samples is required.

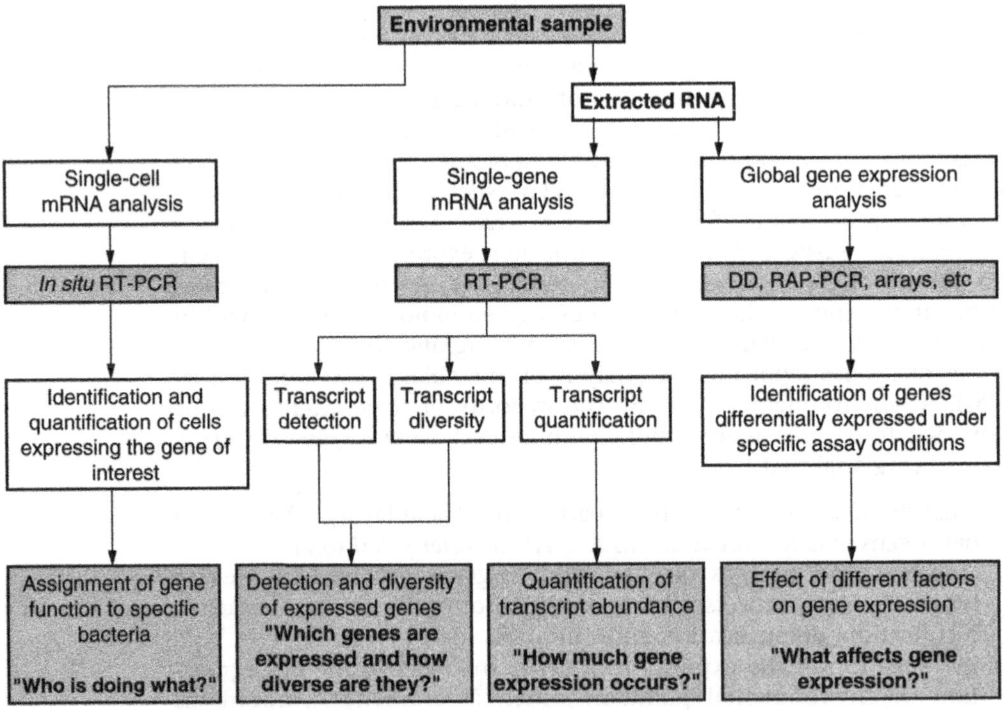

Figure 5.1

Schematic view of the different RT-PCR approaches that can be applied to gene expression analysis in environmental studies.

It must nevertheless be recognized that the analysis of gene transcription by RT-PCR is subject to a number of important methodological limitations. Firstly, prior knowledge of the sequence of genes of interest is a prerequisite of such analysis to enable the design of primers to allow amplification of specific RT-PCR products. Additionally, an exhaustive laboratory analysis of the pattern of expression of the gene of interest, especially in relation to environmentally relevant factors, and if possible using a variety of microorganisms, needs to be undertaken to allow interpretation of the RT-PCR data in an ecological or environmental context. Some of the limitations of RT-PCR techniques are more methodological, such as the quantity, quality and stability of the RNA extract to be used in the reaction. Although the amount of RNA can be low in some cases, RT-PCR amplification from environmental samples would be most probably limited by difficulties encountered during the RNA extraction and by the presence of inhibitory substances in the extracts which will interfere with the RT-PCR reaction. Finally, since RT-PCR is highly sensitive, exhaustive controls to ensure that amplification products are not derived from contaminating DNA need to be performed with every RNA extract used in RT-PCR reactions (18). Typically, RNA extracts are treated with RNase-free DNase and used in PCR reactions without a prior RT reaction (no-RT control).

5.3 The RT-PCR reaction

RT-PCR (19) consist of two sequential steps, namely a reverse transcription reaction (RT) in which a complementary DNA molecule (cDNA) is generated from an RNA template by extension of an oligonucleotide primer due to the action of a reverse transcriptase, and a subsequent PCR amplification reaction in which the cDNA is amplified exponentially by a thermostable DNA polymerase. Both reactions can be performed separately or sequentially in the same tube (see below). Most suppliers of molecular biology products maintain excellent web pages with useful technical information on RT-PCR procedures (*Table 5.1*).

5.3.1 The RNA template

In RT-PCR procedures it is important to work with high quality RNA, i.e. not degraded and free from DNA and ribonuclease contamination, or substances inhibitory to enzymatic reactions. This is critical because prokaryotic mRNA is a labile molecule with a short half-life (4). Therefore, considerable care is required during sample handling to avoid RNA degradation from the time of initial sampling to the isolation of RNA and the subsequent preparation of RT-PCR reactions. RT-PCR amplification of prokaryotic mRNA is usually performed using total RNA extracts that also contain the more abundant RNA fractions: ribosomal (rRNA) and transfer (tRNA) RNA. Prokaryotic mRNAs cannot be reliably enriched by oligo (dT) selection, as is routinely used to isolate eukaryotic mRNA, because although a fraction of mRNA molecules is polyadenylated, poly(A) tails in prokaryotic mRNA are neither as long nor as stable as those from eukaryotes (20).

Table 5.1 Web site details for suppliers of molecular biology products used in mRNA and RT-PCR applications

Company	URL
Ambion Inc., Austin, TX, USA	http://www.ambion.com
Roche Applied Science, F. Hoffmann-LaRoche, Ltd, Switzerland	http://www.roche-applied-science.com
Qiagen GmbH, Hilden, Germany	http://www.qiagen.com
Invitrogen, Carlsbad, CA, USA	http://www.invitrogen.com
Applied Biosystems, Foster City, CA, USA	http://www.appliedbiosystems.com
Promega Corporation, Madison, WI, USA	http://www.promega.com
Stratagene, La Jolla, CA, USA	http://www.stratagene.com

5.3.2 Primers for reverse transcription

Three types of primers can be used in RT reactions:

(i) specific primers that anneal exclusively to the mRNA of interest;
(ii) random hexanucleotides that anneal randomly to any RNA molecule (including rRNA) present in the extract;
(iii) oligo (dT) primers that bind to poly(A) tails at the 3'-end of mRNAs.

However, oligo (dT) primers are seldom used for prokaryotic mRNA amplification and they seem to generate a limited representation of total transcripts (21). The choice between random hexamers or specific primers depends both on the particular research application as well as personal preference. Ideally the objective is to obtain a high proportion of cDNAs complementary to the target RNA and with the maximum possible length. Therefore the preferred choice would in most cases be the use of specific primers for the RT reaction and this has been the method of choice in most studies reporting the amplification of bacterial mRNA. The use of random hexameric primers leads to the transcription of non-coding RNA fractions in addition to the mRNA fraction, although it offers advantages when the product of a single RT reaction is to be used in several PCR reactions using different primer sets.

5.3.3 Reverse transcriptases

In the past few years there has been a revolution in the development of methodologies for RT-PCR, in particular in the identification and/or development of robust and efficient reverse transcriptases (22). *Table 5.2* shows the main features of the reverse transcriptases commercially available. There are two reverse transcriptases of viral origin, the avian myeloblastosis virus reverse transcriptase (AMV) and the Moloney murine leukaemia virus reverse transcriptase (M-MLV), and two of bacterial origin, *C. therm* polymerase (from *Carboxydothermus hydrogenoformans*) and *Tth* DNA polymerase (a recombinant DNA polymerase derived from *Thermus thermophilus*). Both bacterial reverse transcriptases are thermostable enzymes, thereby enabling the reaction to be performed at high temperatures and thus providing optimal conditions for primer annealing and reducing the effects of

Table 5.2. Key features of the different reverse transcriptases commercially available[a]

Enzyme	cDNA length	Temperature optimum (°C)	Ion requirement	RNase H activity	DNA-dependent DNA polymerase	Fidelity in RT	Type of primer
AMV	≤12 kb	37–45	Mg^{2+}	++	–	+	All[b]
AMV RNase H⁻	≤12 kb	50–65	Mg^{2+}	–	–	+	All
M-MLV	≤10 kb	37–42	Mg^{2+}	+	–	+	All
M-MLV RNase H⁻	≤10 kb	50	Mg^{2+}	–	–	+	All
C. therm polymerase	≤4 kb	60–70	Mg^{2+}	–	–	++	All
Tth DNA polymerase	1–2 kb	60–70	Mn^{2+}	–	+[c]	–	Specific

[a]Most suppliers of molecular biology products maintain excellent web pages with technical information about RT-PCR procedures (see *Table 5.1*).
[b]'All' indicates that specific primers, random hexanucleotides and oligo (dT) primers can all be used.
[c]In the presence of Mg^{2+} ions.

secondary structure in the template. Engineered viral reverse transcriptases with higher thermal stability, and lacking RNase H activity that is responsible for the degradation of the RNA strand in an RNA:cDNA hybrid, have also been obtained (AMV RNase H^- and M-MLV RNase H^-). These are very efficient, high-yield enzymes that have been used, as well as *Tth* DNA polymerase, in many RT-PCR experiments with bacterial mRNA in culture-based studies. Bacterial reverse transcriptases have special features that have made them the polymerase of choice for certain applications. *Tth* DNA polymerase has the ability to perform both RT and PCR amplification in the presence of manganese or magnesium ions, respectively (23) and forms the basis of the one-step one-enzyme RT-PCR protocols (see below). *C. therm* polymerase is the only reverse transcriptase with an associated 3' to 5' proofreading activity, which results in an increased fidelity.

5.3.4 One-step and two-step RT-PCR protocols

RT-PCR reactions can be performed following two different approaches, a 'one-step procedure' (also called 'continuous') or a 'two-step procedure' (or 'uncoupled'). In the one-step procedure, the RT and the PCR reactions take place sequentially in the same tube, in the presence of a unique buffer using a specific primer set [random nucleotides or oligo (dT) cannot be used in this procedure], therefore no additional reagents need to be added during the reaction. One-step RT-PCR can be carried out using a single enzyme (*Tth* DNA polymerase), referred to as 'one-enzyme' protocols, or by using commercially available combinations of reverse transcriptase plus thermostable DNA polymerase ('two-enzyme one-step protocols'). The one-step procedure requires less manipulation, reducing pipetting errors, time and minimizing the risk of contamination.

In the two-step procedure, RT and PCR reactions are done sequentially, but separately, in the optimal reaction conditions for each enzyme because a two-buffer system is used. The RT reaction can be performed with specific primers, random hexanucleotides or oligo (dT) (except when using *Tth* polymerase which requires the use of specific primers). After the RT reaction is completed, PCR reagents can be added directly to the tube of the RT reaction ('one-tube system') or alternatively, an aliquot of the RT reaction is transferred to another tube and a PCR reaction with specific primers is performed ('two-tube system'). The two-step procedure, and in particular the two-tube system, provides greatest flexibility since the cDNA can be used in several PCR reactions using different primer sets. It does, however, require more manipulation, increasing the likelihood of contamination. Sensitivity seems to be greatest using the one-step two-enzyme RT-PCR protocol, followed by the two-step two-enzyme protocol with lowest sensitivities reported using the one-step one-enzyme (*Tth* polymerase) protocol (24). Factors other than sensitivity may also dictate the choice of RT-PCR protocol including requirements related to the product yield required, reaction specificity, or the number of samples to be processed.

5.4 Quantitative RT-PCR

Quantitative RT-PCR methods have been developed for the quantification of steady-state mRNA levels (25,26). They are also the subject of a recent review by Sharkey *et al.* (27). Quantification can be performed after the reaction is completed (end-point measurement) or during the course of the reaction (real-time or kinetic). In some cases absolute quantification of the number of target copies per specific unit is performed, but most often quantitative RT-PCR is used to quantify changes in the expression of a specific gene after a treatment by comparison to a reference sample such as an untreated control (relative quantification).

The most frequently used approaches for quantitative RT-PCR are competitive RT-PCR and real-time RT-PCR. Both techniques are reproducible and have comparable sensitivity (7). Competitive RT-PCR is based on the co-amplification of the target RNA and known amounts of a standard RNA (the competitor) that is designed to be amplified with the same primers and with the same efficiency as the target but differs in length or contains a mutation that allows its differentiation from the target (26). In real-time RT-PCR, product synthesis is measured in each cycle during the exponential phase of PCR by measuring an increase in fluorescence as the reaction proceeds (28,29 and Chapter 6). Examples of quantitative analyses of gene expression in environmental samples are the quantification of ribulose-1,5-biphosphate carboxylase/oxygenase, *rbcL*, transcripts in diatom cultures and marine samples (11) and the quantification of ammonia monooxygenase, *amoA*, transcripts in a biofilm (9). Expression of virulence genes has also been quantified in several pathogenic bacteria (6,7).

5.5 Analysis of global gene expression

Genome-wide screening of mRNA transcripts, produced under different environmental conditions, can be compared in order to determine genes that are differentially expressed (induced or repressed) under varying conditions. Several methods are available for analyzing global gene expression in prokaryotes including differential display (DD) (17) and RAP-PCR (16). The basic principle of DD and RAP-PCR is the generation of a collection of cDNAs using short oligonucleotides that in theory covers the whole genome. The cDNAs are subsequently amplified by PCR and the products separated on polyacrylamide or agarose gels. Comparison of the band patterns allows for the detection of differentially expressed genes, which are subsequently confirmed by other techniques such as quantitative RT-PCR or northern blot hybridization analysis, as the frequency of false positives generated by DD or RAP-PCR may be high (30). The DD method uses anchored oligo (dT) primers for the RT reaction, followed by PCR amplification using the same anchored oligo (dT) primers and short arbitrary primers. RAP-PCR on the other hand uses arbitrary primers only, both for the RT and PCR reaction, and as a result has been the method of choice to analyze prokaryotic mRNAs.

In studies of prokaryotic systems, several different strategies with respect to the type of primers used for DD and RAP-PCR have been utilized. Fislage *et al.* (31) designed primers for differential display based on the calculation of oligomer frequency distribution in the coding regions of the genome of *Enterobacteriaceae*. An alternative approach relies on the use of arbitrary primers together with a primer based on Shine–Dalgarno sequences from the 5′-end of bacterial mRNAs (32). Further improvements to this method to decrease the number of false positives caused by rRNA have been described, including the use of antisense probes for 16S and 23S rRNA, which allow discrimination of the fragments amplified from rRNA (33). Another approach uses a large number of arbitrary primers to allow high-density sampling of differentially expressed genes (34). This method is in part designed to avoid annealing of the arbitrary primers to stable RNAs such as rRNA. RAP-PCR has been successfully used for the analysis of differential gene expression in prokaryotes, in particular for the analysis of genes involved in environmentally relevant processes such as response to pollutants (35), symbiotic associations (36) and sulfate respiration (37).

Apart from DD and RAP-PCR there is a variety of related techniques relying on reverse transcription and PCR amplification procedures, including cDNA-amplified fragment length polymorphism (cDNA-AFLP) (38), cDNA representational difference analysis (cDNA-RDA) (39), cDNA-RNA subtractive hybridization (40) and most recently DNA microarrays (41, 42).

5.6 Conclusions

Analysis of functional and global gene expression in the environment has undergone a significant expansion in the past few years, in parallel with the development of robust and reproducible molecular biology tools for RNA analysis. Questions such as which genes are expressed in the environment, how diverse the transcripts are, what are the transcription rates in the environment, how is gene expression affected by changing environmental conditions and which microbes are expressing the genes of interest can now be approached by RT-PCR techniques. The information generated by RT-PCR-based techniques constitutes a solid foundation for the analysis of microbial activity in the environment. Moreover, in combination with data obtained from other molecular biology approaches, from ecological parameters, and from the physiology of isolates, results from RT-PCR studies should significantly contribute to our understanding of the functioning of microbial processes in the environment. However, a key challenge is still to obtain sufficient amounts of high quality RNA from the environment, and in particular from environments in which microbial biomass is limited and/or inhibitory substances (such as humic acids) are present.

References

1. Staley JT, Konopka A (1985) Measurement of *in situ* activities of nonphotosynthetic microorganisms in aquatic and terrestrial habitats. *Annu Rev Microbiol* **39**: 321–346.
2. Spring S, Schulze R, Overmann J, Schleifer KH (2000) Identification and characterization of ecologically significant prokaryotes in the sediment of freshwater lakes: Molecular and cultivation studies. *FEMS Microbiol Rev* **24**: 573–590.
3. Gray ND, Head IM (2001) Linking genetic identity and function in communities of uncultured bacteria. *Environ Microbiol* **3**: 481–492.
4. Grunberg-Manago M (1999) Messenger RNA stability and its role in control of gene expression in bacteria and phages. *Annu Rev Genet* **33**: 193–227.
5. Wagner R (2000) *Transcription Regulation in Prokaryotes*. Oxford University Press, Oxford.
6. Kwon DH, Osato MS, Graham DY, El-Zaatari FAK (2000) Quantitative RT-PCR analysis of multiple genes encoding putative metronidazole nitroreductases from *Helicobacter pylori*. *Int J Antimicrob Agents* **15**: 31–36.
7. Goerke C, Bayer MG, Wolz C (2001) Quantification of bacterial transcripts during infection using competitive reverse transcription-PCR (RT-PCR) and Lightcycler RT-PCR. *Clin Diagn Lab Immunol* **8**: 279–282.
8. Cheng YS, Halsey JL, Fode KA, Remsen CC, Collins MLP (1999) Detection of methanotrophs in groundwater by PCR. *Appl Environ Microbiol* **65**: 648–651.
9. Aoi Y, Shiramasa Y, Tsuneda S, Hirata A, Kitayama A, Nagamune T (2002) Real-time monitoring of ammonia-oxidizing activity in a nitrifying biofilm by *amoA* mRNA analysis. *Water Sci Technol* **46**: 439–442.
10. Nogales B, Timmis KN, Nedwell DB, Osborn AM (2002) Detection and diversity of expressed denitrification genes in estuarine sediments after reverse transcription-PCR amplification from mRNA. *Appl Environ Microbiol* **68**: 5017–5025.
11. Wawrik B, Paul JH, Tabita FR (2002) Real-time PCR quantification of *rbcL* (ribulose-1,5-bisphosphate carboxylase/oxygenase) mRNA in diatoms and pelagophytes. *Appl Environ Microbiol* **68**: 3771–3779.
12. Meckenstock R, Steinle P, van der Meer JR, Snozzi M (1998) Quantification of bacterial mRNA involved in degradation of 1,2,4-trichlorobenzene by *Pseudomonas* sp strain P51 from liquid culture and from river sediment by reverse transcriptase PCR (RT/PCR). *FEMS Microbiol Lett* **167**: 123–129.
13. Wilson MS, Bakermans C, Madsen EL (1999) *In situ*, real-time catabolic gene expression: Extraction and characterization of naphthalene dioxygenase mRNA transcripts from groundwater. *Appl Environ Microbiol* **65**: 80–87.
14. Alfreider A, Vogt C, Barbel W (2003) Expression of chlorocatechol 1,2 dioxygenase and chlorocatechol 1,2 dioxygenase genes in chlorobenzene contaminated subsurface samples. *Appl Environ Microbiol* **69**: 1372–1376.
15. Chen F, Hodson RE (2001) *In situ* PCR/RT-PCR coupled with *in situ* hybridization for detection of functional gene and gene expression in prokaryotic cells. *Meth Microbiol* **30**: 409–424.
16. Welsh J, Chada K, Dalal SS, Cheng R, Ralph D, McClelland M (1992) Arbitrarily primed PCR fingerprinting of RNA. *Nucleic Acids Res* **20**: 4965–4970.
17. Liang P, Pardee AB (1992) Differential display of eukaryotic messenger-RNA by means of the polymerase chain-reaction. *Science* **257**: 967–971.
18. Huang ZQ, Fasco MJ, Kaminsky LS (1996) Optimization of DNase I removal of

contaminating DNA from RNA for use in quantitative RNA-PCR. *Biotechniques* **20**: 1012–1020.
19. Veres G, Gibbs RA, Scherer SE, Caskey CT (1987) The molecular-basis of the sparse fur mouse mutation. *Science* **237**: 415–417.
20. Sarkar N (1997) Polyadenylation of mRNA in prokaryotes. *Annu Rev Biochem* **66**: 173–197.
21. Lakey DL, Zhang Y, Talaat AM, Samten B, DesJardin LE, Eisenach KD, Johnston SA, Barnes PF (2002) Priming reverse transcription with oligo(dT) does not yield representative samples of *Mycobacterium tuberculosis* cDNA. *Microbiology* **14**: 2567–2572.
22. Gerard GF (1998) Reverse transcriptase: a historical perspective. *Focus* **2**: 65–67.
23. Myers TW, Gelfand DH (1991) Reverse transcription and DNA amplification by a *Thermus thermophilus* DNA polymerase. *Biochemistry* **30**: 7661–7666.
24. Sellner LN, Turbett GR (1998) Comparison of three RT-PCR methods. *Biotechniques* **25**: 230–234.
25. Wang AM, Doyle MV, Mark DF (1989) Quantitation of messenger-RNA by the polymerase chain-reaction. *Proc Natl Acad Sci USA* **86**: 9717–9721.
26. Freeman WM, Walker SJ, Vrana KE (1999) Quantitative RT-PCR: Pitfalls and potential. *Biotechniques* **26**: 112–125.
27. Sharkey FH, Banat IM, Marchant R (2004) Detection and quantification of gene expression in environmental bacteriology. *Appl Environ Microbiol* **70**: 3795–3806.
28. Higuchi R, Fockler C, Dollinger G, Watson R (1993) Kinetic PCR analysis - real-time monitoring of DNA amplification reactions. *Bio-Technology* **11**: 1026–1030.
29. Bustin SA (2002) Quantification of mRNA using real-time reverse transcription PCR (RT-PCR): trends and problems. *J Mol Endocrinol* **29**: 23–39.
30. Debouck, C (1995) Differential display or differential dismay. *Curr Opin Biotechnol* **6**: 597–599.
31. Fislage R, Berceanu M, Humboldt Y, Wendt M, Oberender H (1997) Primer design for a prokaryotic differential display RT-PCR. *Nucleic Acids Res* **25**: 1830–1835.
32. Fleming JT, Yao W-H, Sayler GS (1998) Optimization of differential display of prokaryotic mRNA: Application to pure culture and soil microcosms. *Appl Environ Microbiol* **64**: 3698–3706.
33. Nagel AC, Fleming JT, Sayler GS, Beattie KL (2001) Screening for ribosomal-based false positives following prokaryotic mRNA differential display. *Biotechniques* **30**: 988–996.
34. Walters DM, Russ R, Knackmuss HJ, Rouviere PE (2001) High-density sampling of a bacterial operon using mRNA differential display. *Gene* **273**: 305–315.
35. Brzostowicz PC, Walters DM, Thomas SM, Nagarajan V, Rouvière PE (2003) mRNA differential display in a microbial enrichment culture: simultaneous identification of three cyclohexanone monooxygenases from three species. *Appl Environ Microbiol* **69**: 334–342.
36. Cabanes D, Boistard P, Batut J (2000) Identification of *Sinorhizobium meliloti* genes regulated during symbiosis. *J Bacteriol* **182**: 3632–3637.
37. Steger JL, Vincent C, Ballard JD, Krumholz LR (2002) *Desulfovibrio* sp. genes involved in the respiration of sulfate during metabolism of hydrogen and lactate. *Appl Environ Microbiol* **68**: 1932–1937.
38. Noël L, Thieme F, Nennstiel D, Bonas U (2001) cDNA-AFLP analysis unravels a genome-wide *hrpG*-regulon in the plant pathogen *Xanthomonas campestris* pv. *vesicatoria*. *Mol Microbiol* **41**: 1271–1281.
39. Bowler LD, Hubank M, Spratt BG (1999) Representational difference analysis of cDNA for the detection of differential gene expression in bacteria: development using a model of iron-regulated gene expression in *Neisseria meningitidis*. *Microbiology* **145**: 3529–3537.

40. Li MS, Monahan IM, Waddell SJ, Mangan JA, Martin SL, Everett MJ, Butcher PD (2001) cDNA-RNA subtractive hybridization reveals increased expression of mycocerosic acid synthase in intracellular *Mycobacterium bovis* BCG. *Microbiology* **147**: 2293–2305.
41. Small J, Call DR, Brockman FJ, Straub TM, Chandler DP (2001) Direct detection of 16S rRNA in soil extracts by using oligonucleotide microarrays. *Appl Environ Microbiol* **67**: 4708–4716.
42. Wu L, Thompson DK, Li G, Hurt RA, Tiedje JM, Zhou J (2001) Development and evaluation of functional gene arrays for detection of selected genes in the environment. *Appl Environ Microbiol* **67**: 5780–5790.

Protocol 5.1: RT-PCR

This is a standard protocol for amplification of functional genes from environmental samples using RT-PCR. A two-step two-enzyme protocol is detailed (using a viral reverse transcriptase and *Taq* DNA polymerase), although protocols using *Tth* DNA polymerase and two-enzyme one-step protocols have also been employed successfully for the amplification of environmental mRNA.

MATERIALS

Reagents

RT reagents:	RT buffer 5 × (containing 250 mM Tris-HCl pH 8.3, 375 mM KCl, 15 mM $MgCl_2$)
	0.1 M dithiothreitol (DTT)
	dNTP mix (10 mM each dATP, dCTP, dGTP and dTTP)
	Sequence-specific reverse primer complementary to mRNA or random hexamers
	RNA sample (full methods for isolation of total RNA from environmental samples are given in Chapter 1).
PCR reagents:	Superscript II reverse transcriptase (Invitrogen, Carlsbad, CA, USA)
	PCR buffer 10 × (composition depends on the supplier)
	$MgCl_2$ 25 mM
	dNTP mix
	Forward and reverse sequence-specific primers
Additional reagents:	*Taq* polymerase
	TAE buffer (0.04 M Tris-acetate, 1 mM EDTA, pH 8.0); loading dye (0.25% bromophenol blue, 0.25% xylene cyanol, 15% Ficoll type 400); ethidium bromide solution (0.5 μg ml^{-1}); diethyl pyrocarbonate (DEPC); agarose; RNase inhibitors
Equipment:	Thermal cycler
	Agarose gel electrophoresis apparatus and power supply
	UV transilluminator

METHODS

RT reaction:

1. In a ribonuclease-free microcentrifuge tube, mix the RNA extract (1 ng to 5 μg for total RNA) with 1 μl of dNTP mix, the primer (2 pmol for specific primers and 50–250 ng for hexamer primers) and add autoclaved milliQ water (DEPC-treated water or ribonuclease inhibitors are frequently used for ribonuclease inactivation) to give a total volume of 12 μl. Prepare a premix of reagents whenever possible.
2. Heat the mixture at 65°C for 5 min and place immediately on ice.
3. Add 4 μl of 5 × RT-buffer and 2 μl of DTT to each tube and incubate at 42°C for 2 min. If random hexamers are used, a prior incubation for 10 min at 25°C is required.
4. Add 1 μl of reverse transcriptase (200 IU) and incubate at 42°C for 50 min.
5. Heat to 70°C for 10 min to inactivate the reverse transcriptase. The cDNA can be use immediately for a PCR reaction or stored at –20°C for later use.

PCR reaction:

1. Prepare a reagent premix with 10 × PCR buffer (adjust $MgCl_2$ as appropriate), dNTPs, forward and reverse specific primers and *Taq* DNA polymerase at the concentrations defined for optimal PCR amplification of the target of interest. Aliquot the premix into PCR microcentrifuge tubes.
2. Add 2 μl of cDNA from the RT reaction
3. Perform the PCR reaction at the cycling conditions optimized for the amplification of the target.

Controls for RT-PCR reaction:

1. Control for contamination of the reagents used in the RT and PCR reactions: perform RT and PCR reactions with all the reagents but without RNA or cDNA template, respectively. No amplification products should be obtained in these reactions.
2. Control for DNA contamination of the RNA extract: perform a PCR reaction with the RNA sample directly, without a prior reverse transcription (no RT control). No amplification products should be obtained in this reaction.

3. Control for the presence of inhibitors of the RT-PCR in the RNA extract. A positive control to confirm the quality of the RNA template can be done by running RT-PCR reactions for RNA molecules known to be present in a total RNA extract, such as 16S ribosomal RNA (rRNA) or mRNAs for housekeeping genes. The generation of RT-PCR products from these controls indicates the absence of inhibitors in the sample.

Agarose gel electrophoresis:

1. Prepare an agarose gel (concentration of agarose depending on the size of the amplified product) in TAE buffer.
2. Load the RT-PCR products and a molecular weight marker of the appropriate size range.
3. Run the electrophoresis at 80–100 V
4. Stain the RT-PCR products by submerging the gel in solution containing ethidium bromide. Destain the gel by soaking in water.
5. Place the gel on the transilluminator and visualize the RT-PCR products under UV light (302 nm).

Additional product detection methods:

Some simple alternatives are:

1. Staining of the agarose gels with SYBR Green instead of ethidium bromide to increase sensitivity of detection.
2. Generation of labeled RT-PCR products by using digoxigenin (DIG)-labeled dUTP during the PCR reaction.
3. Transfer of the RT-PCR products to nylon membranes and detection by Southern blot hybridization with DIG-labeled polynucleotide probes (see 10).

Protocol 5.2: RAP-PCR

MATERIALS

(Additional reagents and material only)

RT reagents:	Arbitrary primers
PCR reagents:	1–2 μCi [α-^{33}P]dCTP (only in case of synthesis of radiolabeled products)
Reagents for polyacrylamide gel electrophoresis:	Three different possibilities: 1. Pre-cast gels (commercially available) 2. Non-denaturing 6% acrylamide gels: 29:1 acrylamide-bisacrylamide, in 1× TBE (100 mM Tris, 90 mM Borate, 1 mM EDTA (pH 8.3) 3. Conventional acrylamide-urea sequencing gels (denaturing gel) 4. 10 mM Tris, 10 mM KCl (pH 8.3)
Additional reagents:	Silver staining kit, TBE buffer, autoradiography films (only for radiolabeled products)
Equipment:	1. Polyacrylamide gel electrophoresis equipment and power supply 2. DNA sequencer

METHODS

RT-PCR reaction:	1. Perform the RT-PCR reaction as detailed above using each one of the arbitrary primers in parallel reactions. Adjust dNTPs, primer and $MgCl_2$ concentrations to meet the optimal conditions described for the type or arbitrary primers used. If using radiolabeling of products add [α-^{33}P] dCTP to the PCR reaction.
Electrophoretic separation of the products:	1. Run the samples in a polyacrylamide gel. Electrophoresis voltage and run-time would depend on the type of electrophoretic system used and the length of the gel. 2. Stain the gels with silver nitrate to visualize the RT-PCR products (if radiolabeled products

have been generated, expose autoradiography films to gels)

Confirmation of products from differentially expressed genes:

1. Excise the bands from the gel that correspond to differentially expressed genes and elute the DNA in 10 mM Tris-HCl, 10 mM KCl (pH 8.3) at 95°C for 20–60 min.
2. Re-amplify the RT-PCR products using the same arbitrary primer as the one used to generate the electrophoretic pattern (omit radiolabeled dCTP).
3. Clone the re-amplified products and sequence them to identify the gene differentially expressed.
4. Confirm the differential expression of the gene by northern blot analysis or quantitative RT-PCR with primers specific for the gene of interest.

Quantitative real-time PCR

6

Cindy J. Smith

6.1 Introduction

The development of molecular methods to detect and analyze microorganisms from the natural environment without prior culture has provided fascinating insights into microbial diversity and function. Much of this work has been facilitated by the use of the polymerase chain reaction (PCR) to amplify sequences from DNA and RNA extracted directly from environmental samples (see Chapters 1, 2 and 5) (1,2). However, a major drawback of microbial ecological studies based on classical end-point PCR is that the process inherently introduces biases as the mixed environmental target template is amplified. However, the resulting PCR amplicons, although providing a rich source of information relating to previously uncharacterized genes, cannot be used accurately to quantify numbers or proportions of specific genes or phylotypes within natural environments.

End-point PCR works by exponentially amplifying the target gene over a number of cycles to result in detectable levels (usually visualized via agarose gel electrophoresis) of the template gene. End-point PCR-based community analysis assumes that all genes within a mixed community are amplified at equal efficiencies by the primer set used, and that the yields of specific products are not influenced by either PCR conditions or cycle number. However, several aspects of the PCR can lead to the preferential amplification of certain templates (3,4,5). In particular, PCR drift, a stochastic effect of the PCR cycle itself, can result in end-point ratios of target genes that differ significantly from the proportions of the initial templates (6). Initially in a PCR amplification, primers are present as an excess in proportion to the template concentration. At some point in the reaction the template concentration increases to a critical amount that makes it as likely to bind to itself rather than to anneal with the primers. This results in competition events between template–template and template–primer hybridizations (the so called C_0t effect) (7). Therefore the amplification of that particular template is now self-limiting. Concurrently template that was originally less abundant is still being exponentially amplified by the primers without self-inhibition. Theoretically this can result in equal end-point ratios of template from unequal starting template ratios. Therefore with end-point PCR, the number of products generated from numerically dominant bacteria in the environment may not necessarily correspond to the numerically dominant members of the original community.

Prior to the development of real-time PCR, the other strategies available for quantification were competitive PCR (8) and limiting dilution PCR (9). These methods are both time- and resource-consuming and require post-PCR analysis. However, quantification of genes can be achieved using quantitative real-time PCR (qPCR). Unlike end-point PCR, the data is collected as the genes are amplified in 'real time'. Gene quantification is therefore carried out during the initial cycles of PCR, before PCR kinetics bias the reaction. Since product yields at this stage are below the detection limits of ethidium bromide, fluorescence-based labeling is used, whereby a fluorescent dye is detected that accumulates in direct proportion to the yield of the amplified PCR product. As a result, template amplification is recorded for every cycle via the corresponding increase in fluorescence. Quantification of the starting template is achieved by determining the threshold cycle (Ct) of the unknown samples and of a range of known standards. The Ct value is defined as the point at which the accumulation of amplicons, as measured by an increase in fluorescence, is significantly above the background levels of fluorescence (10) (*Figure 6.1*). Yields during the very first cycles of amplification cannot be determined as fluorescence levels are indistinguishable from background levels. Once the Ct value is exceeded, the exponential accumulation of product can be measured by the increase in fluorescence (*Figure 6.1*). Since the initial stages of amplification are exponential, amplification is not affected by PCR drift as product renaturation is not competing with primer annealing. At this point in the PCR, amplified gene copy numbers are proportional to those of the initial template. The greater the initial template concentration, the earlier the Ct value is reached, whilst reactions containing lower concentrations of template will require a greater number of amplification cycles before the fluorescence level is greater than the background level. Additionally, using the Ct method to identify starting template results in a greater dynamic range of detection than is found for end-point PCR (11) (*Figure 6.1A*).

6.2 Quantification

The actual quantification of the target gene in environmental samples is calculated using standard curves constructed from known amounts of the target gene. Standard curves can be created from genomic DNA/RNA (11–13), plasmid DNA (14) or from a PCR product of the target gene. For accurate quantification the range of concentrations of the standard template DNA should span the expected amount of the unknown samples. The known standards should be amplified using the same PCR conditions as used for the unknown samples and Ct values are determined for each. A linear regression line is constructed from the Ct values of the standards plotted against the log of their starting concentration. Unknown samples are then quantified by comparing their Ct values to the standard curve. The value reported for the unknown sample is relative to the standards used, for example copies of gene per milliliter, genomes per gram of soil etc. Determination of Ct values, standard curve construction and unknown gene quantification is carried out using the software accompanying the real-time PCR system in use.

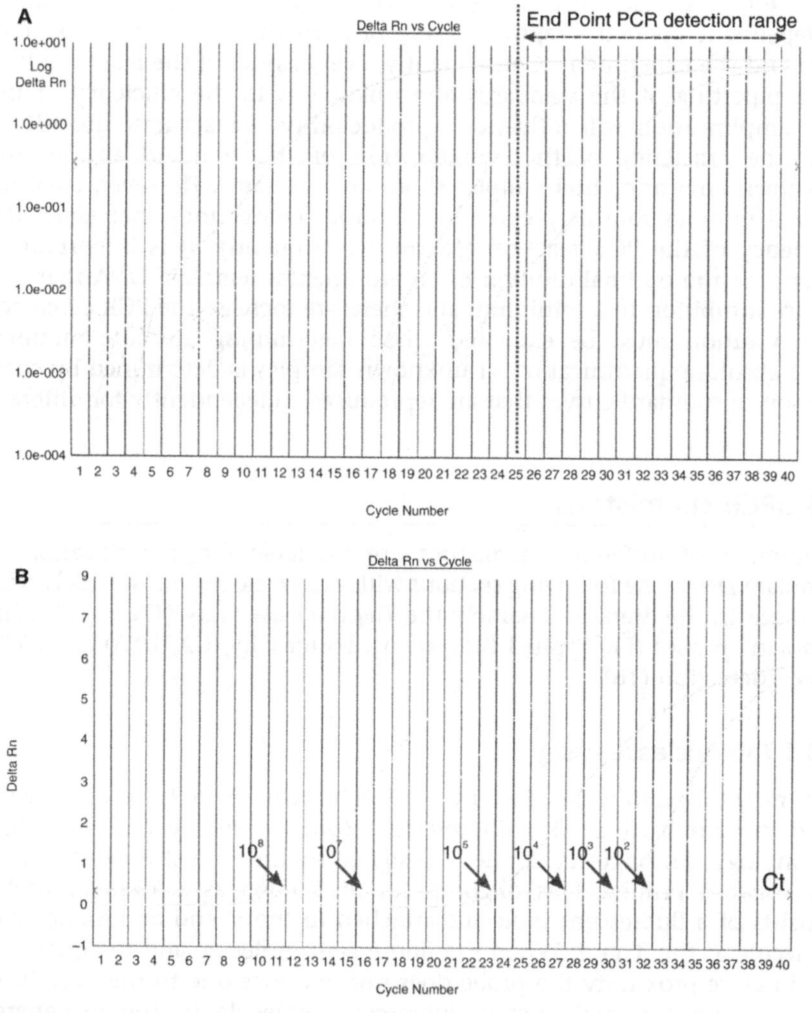

Figure 6.1

(A) Typical PCR amplification plot generated by real-time PCR. Cycle number is plotted against the log increase (Delta Rn) in fluorescence. Delta Rn is the increase in fluorescence from the reporter signal normalized against the internal instrument fluorescent dye ROX = (6-carboxy-x-rhodamine). Quantification of starting template by real-time PCR occurs during the linear amplification stage. The position of the horizontal arrow indicates the point at which the starting template from all samples is still undergoing linear amplification. During this phase the number of amplified products is directly proportional to the initial number of original DNA target templates in the reaction. The detection range for end-point PCR is indicated by the vertical dotted line, and the end-point PCR detection range by the dashed horizontal arrow.
(B) Determination of Ct value. Linear amplification plot of increasing fluorescence (Delta Rn) vs. cycle number. The Ct value (horizontal arrow) is determined as the point at which amplification of template is above background fluorescence. The arrows indicate the Ct values from a range of starting template concentrations of environmental DNA. The greater the amount of starting template the earlier the Ct cycle threshold is reached. Standard curves are constructed from the Ct values of known amounts of starting template.

Since the standard curve is constructed for each analysis, quantification is dependent upon the reproducibility of both the real-time PCR system used and the experiment. Consequently, every aspect of the real-time PCR from pipetting of the standards and unknowns to the efficiency of the PCR amplification will influence reproducibility. For accurate quantification the efficiency of the amplification of the standard and of the unknown environmental samples should be the same. However, environmental samples routinely contain inhibitory compounds that affect the efficiency of the PCR reaction (15,16). Consequently, it is important to determine the optimal dilution of environmental template DNA that will reduce inhibition to a minimum and therefore increase the PCR efficiency (13). Caution must be exercised when determining absolute numbers since absolute quantification of unknown samples is determined by comparison to standard curves that are reproduced independently for different runs.

6.3 qPCR chemistries

A number of different chemistries are available for quantification of template DNA. The following sections will review the two most widely used methods in the literature, namely the *Taq* nuclease assay (TNA) (17) using *Taq*Man® probes (PE Applied Biosystems, Foster City, CA, USA) and SYBR Green detection (18).

6.3.1 *Taq* nuclease assay

This method exploits the 5′ exonuclease activity of the polymerase enzyme used in combination with fluorescent resonance energy transfer (FRET) technology. An oligonucleotide is designed to anneal just downstream of the forward primer. This oligonucleotide, known as a *Taq*Man probe, consists of a fluorescent molecule attached to the 5' end and a quencher molecule at the 3' end. When the fluorescent molecule and the quencher are in close proximity the probe does not fluoresce due to the transfer of energy from the high-energy fluorescent molecule to the low-energy quencher (FRET). During the elongation step, the 5' exonuclease activity of the *Taq* polymerase cleaves nucleotides from the probe molecule; as this occurs the fluorescent molecule bound to the 5' end of the *Taq*Man probe is removed from the path of the newly forming amplicon (17). Once the fluorescent molecule and quencher are no longer in close proximity, fluorescence is emitted and can then be detected (*Figure 6.2A*). The fluorescence detected is therefore a direct measure of the amount of amplified target template. A major advantage of this method is that fluorescence is only generated by cleavage of the sequence-specific probe during amplification of the target template, and hence non-specific amplification products are not detected. However, design of primers and probes for *Taq*Man assay is restricted by the requirement of an additional conserved site necessary for probe hybridization.

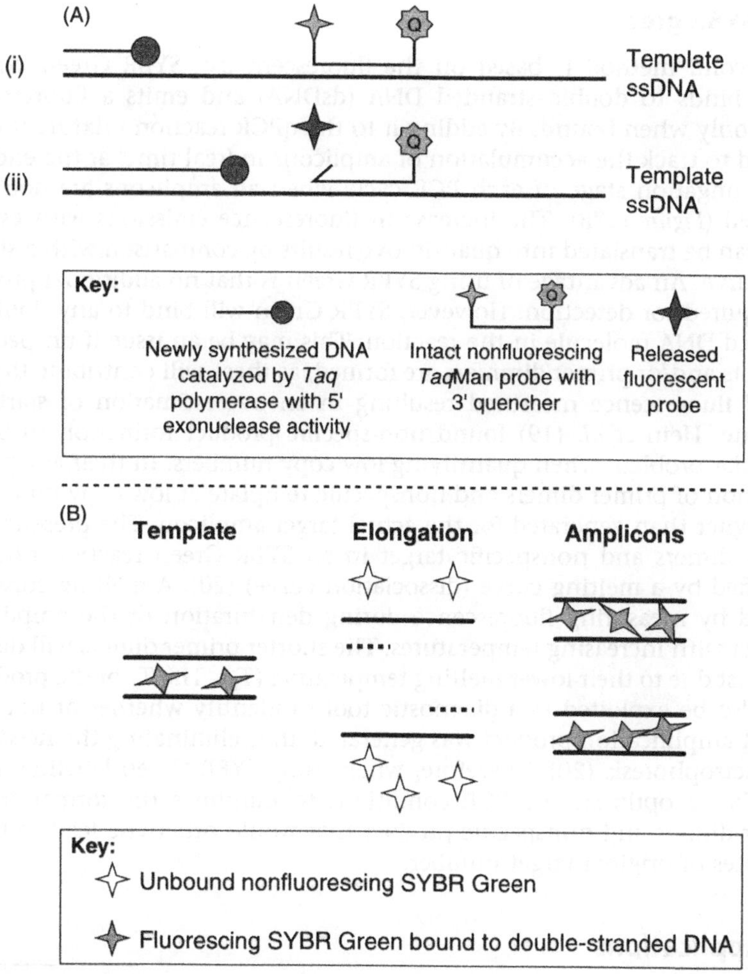

Figure 6.2

Real-time PCR chemistries. **(A)** *Taq* nuclease assay (TNA) using *Taq*Man® probes. During annealing the *Taq*Man probe and primers bind to the template. When the *Taq*Man probe is intact (i), energy is transferred between the reporter and the quencher; as a result, no fluorescent signal is detected. As the new strand is synthesized by *Taq* polymerase the 5′ exonuclease activity of the enzyme cleaves the labeled 5′ nucleotide of the probe, releasing the reporter from the probe (ii). Once it is no longer in close proximity the fluorescent signal from the probe is detected and template amplification is recorded by the corresponding increase in fluorescence.
(B) SYBR Green Detection. SYBR Green binds to all double-stranded DNA and emits a fluorescent signal. In its unbound state, SYBR Green does not fluoresce. Template amplification is therefore measured by the increase in fluorescence.

6.3.2 SYBR green

The second method is based on the fluorescent dye SYBR Green. SYBR Green binds to double stranded DNA (dsDNA) and emits a fluorescent signal only when bound. By adding it to the qPCR reaction mixture it can be used to track the accumulation of amplicons in 'real time' at the end of each elongation stage of each PCR cycle when all amplicons are double-stranded (*Figure 6.2B*). The increase in fluorescence emissions with every cycle can be translated into quantitative results by comparison with a standard curve. An advantage of using SYBR Green is that no additional probes are required for detection. However, SYBR Green will bind to any double-stranded DNA molecule in the reaction. This may be an issue if unspecific products and/or primer dimmers are formed, as these will contribute to the overall fluorescence measured resulting in an overestimation of starting template. Hein *et al.* (19) found non-specific product formation to be a particular problem when quantifying low copy numbers. In their assay the formation of primer dimers and nonspecific template at low copy numbers was higher than generated for the actual target amplicon. The presence of primer dimers and nonspecific target in an SYBR Green reaction can be identified by a melting curve (dissociation curve) (20). A melting curve is created by measuring fluorescence during denaturation of the amplified product with increasing temperatures. The shorter primer dimers will dissociate first due to their lower melting temperature (T_m). The T_m of the product may also be exploited as a diagnostic tool to identify whether or not the correct amplification product was generated, thus eliminating the need for gel electrophoresis (20). Therefore, when using SYBR Green labeling, it is essential to optimize the PCR conditions to minimize the formation of primer dimers and nonspecific product that would otherwise lead to false estimates of original target number.

6.4 Applications

qPCR has proven to be a valuable tool in microbial ecology studies. To date it has been used to examine total microbial communities and the relative proportions of specific phylotypes within a number of different environments (4,11,13,14,21–24) using the 16S rRNA gene. Specific biological functions have also been addressed by the targeting of functional genes (12,25). The following section reviews significant applications of qPCR from the current literature.

6.4.1 qPCR analysis of total microbial communities via analysis of amplified ribosomal RNA genes

Aquatic microbial communities were one of the first to be investigated by qPCR using *Taq*Man probes (11,14,22). Suzuki *et al.* (14) developed and compared a number of primers and *Taq*Man probes for the domains Bacteria and Archaea together with narrower phylogenetic-specific primers and probes for *Synechococcus* and *Prochlorococcus*. Using standard curves constructed from linearized plasmid, the relative proportion of these groups compared to the overall bacterial numbers was determined over a 200 m depth profile in seawater from the Californian coast. The ability of *Taq*Man probes to quantify target genes of a specific *Synechococcus* sp. from among other phylogenetically closely related *Synechococcus* spp. was examined by Becker *et al.* (11). A subsequent study (22) determined the contribution of *Synechococcus* spp. to the overall cyanobacterial population in Lake Constance. The use of qPCR allowed the dynamics of *Synechococcus* populations and subpopulations to be followed over a 1 year period. The SYBR Green method has been used to quantify *Pseudoaltermonas* species relative to total eubacterial abundance in a range of different marine habitats (24).

In general, aquatic environments are relatively free from compounds that are co-purified with nucleic acids and then inhibit the polymerase enzyme activity. Indeed, Becker *et al.* (22) found that serial dilutions of the genomic DNA extracted from lake water did not indicate the presence of any inhibitory substances. However, the presence of inhibitors in soil and sediment extractions has been well documented (26). qPCR has been successfully used to quantify bacteria present in environments that contain compounds inhibitory to PCR, for example ammonia-oxidizing bacteria (AOB) in arable soil (27) and in a municipal waste-water treatment plant (28). Archaeal abundance in a range of temperate environments such as different soil types, sediment and microbial mats (29) and archaea inhabiting the extreme environments of hot spring water, surrounding anoxic sediment and hydrothermal vent effluent (30) have also been quantified. The effects of inhibitors on real-time PCR was examined by Stults *et al.* (13) using *Taq*Man qPCR to quantify *Geobacter* spp. present in aquifer sediment. They noted that humic acids do indeed reduce the activity of the polymerase but also found that the presence of inhibitors can additionally quench fluorescence, resulting in higher Ct values and an underestimation of the target gene. To reduce this effect, they recommend running a series of template dilutions for all environmental samples.

6.4.2 Functional ecology

Not only can valuable insights into community structure be gained from qPCR but important questions addressing the functional roles of microorganisms in the environment can also be investigated. Primers and probes (if using the *Taq* nuclease assay), can be designed to target conserved sequences of a functional gene. The target can be either DNA to quantify gene abundance, or alternatively mRNA for gene expression studies. Standard curves are constructed using a target sequence. If mRNA levels are being examined then the standard curve should be constructed from cDNA and not double-stranded DNA, as otherwise there is no control for the efficiency of the RT reaction. The cDNA standard curve can be constructed from either a known amount of RNA that has been reverse-transcribed into cDNA and then serially diluted or alternatively from a serial dilution of known quantities of RNA each of which are then individually reverse-transcribed into cDNA and subsequently used in the qPCR reaction to construct the standard curve.

A number of functional gene studies have used SYBR Green to quantify PCR product yield. Whilst this circumvents the problem of identifying a third conserved region for a probe within the short amplicon, the use of

SYBR Green detection may be more problematic for amplification of functional genes where non-specific products are more often generated. However, *Taq*Man probes may provide their own limitation in that it is difficult to design satisfactory primer–probe combinations for functional genes that exhibit considerable sequence variation.

Using DNA as the target, genes that play key roles in important biogeochemical cycles have been quantified: denitrification processes have been investigated via the nitrite reductase gene *nirS* (12); methane oxidation by methanotrophs in soil via the *pmoA* gene (31); ammonium oxidation by AOB (32). Other key environmental processes carried out by microorganisms examined by qPCR include the quantification of key genes involved the biodegradation of aromatic pollutants (25).

Quantification of mRNA (see Chapter 5 for details of RT-PCR amplification of environmental mRNA) can provide further insights into microbial activity in the environment. Detection and quantification of mRNA from environmental samples by qPCR has been reported to be three orders of magnitude more sensitive than hybridization techniques (33). To date, RT-qPCR has not been directly applied to environmental samples. However, a few studies have used it to monitor genes involved in environmental processes in pure cultures (33,34). Wawrik *et al.* (33) used RT-qPCR to measure carbon fixation in diatoms and pelagophytes by targeting *rbcL*, the gene encoding the large subunit of ribulose-1,5-bisphosphate carboxylase/oxygenase (RubisCO). Similarly dissimilatory (bi)sulphate reductase (DSR) gene expression was quantified during the growth of *Desulfobacterium autotrophicum* under a variety of different conditions by Neretin *et al.* (34). The application of RT-qPCR to quantify gene expression in environmental samples will undoubtedly further our understanding of many important processes that are mediated by microorganisms.

References

1. Ward DM, Weller R, Bateson MM (1990) 16S rRNA sequences reveal numerous uncultured microorganisms in a natural community. *Nature* **345**: 63–65.
2. Suzuki M, Rappe M, Haimberger ZW, Winfield H, Adair N, Strobel J, Giovannoni SJ (1997) Bacterial diversity among small-subunit rRNA gene clones and cellular isolates from the same seawater sample. *Appl Environ Microbiol* **63**: 983–989.
3. Polz MF, Cavanaugh CM (1998) Bias in template-to-product ratios in multi template PCR. *Appl Environ Microbiol* **64**: 3724–3730.
4. Reysenbach AL, Giver LJ, Wickman GS, Pace NR (1992) Differential amplification of rRNA genes by polymerase chain reaction. *Appl Environ Microbiol* **58**: 3417–3418.
5. Wintzingerode FV, Gobel UB, Stackebrandt E (1997) Determination of microbial diversity in environmental samples: pitfalls of PCR-based rRNA analysis. *FEMS Microbiol Rev* **21**: 213–229.
6. Suzuki M, Rappe MS, Giovannoni SJ (1998) Kinetic bias in estimates of coastal picoplankton community structure obtained by measurements of small-subunit rRNA amplicon length heterogeneity. *Appl Environ Microbiol* **64**: 4522–4529.
7. Mathieu-Daudé F, Welsh J, Vogt T, McClelland M (1996) DNA rehybridization during PCR: the 'C_0t effect' and its consequences. *Nucleic Acids Res* **24**: 2080–2086.
8. Diviacco S, Norio P, Zentilin L, Menzo S, Clementi M, Biamonti G, Riva S,

Falaschi A, Giacca M (1992) A novel procedure for quantitative polymerase chain reaction by coamplification of competitive templates. *Gene* **122**: 313–320.

9. Skyes PJ, Neoh SH, Brisco MJ, Hughes E, Condon J, Morley AA (1992) Quantitation of targets for PCR by use of limiting dilutions. *Biotechniques* **13**: 444–449.
10. Heid CA, Stevens J, Livak KJ, Williams PM (1996) Real time quantitative PCR. *Genome Res* **6**: 986–994.
11. Becker S, Boger P, Oehlmann R, Ernst A (2000) PCR bias in ecological analysis: a case study for quantitative Taq nuclease assays in analyses of microbial communities. *Appl Environ Microbiol* **66**: 4945–4953.
12. Gruntzig V, Nold SC, Zhou J, Tiedje JM (2001) *Pseudomonas stutzeri* nitrite reductase gene abundance in environmental samples measured by real-time PCR. *Appl Environ Microbiol* **67**: 760–768.
13. Stults JR, Snoeyenbos-West O, Methe B, Lovley DR, Chandler DP (2001) Application of the 5′ fluorogenic exonuclease assay (TaqMan) for quantitative ribosomal DNA and rRNA analysis in sediments. *Appl Environ Microbiol* **67**: 2781–2789.
14. Suzuki MT, Taylor LT, DeLong EF (2000) Quantitative analysis of small-subunit rRNA genes in mixed microbial populations via 5′-nuclease assays. *Appl Environ Microbiol* **66**: 4605–4614.
15. Kreader CA (1996) Relief of amplification inhibition in PCR with Bovine Serum Albumin or T4 gene 32 protein. *Appl Environ Microbiol* **62**: 1102–1106.
16. Monteiro L, Bonnemaison D, Vekris A, Petry KG, Bonnet J, Vidal R, Cabrita J, Megraud F (1997) Complex polysaccharides as PCR inhibitors in feces: *Helicobacter pylori* model. *J Clin Microbiol* **35**: 995–998.
17. Holland PM, Abramson RD, Watson R, Gelfand DH (1991) Detection of specific polymerase chain reaction product by utilizing the 5′→3′ exonuclease activity of *Thermus aquaticus* DNA polymerase. *Proc Natl Acad Sci USA* **88**: 7276–7280.
18. Wittwer CT, Herrmann MG, Moss AA, Rasmussen RP (1997) Continuous fluorescence monitoring of rapid cycle DNA amplification. *Biotechniques* **22**: 130–138.
19. Hein I, Lehner A, Rieck P, Klein K, Brandl E, Wagner, M (2001) Comparison of different approaches to quantify *Staphylococcus aureus* cells by real-time quantitative PCR and application of this technique for examination of cheese. *Appl Environ Microbiol* **67**: 3122–3126.
20. Ririe KM, Rasmussen RP, Wittwer CP (1997) Product differentiation by analysis of DNA melting curves during the polymerase chain reaction. *Anal Biochem* **245**: 154–160.
21. Suzuki MT, Preston CM, Chavez FP, DeLong EF (2001) Quantitative mapping of bacterioplankton populations in seawater: field tests across an upwelling plume in Monterey Bay. *Aquatic Microb Ecol* **24**: 117–127.
22. Becker S, Fahrbach M, Boger P, Ernst A (2002) Quantitative tracing, by *Taq* nuclease assays, of a *Synechococcus* ecotype in a highly diversified natural population. *Appl Environ Microbiol* **68**: 4486–4494.
23. Hristova K, Gebreyesus B, Mackay D, Scow KM (2003) Naturally occurring bacteria similar to the methyl tert-butyl ether (MTBE)-degrading strain PM1 are present in MTBE-contaminated groundwater. *Appl Environ Microbiol* **69**: 2616–2623.
24. Skovhus TL, Ramsing NB, Holmstrom C, Kjelleberg S, Dahllof I (2004) Real-time quantitative PCR for assessment of abundance of *Pseudoalteromonas* species in marine samples. *Appl Environ Microbiol* **70**: 2373–2382.
25. Baldwin BR, Nakatsu CH, Nies L (2003) Detection and enumeration of aromatic oxygenase genes by multiplex and real-time PCR. *Appl Environ Microbiol* **69**: 3350–3358.

26. Tebbe CC, Vahjen W (1993) Interference of humic acids and DNA extracted directly from soil in detection and transformation of recombinant DNA from bacteria and a yeast. *Appl Environ Microbiol* **59**: 2657–2665.
27. Hermansson A, Lindgren PE (2001) Quantification of ammonia-oxidizing bacteria in arable soil by real-time PCR. *Appl Environ Microbiol* **67**: 972–976.
28. Harms G, Layton AC, Dionisi HM, Gregory IR, Garrett VM, Hawkins SA, Robinson KG, Sayler GS (2003) Real time PCR quantification of nitrifying bacteria in a municipal wastewater treatment plant. *Environ Sci Technol* **37**: 343–351.
29. Ochsenreiter T, Selezi D, Quaiser A, Bonch-Osmolovskaya L, Schleper C (2003) Diversity and abundance of Crenarchaeota in terrestrial habitats studied by 16S RNA surveys and real time PCR. *Environ Microbiol* **5**: 787–797.
30. Takai K, Horikoshi K (2000) Rapid detection and quantification of members of the Archaeal community by quantitative PCR using fluorogenic probes. *Appl Environ Microbiol* **66**: 5066–5072.
31. Kolb S, Knief C, Stubner S, Conrad R (2003) Quantitative detection of methanotrophs in soil by novel *pmoA*-targeted real-time PCR assays. *Appl Environ Microbiol* **69**: 2423–2429.
32. Okano Y, Hristova KR, Leutenegger CM, Jackson LR, Denison RF, Gebreyesus B, Lebauer D, Scow KM (2004) Application of Real-time PCR to study effects of ammonium on population size of ammonia-oxidizing bacteria in soil. *Appl Environ Microbiol* **70**: 1008–1016.
33. Wawrik B, Paul JH, Tabita FR (2002) Real-time PCR quantification of *rbcL* (ribulose-1,5-bisphosphate carboxylase/oxygenase) mRNA in diatoms and pelagophytes. *Appl Environ Microbiol* **68**: 3771–3779.
34. Neretin LN, Schippers A, Pernthaler A, Hamann K, Amann R, Jorgensen BB (2003) Quantification of dissimilatory (bi)sulphite reductase gene expression in *Desulfobacterium autotrophicum* using real time PCR. *Environ Microbiol* **5**: 660–671.

Protocol 6.1

The following section outlines the steps involved in quantifying the total bacterial community from an environmental sample by targeting the 16S rRNA gene using an ABI Prism 7000 sequence detection system (Applied BioSystems). Recommendations for: (i) primer and probe design; (ii) construction of a standard curve; and (iii) protocols for *Taq*Man and SYBR Green assays are given and can be applied for use with any qPCR system on the market.

MATERIALS

Reagents

SYBR Green or *Taq*Man PCR Mastermix (Applied Biosystems)

Primers: BACT1369F 5'-CGGTGAATACGTTCYCGG-3'

PROYK1492R 5'-GGWTACCTTGTTACGACTT -3'

*Taq*Man probe: TM1389F 5'-CTTGTACACACCGCCCGTA-3'

Primers and probes are detailed in Suzuki *et al.* (14)

Sterile milliQ water

Equipment

Real-time PCR thermal cycler

Conventional PCR thermal cycler

Spectrophotometer

Microcentrifuge

Benchtop centrifuge with rotor for centrifuging microtitre plates

Additional materials

Environmental DNA (see Chapter 1 for typical extraction procedures)

Sequenced 16S rRNA gene containing conserved sites for primers and probe to be used for construction of standard curve

Thermal cycler and reagents for endpoint PCR (see Chapter 2 for further details)

Cloning kit for cloning amplicon standard

Plasmid isolation kit

Restriction enzyme and buffer to linearize plasmid

Agarose gel electrophoresis equipment

Eppendorf tubes (0.5 ml)

96-well optical plates (Applied Biosystems)

METHODS

Primer and probe design

General considerations:

For both methods of detection, a short amplicon of ~50–150 bp should be targeted.

G/C content of the primers should be between 20 and 80%. This is particularly important for SYBR Green assays in order to minimize non-specific products.

Avoid runs of identical nucleotides.

Probe (*Taq*Man assays only):

Should have a T_m 8–10°C higher than those of the primers.

Should be located as close as possible to the forward primer without overlapping with it.

Avoid G residues at the 5′ end. Guanine residues are natural quenchers.

There should be more Gs than Cs in the probe for the same reason.

Standard curve construction from plasmid DNA

1. PCR amplify a 16S rRNA sequence from a pure culture (or use a clone from a previously constructed library) using primers F1369 and R1492 and clone product into an appropriate vector (e.g. p-GEM T vector, Promega).
2. Isolate and purify the plasmids using a plasmid isolation kit such as Plasmid Midi kit (Qiagen). Linearize the plasmid using an appropriate restriction enzyme (following the manufacturer's instructions), choosing an enzyme that will cut the plasmid only once.
3. Determine the purity of the linearized plasmid by measuring the absorption ratio A_{260}/A_{280}. A

ratio of 1.8 indicates pure DNA. Any value below this indicates that the plasmid preparation may contain contaminants such as proteins, residual phenol, etc. If this is the case, re-extract the linearized plasmid with phenol-chloroform. Determine the concentration of plasmid DNA by measuring the A_{260}. DNA with an absorbance of 1 has a concentration of 50 μg ml^{-1}.

4. From the concentration of the linearized plasmid and the number of base pairs in the target amplicon, the gene target copy number per μl of plasmid can be determined. Using the primer pair F1396-R1492 a 123 bp product is targeted. Calculate the proportion (as a percentage) of the recombinant plasmid that consists of the target amplicon by dividing the size of the target amplicon by the total size of the recombinant plasmid (vector + insert); e.g. for a 123 bp target amplicon cloned into a 3000 bp vector the proportion of target DNA would be:

   (123/3123 bp) × 100 = 3.9% insert DNA.

   If the concentration of the plasmid was determined to be 500 ng/μl then the concentration of the target insert would therefore represent 3.9% of the total, i.e. 19.5 ng/μl. The concentration of the target (insert) template can be converted to copies/μl of plasmid using the following equation:

$$\frac{{}^{*}6.023 \times 10^{23} \text{ (copies/mol)} \times \text{concentration of plasmid (g/μl)}}{\text{relative molecular mass (g/mol)}}$$

   *Avogadro's number.

   The molecular weight of the target insert is determined by multiplying the number of base pairs by the average molecular weight of dsDNA which is 660 Da per base pair.

   For the example plasmid for which the 123 bp cloned amplicon would have a mass of 19.5 ng, the number of copies/μl of plasmid would be calculated as follows:

   relative molecular mass = 123 bp × 660 Da

   = 81 180 Da

Convert formula to ng and insert values

$$\frac{*6.023 \times 10^{14}\ (\text{Da/ng}) \times 19.5\ \text{ng/µl}}{81\,180\ \text{Da}}$$

$= 1.4 \times 10^{11}$ copies of target plasmid/µl.

5. Prepare a range of dilutions representing 10^{10} to 10^{1} copies of target DNA. Careful preparation of the standard curve is vital for accurate quantification. Insure each dilution is well mixed and spun down briefly before making the next dilution.

q-PCR reaction

The following protocol is for use with the ABI Prism 7000 detection system, but should be equally applicable for other systems. The same DNA standard can be used for constructing qPCR standard curves using *Taq*Man Probes or SYBR Green. Every run should include in triplicate the following reactions: template DNA to construct the standard curve, the unknown sample DNA, and a no template control (NTC). All reactions should be run in triplicate on the same 96-well plate.

*Taq*Man Probe Reaction:

1. Thaw all samples on ice. Protect the *Taq*Man Probe from light.
2. Prepare the dilution range for the standard as outlined above. Prepare the dilution range for the unknown samples, e.g. undiluted, 10^{-1}, 10^{-2}.
3. For an initial reaction, a 'no optimization' reaction containing 900 nM of each primer and 200 nM of probe can be used (Applied Biosystems).
4. Make up a master mix for *n* + 10% reactions (e.g. for 30 reactions make up a master mix for 33) with each individual reaction to contain:

Universal Master Mix	12.5 µl
Forward primer (10 µM)	2.25 µl
Reverse primer (10 µM)	2.25 µl
Probe (10 µM)	0.50 µl

Template	1.0 µl
Water	6.5 µl
Total	25 µl

Mix reaction and spin down.

5. Aliquot 24 µl of Master Mix for each reaction into 0.5 ml PCR tubes. For each template (standard, unknown and NTC) the reaction is carried out in triplicate. Add 1 µl of the template to each of three individual reactions. Mix all reactions well, spin down briefly and keep on ice until all reactions are set up.
6. Add each reaction to the 96-well optical plate. Cover with the optical cover. Centrifuge plate briefly to bring down samples.
7. Load onto ABI Prism 7000 Sequence Detection System.
8. Thermal cycling program: 2 min at 50°C*, 10 min at 95°C, then 40 cycles of 15 s at 95°C, 1 min at 60°C; *the first 2 mins at 50°C are only required if using AmpErase UNG.
9. Ct values and standard curve construction is carried out using the ABI 7000 sequence detection system software. The number of original templates in the qPCR of the unknownl sample is then calculated from the standard curve.

qPCR using SYBR Green:

1. A standard should be prepared as described above.
2. Optimizing the primer concentration for a SYBR Green reaction is more complicated than for the *Taq*Man probe as non-specific priming and primer dimers can be problematic. In general a much lower concentration of primer is required. Applied Biosystems and Roche both recommend starting with 50 nM although this is not universal. Primer conditions will have to be optimized empirically. Include the NTC in any optimization
3. Again make up master mix for *n* + 10% reactions with each individual reaction to contain:

Universal SYBR Green Master Mix	12.5 µl
Forward primer (10 µM)	1.25 µl
Reverse primer (10 µM)	1.25 µl
Template	1.0 µl
Water	9.0 µl
Total	25 µl

4. Aliquot 24 µl of master mix for each reaction into 0.5 ml PCR tubes. For each template (standard, unknown and NTC) is carried out in triplicate. Add 1 µl of the template to each of three individual reactions. Mix all reactions well, spin down briefly and keep on ice until all reactions are set up.
5. Add each reaction to the 96-well optical plate. Cover with the optical cover. Spin down plate briefly.
6. Load onto ABI Prism 7000 Sequence Detection System.
7. Thermal cycling program: 2 min at 50°C*, 10 min at 95°C, then 40 cycles of 15 sec at 95°C, 1 min at 60°C; *the first 2 min at 50°C are only required if using AmpErase UNG.
8. After the reaction is complete, run a melting curve analysis to determine if primer dimers/nonspecific products are present.
9. Ct values and standard curve construction is carried out using the ABI 7000 sequence detection system software. The number of original templates in the qPCR of the unknownl sample is then calculated from the standard curve.

7 Stable-isotope probing

Stefan Radajewski and J. Colin Murrell

7.1 Introduction

Determining the metabolic function of individual taxa within microbial communities represents a major challenge in microbial ecology. One approach to address this question has first involved the isolation, identification and characterization of microorganisms to which a particular function can be attributed. A functional group can sometimes be defined by small subunit rRNA gene similarities, enabling the subsequent use of molecular biological techniques to investigate these closely related populations *in situ* (1–3). An analogous approach has defined functional groups on the basis of similarities between genes that encode key enzymes in metabolic pathways, 'functional genes' (4–6). It is likely that many microorganisms will share metabolic functions, and therefore some uncultivated taxa will be detected using these molecular approaches. However, not all uncultivated taxa will necessarily share the genetic similarities used to define an individual functional group, and thus the metabolic function and identity of these organisms will remain unclear.

An alternative approach to link metabolic function with taxonomic identity is first to establish the function of uncultivated microbial populations and then determine their identity using molecular biological techniques. Several techniques involving the use of substrates labeled with radioisotopes or stable-isotopes can achieve this goal and simultaneously link identity, activity and function under conditions which approach those occurring *in situ*. The technique of microautoradiography, developed for microscopic observation of microorganisms involved in uptake of radiolabeled substrates (7), has recently been combined with molecular identification using 16S rRNA probes and fluorescent *in situ* hybridization (8,9; see Chapter 8). Substrates labeled with ^{14}C or ^{13}C have also been added to environmental samples and recovered as labeled lipid fractions that can be compared with the lipid fractions of cultivated strains (10,11). More recently, the technique of stable-isotope probing (SIP) was described, which used substrates highly enriched with ^{13}C to selectively recover the DNA of functional groups of microorganisms, enabling subsequent identification with molecular biological techniques (12). In this chapter, we will introduce the basis of SIP, outline technical considerations for the use of stable-isotopes, provide examples of its application and detail a general procedure suitable for recovering the labeled DNA fraction from a functional group of microorganisms.

7.2 Stable-isotope labeling of DNA

Stable-isotope labeling of biomarkers exploits intrinsic physical properties of atoms. These include the low natural abundance of certain stable-isotopes: the stable carbon isotopes are ^{12}C (98.9%) and ^{13}C (1.1%), and stable nitrogen isotopes are ^{14}N (99.63%) and ^{15}N (0.37%). Therefore, commercially available substrates that are highly enriched in the rare stable-isotopes (e.g. >99%, ^{13}C or ^{15}N) can be added to complex environments, permitting the labeled isotopes to be tracked using techniques that detect the mass increase due to the single additional neutron.

SIP relies on the fact that DNA synthesized during microbial growth on a substrate enriched with a 'heavy' stable-isotope becomes labeled sufficiently to be resolved from unlabeled DNA by equilibrium centrifugation in a CsCl density-gradient. The classical experiments of Meselson and Stahl (13) demonstrated this principle with *Escherichia coli* grown on $^{15}NH_4^+$. Although the buoyant density of DNA varies with its guanine–cytosine content, the incorporation of a high proportion of a naturally rare stable-isotope (^2H, ^{15}N or ^{13}C) into DNA enhances greatly the density difference between labeled and unlabeled DNA fractions (14,15). This principle has recently been demonstrated with bacterial cultures grown on $^{13}CH_3OH$ (12; see *Figure 7.1*) and $^{13}CO_2$ (16) as a carbon source, and has been applied to identify microorganisms in soil that actively assimilated methanol (12) and methane (17). For the purposes of this chapter, we will focus upon applications using the ^{13}C isotope.

A principal consideration for determining if SIP is a suitable technique to employ is whether each DNA molecule in the target microorganisms will

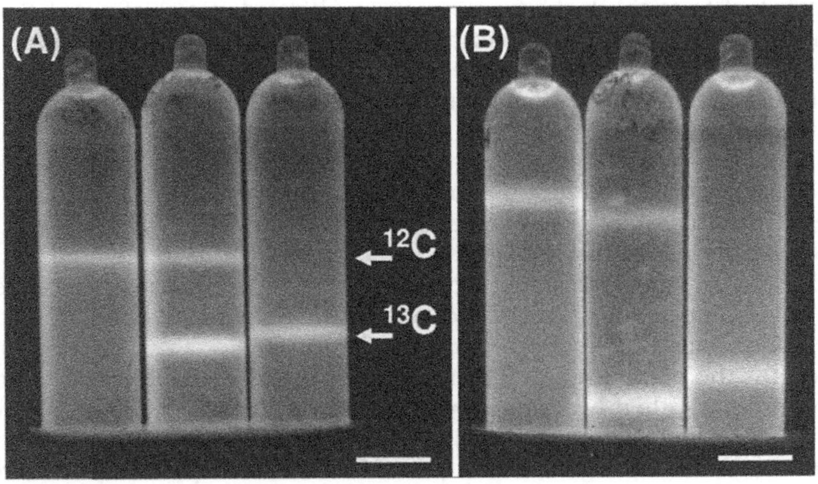

Figure 7.1

CsCl/ethidium bromide density gradients of DNA fractions extracted from *Methylobacterium extorquens* grown on either [^{12}C]- or [^{13}C]methanol as the carbon source. Visualization of ^{13}C- and ^{12}C-labeled DNA bands (arrows) with long-wavelength UV light following equilibrium ultracentrifugation at **(A)** 265 000 ***g*** for 16 h at 20°C and **(B)** 140 000 ***g*** for 60 h at 20°C. Bar = 1 cm.

contain a sufficient proportion of stable-isotope (^{13}C) so as to permit collection of a ^{13}C-labeled 'heavy' DNA fraction ([^{13}C]DNA). Factors including ^{13}C dilution due to the simultaneous assimilation of naturally occurring carbon substrates (i.e. ^{12}C-labeled), ^{13}C turnover due to substrate co-oxidation or predation of the target microorganisms, or ^{13}C assimilation without DNA replication, might influence the proportion of a microbial genome that will become ^{13}C-labeled. To date, SIP has been used to target metabolically restricted groups [methylotrophs and ammonia-oxidizing bacteria (AOB)] that grow in the presence of high concentrations of labeled substrate.

7.3 Application of stable-isotope probing

The availability of PCR primers that are universal for the small-subunit rRNA genes of Bacteria, Archaea (Chapter 2 and 11) and Eukarya (Chapter 12) is integral to the use of SIP for identification, *a priori*, of microorganisms involved in a specific function. More selective PCR primers, such as those targeting functional genes (Chapter 11), can also be applied to study defined populations that are known to be involved in specific processes.

7.3.1 Methylotroph populations

Methylotrophs are microorganisms that can use reduced one-carbon compounds as a sole source of carbon and energy (18). Although the known methylotrophs include a variety of Bacteria, Archaea and Eukarya, most extant aerobic strains are Bacteria belonging to the class *Proteobacteria*. A specialized subgroup of these methylotrophs is the methane oxidizing bacteria (methanotrophs). Characterization of proteobacterial methylotrophs has identified certain common features of their biochemistry, which has permitted the design of PCR primers that target key functional genes of methylotrophs and methanotrophs; those encoding the active-site subunits of methanol dehydrogenase (MDH) and methane monooxygenase (MMO), respectively (19,20).

7.3.1.1 Methanol assimilation

Stable-isotope probing was first applied to identify the active methanol-assimilating microorganisms in an acidic forest soil (12). Soil in a microcosm (small-scale experimental chamber that attempts to mimic environmental conditions; see Protocol section for further details) was exposed to $^{13}CH_3OH$ (0.5% v/w) for 44 days, after which a distinct [^{13}C]DNA fraction was resolved from total community DNA using a CsCl density-gradient. Domain level PCR primers only detected bacterial sequences in the [^{13}C]DNA fraction. Analysis of 16S rDNA sequences identified that three closely related genera within the α-*Proteobacteria* had assimilated the [^{13}C]methanol, which was reflected in a parallel analysis of genes encoding MDH. Other 16S rDNA sequences retrieved from the [^{13}C]DNA grouped with the *Acidobacterium* division, which is poorly represented by cultivated strains (21). Association of these bacteria with assimilation of methanol or the by-products of methylotrophic carbon metabolism thus provides insight into the metabolic function of a diverse, poorly studied and potentially important group of bacteria.

7.3.1.2 Methane assimilation

The active population of methanotrophs in a peat soil microcosm with a gas headspace containing $^{13}CH_4$ (8% v/v) was characterized following recovery of a 'heavy' [^{13}C]DNA fraction (17). In contrast to the discrete [^{13}C]DNA band observed in the previous methanol experiment, the [^{13}C]DNA fraction was observed as a 'smear', indicating intermediate density DNA species that probably resulted from the growth of methanotrophs and additionally of bacteria using ^{13}C-labeled intermediates/by-products of methanotroph metabolism. PCR amplification products of 16S rRNA genes and functional genes for MDH and MMO identified many sequences related to those of extant methanotrophs, demonstrating the activity of these bacteria *in situ* and reinforcing that SIP can identify the target group of microorganisms *a priori*. The [^{13}C]DNA fraction also contained a large proportion of 16S rDNA sequences of bacteria not recognized as methanotrophs or methylotrophs, suggesting that other groups of bacteria are also involved in cycling the carbon derived from CH_4 (possibly in the form of ^{13}C-labeled metabolites or biomass) under these experimental conditions. SIP can thus identify the microbial population involved in the cycling of a specific compound, even though the exact function of individual members of the community may be unclear.

7.3.2 Ammonia-oxidizing populations

Autotrophic ammonia-oxidizing bacteria (AOB) are a specialized group of bacteria that are slow-growing and relatively difficult to study in culture, but are important in the global cycling of nitrogen. Phylogenetic analysis of rRNA gene sequences places nearly all AOB in a monophyletic group within the β-*Proteobacteria* (22), which has resulted in the wide use of selective 16S rDNA PCR primers to study their ecology. SIP was recently used to identify the active $^{13}CO_2$-assimilating species of AOB in enrichment cultures inoculated with a fresh-water sediment (16). Although several types of AOB were detected in total DNA extracted from the enrichment cultures, only some subgroups of AOB were present in the [^{13}C]DNA fraction. These results not only support previous observations that certain subgroups of AOB are outcompeted in laboratory culture, but also illustrate that 'heavy' isotope-labeled DNA can be used to identify which members of a metabolically defined population are active under a specific set of conditions.

7.4 Future prospects

SIP is able to enrich and isolate the combined 'genome' of a microbial population that is involved in a specific function, thereby providing a rational basis for investigating further the ecology of these microorganisms *in situ* using a variety of molecular biological techniques. One attractive option is the use of cloning and hybridization techniques to retrieve entire gene clusters from uncultivated bacteria whose function has been defined. In this manner, biases implicit in the use of selective PCR primers may be circumvented and improved PCR primers could be designed (see Chapter 2). A further development which could be envisaged is the targeting of molecules that do not rely on replication of the chromosome for ^{13}C-labeling (e.g. rRNA; 23). As ribosomes are naturally amplified in active cells, this would improve the sensitivity of SIP by reducing the amount of label required for linking identity with function.

References

1. Devereux R, Kane MD, Winfrey J, Stahl DA (1992) Genus- and group-specific hybridization probes for determinative and environmental studies of sulfate-reducing bacteria. *Syst Appl Microbiol* **15**: 601–609.
2. McCaig AE, Embley TM, Prosser JI (1994) Molecular analysis of enrichment cultures of marine ammonia oxidisers. *FEMS Microbiol Lett* **120**: 363–367.
3. Gulledge J, Ahmad A, Steudler PA, Pomerantz WJ, Cavanaugh CM (2001) Family- and genus-level 16S rRNA-targeted oligonucleotide probes for ecological studies of methanotrophic bacteria. *Appl Environ Microbiol* **67**: 4726–4733.
4. McDonald IR, Murrell JC (1997) The methanol dehydrogenase structural gene *mxaF* and its use as a functional gene probe for methanotrophs and methylotrophs. *Appl Environ Microbiol* **63**: 3218–3224.
5. Rotthauwe J-H, Witzel K-P, Liesack W (1997) The ammonia monooxygenase structural gene *amoA* as a functional marker: molecular fine-scale analysis of natural ammonia-oxidizing populations. *Appl Environ Microbiol* **63**: 4704–4712.
6. Wagner M, Roger AJ, Flax JL, Brusseau GA, Stahl DA (1998) Phylogeny of dissimilatory sulfite reductases supports an early origin of sulfate respiration. *J Bacteriol* **180**: 2975–2982.
7. Brock TD, Brock ML (1966) Autoradiography as a tool in microbial ecology. *Nature* **209**: 734–736.
8. Lee N, Nielsen PH, Andreasen KH, Juretschko S, Nielsen JL, Schleifer K-H, Wagner M (1999) Combination of fluorescent in situ hybridization and microautoradiography—a new tool for structure–function analyses in microbial ecology *Appl Environ Microbiol* **65**: 1289–1297.
9. Ouverney CC, Fuhrman JA (1999) Combined microautoradiography-16S rRNA probe technique for determination of radioisotope uptake by specific microbial cell types in situ. *Appl Environ Microbiol* **65**: 1746–1752.
10. Boschker HTS, Nold SC, Wellsbury P, Bos D, de Graaf W, Pel R, Parkes RJ, Cappenberg TE (1998) Direct linking of microbial populations to specific biogeochemical processes by ^{13}C-labelling of biomarkers. *Nature* **392**: 801–805.
11. Roslev P, Iversen N (1999) Radioactive fingerprinting of microorganisms that oxidize atmospheric methane in different soils. *Appl Environ Microbiol* **65**: 4064–4070.
12. Radajewski S, Ineson P, Parekh NR, Murrell JC (2000) Stable-isotope probing as a tool in microbial ecology. *Nature* **403**: 646–649.
13. Meselson M, Stahl FW (1958) The replication of DNA in *Escherichia coli*. *Proc Natl Acad Sci USA* **44**: 671–682.
14. Rolfe R, Meselson M (1959) The relative homogeneity of microbial DNA. *Proc Nat Acad Sci USA* **45**: 1039–1043.
15. Vinograd J (1963) Sedimentation equilibrium in a buoyant density gradient. In: Colowicker SP, Kaplan NO (eds) *Methods in Enzymology*, Vol. VI, pp. 854–870. Academic Press, London.
16. Whitby CB, Hall G, Pickup R, Saunders JR, Ineson P, Parekh NR, McCarthy A (2001) ^{13}C incorporation into DNA as a means of identifying the active components of ammonia-oxidizer populations. *Lett Appl Microbiol* **32**: 398–401.

17. Morris SA, Radajewski S, Willison TW, Murrell JC (2002) Identification of the functionally active methanotroph population in a peat soil microcosm by stable-isotope probing. *Appl Environ Microbiol* **68**: 1446–1453.
18. Lidstrom ME (1992) The aerobic methylotrophic bacteria. In: Balows A, Trüper HG, Dworkin M, Harder W, Schleifer K-H (eds) *The Prokaryotes*, 2nd edn, Vol. 1, pp. 431–445. Springer-Verlag, New York.
19. McDonald IR, Kenna EM, Murrell JC (1995) Detection of methanotrophic bacteria in environmental samples with the PCR. *Appl Environ Microbiol* **61**: 116–121.
20. Holmes AJ, Costello A, Lidstrom ME, Murrell JC (1995) Evidence that particulate methane monooxygenase and ammonia monooxygenase may be evolutionarily related. *FEMS Microbiol Lett* **132**: 203–208.
21. Barns SM, Takala SL, Kuske CR (1999) Wide distribution and diversity of members of the bacterial kingdom *Acidobacterium* in the environment. *Appl Environ Microbiol* **65**: 1731–1737.
22. Head IM, Hiorns WD, Embley TM, McCarthy AJ, Saunders JR (1993) The phylogeny of autotrophic ammonia-oxidizing bacteria as determined by analysis of 16S ribosomal-RNA gene-sequences. *J Gen Microbiol* **139**: 1147–1153.
23. Gray ND, Head IM (2001) Linking genetic identity and function in communities of uncultured bacteria. *Environ Microbiol* **3**: 481–492.

Protocol 7.1

Three steps are involved in SIP:

1. Incubation of an environmental sample with a stable-isotope-enriched substrate so as to achieve sufficient incorporation of the isotope into the target population.
2. Collection of [^{13}C]DNA following CsCl density-gradient centrifugation.
3. Identification of the target population by analysis of [^{13}C]DNA using molecular biological techniques.

The following protocols describe the application of SIP to detect aerobic methylotrophs. This method has been generally successful for labeling methylotroph DNA, although some optimization may be required for samples from certain environments. Alternative labeled substrates will require individual modifications to the Materials and Methods.

MATERIALS

Reagents

$^{13}CH_3OH$ or $^{13}CH_4$ (99% ^{13}C; Sigma–Aldrich)

TE buffer; 10 mM Tris-HCl, pH 7.6, 1 mM disodium EDTA

Ethidium bromide solution: 10 mg ml^{-1} in water

CsCl solution: prepared by dissolving 10 g CsCl in 10 ml TE

1-Butanol saturated with TE buffer

Ammonium acetate: 10 M, dissolved in water and filter-sterilized

Ethanol

Equipment

Crimp-sealed serum vials (125 ml), chlorobutyl rubber seals and aluminium crimps

Gas chromatograph equipped with a flame ionization detector

Ultracentrifuge equipped with a VTi 65 or VTi 65.2 rotor (Beckman)

Polyallomer Quick-Seal ultracentrifuge tubes (13 mm × 51 mm; Beckman)

1 ml syringe

19 gauge hypodermic needles

Long-wavelength UV light (365 nm)

Retort stand with clamps

Dialysis clips

Additional materials

DNA extraction procedure (see Chapter 1 for further details)

Dialysis membrane tubing is prepared by boiling lengths (~10–20 cm) for 10 min in a large volume of 2% (w/v) $NaHCO_3$ and 1 mM EDTA (pH 8.0); tubing is rinsed thoroughly in distilled water and boiled for 10 min in 1 mM EDTA (pH 8.0) and stored at 4°C

Thermal cycler, *Taq* DNA polymerase, dNTPs, primers and reaction buffers for PCR (see Chapter 2 for further details)

Kit for cloning PCR products

Restriction endonucleases

METHODS

Incubation with ^{13}C

Incorporation of sufficient ^{13}C into the DNA of the active methylotrophs is critical for the success of SIP. As effective separation of DNA requires preferential use of the ^{13}C-labeled substrate instead of 'natural' ^{12}C-labeled compounds, SIP is best suited to higher substrate concentrations (e.g. >1% v/v $^{13}CH_4$). Nevertheless, in some soil samples we have been unable to observe a 'heavy' ^{13}C-labeled DNA fraction, despite rapid $^{13}CH_4$ oxidation, possibly due to some of the reasons discussed earlier. The addition of an equal volume of dilute minimal medium containing N and P to this soil sample has subsequently yielded a [^{13}C]DNA fraction.

1. SIP is performed in microcosms consisting of an environmental sample (e.g. soil, 10 g; aquatic, 10 ml) in a crimp-sealed serum vial.
2. ^{13}C-labeled substrate is added (e.g. 12 ml of $^{13}CH_4$ is injected into the headspace of a sealed serum vial to achieve a concentration of 10% v/v; or $^{13}CH_3OH$ is added drop-wise to soil to achieve a concentration of 0.5% v/w) and the microcosm incubated appropriately (e.g. 20°C in the dark). Note that ^{13}C-labeled gases (e.g. $^{13}CH_4$) are often supplied as small volumes (500 ml) in steel lecture bottles and are more conveniently handled in sealed serum vials. Gas can be transferred from the lecture bottle into a serum vial that has been filled with water, crimp-sealed, inverted and clamped in a retort stand. Two needles are inserted through the seal of the serum vial; one connected to the gas source and the other acting as a drain for the water. Slow release of the gas from the bottle will displace the water, after which the vial will only contain gas at atmospheric pressure that can be stored for later use. Gas samples can then be withdrawn as required and replaced by the equivalent volume of water to maintain 1 atmosphere of pressure in the serum vial.
3. The disappearance of substrate is monitored at regular intervals by gas chromatography. After >90% of the ^{13}C-substrate has been oxidized, the vials are opened and flushed with air (~500 ml) to keep the headspace gas aerobic and remove $^{13}CO_2$ produced during the oxidation of the substrate.
4. Results from several different experiments indicate that for [^{13}C]DNA to be visible in a CsCl density gradient, a minimum of 0.2 mmol ^{13}C must be assimilated (5 ml $^{13}CH_4$). Higher yields of [^{13}C]DNA are obtained if the vial is re-sealed and the incubation continued until 1.0–2.0 mmol ^{13}C has been consumed.

Separation of [^{13}C]DNA

1. Total DNA is extracted from the microcosm sample using methods preferred by the individual laboratory (see Chapter 1). When choosing the DNA extraction method it is important to consider that environmental

DNA samples containing large amounts of discoloration (e.g. humic material) can be difficult to see in a CsCl gradient. For SIP, the method should be suitable for extracting DNA from up to 10 g soil, and the resulting DNA pellet should be dissolved in <4 ml of TE.

2. A subsample of the total DNA can be reserved for later comparison with the [^{13}C]DNA.
3. The solution of total DNA is adjusted to the density appropriate for ultracentrifugation by dissolving 1 g solid CsCl in each 1 ml DNA solution.
4. The CsCl-containing DNA solution and ethidium bromide solution (100 µl) is added to an ultracentrifuge tube, which is then topped-up with CsCl solution (1 g ml^{-1}) and sealed.
5. Tubes are balanced and centrifuged at 265 000 ***g*** (rotor VTi 65; 55 000 rpm) for 14–16 h at 20°C
6. DNA bands are visualized by illumination with long-wavelength (365 nm) UV light. Typical gradients of DNA from a methylotroph grown on $^{12}CH_3OH$ or $^{13}CH_3OH$ are shown in *Figure 7.1a*.
   The ^{13}C-labeled DNA fraction appears as a distinct dense ('heavy') band below the [^{12}C]DNA.
7. When labeling environmental samples, [^{13}C]- and [^{12}C]DNA fractions are not always separated as distinctly as is observed with pure cultures. In such cases, the most labeled fraction can be collected more easily following ultracentrifugation at slower speed [140 000 ***g*** (rotor VTi 65; 40 000 rpm), 60 h, 20°C], which increases the separation between DNA fractions at the expense of less sharp band formation (*Figure 7.1b*).

Collection and purification of [^{13}C]DNA

1. Following ultracentrifugation and visualization of the [^{13}C]DNA fraction, the top of the centrifuge tube is pierced with a 19 gauge needle to provide an air vent.

2. The tube is clamped firmly in a retort stand, illuminated with a long-wavelength UV light source and pierced 2 mm below the most dense DNA fraction with a hypodermic needle (19 gauge) attached to a syringe (1 ml).
3. The tip of the needle is located in the centre of the tube and the [^{13}C]DNA fraction is gently withdrawn (~0.5 ml) and transferred to a microcentrifuge tube.
4. A second ultracentrifugation step is advisable to purify the ^{13}C-labeled fraction from the small proportion of [^{12}C]DNA that can be inadvertently collected during the primary extraction. The ^{13}C-labeled DNA fraction is transferred to a new ultracentrifuge tube, which is filled, sealed and centrifuged as described previously.
5. Only a single 'heavy' DNA fraction should be visible in the second gradient. Labeled DNA is collected as described previously, attempting to collect the minimum volume possible (<0.5 ml).
6. Ethidium bromide is extracted from DNA by the addition of an equal volume of 1-butanol saturated with TE, followed by gentle mixing and brief centrifugation at 13 000 ***g***. The organic (upper) layer is discarded and the extraction repeated until the pink colour has been removed from the aqueous (lower) phase.
7. DNA is transferred to the prepared dialysis tubing, which is sealed with clips and dialysed against a large volume (2 l) of TE buffer for 2 h at 4°C. The buffer is replaced with fresh TE buffer and the dialysis repeated three times in total.
8. DNA is transferred to a microcentrifuge tube and precipitated overnight at –20°C by addition of 1/3× vol of ammonium acetate and 2× vol of 100% ethanol.
9. DNA is recovered by centrifugation at 13 000 ***g*** for 20 min. The DNA pellet is washed with 1 ml of 70% ethanol, centrifuged at 13 000 ***g*** for 10 min, dissolved in 50–100 µl TE and stored at –70°C.

Molecular analysis of [^{13}C]DNA for identification of methylotrophs

Molecular analysis of [^{13}C]DNA uses the established technology of the polymerase chain reaction (PCR) to target both small-subunit rRNA and functional genes. The total community DNA sample reserved prior to ultracentrifugation can be processed identically and the populations compared. A general strategy is provided here; specific details of PCR-based approaches and analysis methods are provided in Chapter 2.

1. Total diversity within the [^{13}C]DNA is assessed with domain level PCR amplification of small-subunit rRNA genes (Bacteria, Archaea and Eukarya). Group-specific primers can be used to target specific genes within the [^{13}C]DNA.
2. Clone libraries are constructed from amplification products using a suitable cloning kit. Libraries should include as many clones as is feasible to screen, especially for small-subunit rRNA genes (>50 clones).
3. Clone inserts are screened by digestion with restriction endonucleases. Fragments are resolved by agarose gel (2–3% w/v) electrophoresis and clones grouped according to the restriction pattern.
4. The DNA sequence of representative clones is obtained and compared with publicly available sequence databases (GenBank; http://www.ncbi.nlm.nih.gov), and subsequent sequence alignments and phylogenies constructed (see Chapter 2).

Applications of nucleic acid hybridization in microbial ecology

8

A. Mark Osborn, Vivien Prior and Konstantinos Damianakis

8.1 Introduction

Nucleic acid hybridization can be defined as the complementary base pairing between two nucleotide strands mediated by Watson–Crick hydrogen bond formation between individual nucleotides. As such, hybridization is central to biology, and is responsible for genomic rearrangements via both homologous and site-specific recombination. Nucleic acid hybridization is also central to molecular biology, both in its direct application as a tool to detect specific DNA sequences (1), and also as an integral process during the polymerase chain reaction (PCR) and DNA sequencing during the hybridization (annealing) of oligonucleotide primers to complementary single-stranded DNA templates.

In microbial ecology, as in all areas of biology, a tremendous assortment of nucleic acid hybridization methodologies has been developed, beginning with an initial emphasis on applying DNA probes to confirm the presence of particular genes in individual microorganisms and/or recombinant DNA constructs, then subsequently in the specific detection of particular organisms or genes within environmental samples, and also as a tool to estimate the complexity in terms of species diversity of microbial communities. One major application of nucleic acid hybridization has been the development of fluorescently *in situ* hybridization (FISH) to detect and enumerate specific microbial taxa within environmental systems, using oligonucleotide probes. This approach is of considerable importance for investigating spatial distribution of microbial communities, and is discussed at length in Chapter 9.

This chapter will thus focus on *in vitro* applications of nucleic acid hybridization, defined here as those hybridization experiments that are conducted using nucleic acids isolated either from individual microorganisms, or directly from environmental samples. We will begin by briefly introducing key methodological concepts in nucleic acid hybridization, prior to illustrating how such approaches have been applied to screen for the presence of genes in cultured bacteria, and subsequently in environmental samples following isolation of environmental nucleic acids, and we then introduce one of the most exciting recent developments in microbial ecology, namely the application of microarrays to investigate gene distribution, diversity and expression both in cultured microorganisms, and in the environment. The chapter concludes with the Protocols section providing coverage of some of the more fundamental applications of DNA hybridization to microbial ecology.

8.2 Fundamentals of DNA hybridization

Nucleic acid hybridization is theoretically an extremely simple molecular methodology. At its simplest level it involves the generation of a single-stranded nucleic acid probe (most typically ssDNA) that is labeled for example with a radioisotope that will enable subsequent detection when the labeled probe binds to a single-stranded nucleic acid molecule (the target) which has usually first been immobilized on a solid matrix, e.g. a nylon membrane. If the probe and the target nucleic acid show complementarity, i.e. significant similarity at the nucleotide level that will allow a double-stranded hybrid molecule to be formed, then the probe will anneal to the target nucleic acid and can be detected using autoradiography or other visualization approaches. However, if the probe sequence is not similar to the target nucleic acid then hybridization (i.e. Watson–Crick base pairing) will not occur, the probe will not bind and no hybridization signal will be detected. In principle, therefore, hybridization will enable detection of a specific DNA sequence from a mixed nucleic acid sample (e.g. chromosomal DNA, or total environmental DNA). All hybridization methods rely on the above principles but vary considerably with respect to the type of nucleic acid being screened for the sequence of interest, the extent to which a probe will provide unambiguous results, i.e. probe specificity, and indeed the type of probe that will be utilized in the hybridization experiment. In this section, fundamental considerations for the design of effective hybridization experiments will be addressed. The reader is also referred to the textbook by Anderson (2) for more detailed coverage of the principles of nucleic acid hybridization.

8.2.1 Probe design

When designing nucleic acid hybridization experiments, the choice and/or design of the probe is a key consideration—see Sambrook and Russell (3) for more details on the generation and application of nucleic acid probes. In most applications, the probe will be a DNA molecule and these are of two main types: fragment probes or oligonucleotide probes. Fragment probes usually consist of a double-stranded nucleic acid molecule, for example a restriction fragment or a PCR amplification product, and such probes are typically ≥200 bp in length, but may be several kilobases in size. Prior to hybridization, the double-stranded probe will be denatured typically via boiling the probe for a few minutes. In contrast, oligonucleotide probes typically consist of single-stranded DNA molecules of ~20 nucleotides in length, with denaturation of the probe maintained by inclusion of formamide in the hybridization reaction. A third, less commonly used type of probe, is the polynucleotide probe (4,5) that consist of between ~60 and 350 nucleotides, either following chemical synthesis for shorter polynucleotides, or as RNA transcripts for larger probes (5). The choice of whether to use fragment or oligonucleotide probes will be influenced in particular by whether the researcher is interested in detecting or identifying sequences that share similarity to the sequence of interest or near or absolute identity. Thus, for those applications in which similar but not necessarily identical sequences are sought, larger fragment probes are utilized, whereas for very specific detection, oligonucleotide probes are used. The rationale for this is seen in *Figure 8.1*: decreasing similarities between the probe and target sequences will result in the hybrid molecule becoming unstable. For oligonucleotide probes, single base changes can result in release of the probe from the target, whilst for fragment probes, much lower levels of similarity are permissible. Additionally, the stringency (i.e. the temperature or salt conditions under which the hybridization experiment is conducted) will also result in different behaviors for fragment and oligonucleotide probes, with oligonucleotide probes again showing a far higher degree of specificity than fragment probes. Thus fragment probes should typically be chosen for hybridization experiments in which the detection of sequences that are divergent (e.g. up to ~70% identity) would be of interest (e.g. for detection of functional genes, for which there may be considerable divergence between different bacterial species). By contrast, for detection of

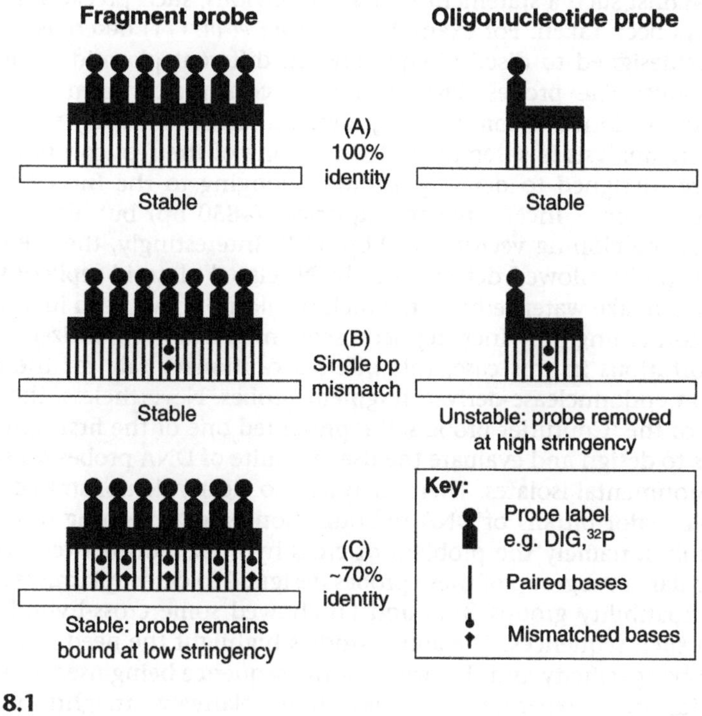

Figure 8.1

Schematic representation of DNA–DNA hybridization using fragment and oligonucleotide probes. (A–C) Results of hybridization reactions in which the template DNA attached to the membrane (white box) shows decreasing similarity to the DNA.

specific sequences, e.g. a particular 16S rRNA sequence representing a species, oligonucleotide probes are preferable. In polynucleotide probes, single mismatches between the probe and target sequence will not result in dissociation of the probe except under very high stringency conditions (6), and therefore such probes can be used for experiments where high but not complete specificity is required. This has made such probes ideal for use in group-specific detection of bacterial genera (5). More recently polynucleotide probes have been adopted for use in microarray experiments by some commercial array suppliers such as Agilent (7). A comprehensive review of technical considerations for the design of optimal hybridization conditions for experiments using oligonucleotide and polynucleotide probes is given by Wetmur (8).

The choice of probe is absolutely critical to the design of hybridization experiments. Historically, most probes were generated from restriction fragments of the cloned gene or operon of interest. For example, Barkay *et al.* (9) used a series of restriction fragments (0.35–6.4 kb) containing fragments of cloned mercury resistance (*mer*) operons from Gram-negative and Gram-positive bacteria to screen for the presence of these *mer* sequences in whole community DNA from mercury-polluted estuarine waters. In an earlier study, Barkay *et al.* (10) had stressed the importance of choosing restriction fragments for probes that excluded any sequences that may have been derived from the cloning vector into which the gene of interest has been cloned. Whilst such a statement may seem obvious, such precautions have not always been taken. For example Couturier *et al.* (11) described a series of probes designed to discriminate between different plasmid incompatibility groups. The probes developed were constructed from restriction fragments of cloned incompatibility determinants. However, subsequent sequencing analysis has demonstrated that one of the replicon probes (~1 kb in size) designed to detect plasmids belonging to the IncN family is comprised of both IncN replicon sequences (~850 bp) but also adjacent regions of the cloning vector (~150 bp) (12). Interestingly, the use of this particular probe allowed detection of IncN-related plasmid replicons from isolates from lake water sediment, which would not have been identified if a probe containing only IncN replicon sequences had been utilized. Whilst rather fortuitous in this case, this offers a cautionary tale for the use of restriction endonuclease-derived fragment probes. Nevertheless, the development of the 'Couturier probe set' represented one of the first significant attempts to design and evaluate the use of a suite of DNA probes for screening environmental isolates. Their analysis also clearly demonstrated one of the other major pitfalls of DNA hybridization-based screening using fragment probes, namely the problem of cross-hybridization between probes. In particular, a subgroup of these probes designed to detect plasmids related to incompatibility groups B, FI and FII showed some cross-hybridization between such sequences. The above studies highlight the need to consider both probe specificity and the origin of the sequence being used as a probe in hybridization experiments. Today, it is relatively straightforward to determine the sequence of any particular probe of interest. Having the DNA sequence enables an *in silico* hybridization experiment to be conducted by doing a FASTA (Fast All) or BLAST (Basic Local Alignment Search Tool) comparison with the GenBank databases, whereby sequences showing high identity to the probe sequence can be identified. Whilst such analyses can provide an indication of the likely specificity of any given probe, it is advised that empirical testing via actual hybridization experiments using the probe with a number of sequences showing varying degrees of identity should be performed (10,11).

Another major consideration in the design of fragment probes is the length of the probe being used. For larger probes, defined here as those >1 kb in length, it is entirely possible that a significant proportion of the probe will hybridize to a target sequence, but that in other regions of the probe no homology will be found to the target DNA. This may mean that under stringent conditions, the probe may be removed during washing stages in the hybridization protocol that remove unbound or partially bound probes. This problem may be particularly true for probes that span more than one gene. Thus to increase specificity in hybridization reactions, the use of shorter probes (<1 kb) is recommended, and where the researcher is interested in detecting multiple genes a suite of probes can be utilized to cover multiple genes in an operon (13). As a simple alternative to the use of restriction fragment-derived probes, the polymerase chain reaction can amplify the sequence to be used as a probe (14,15). This has the obvious advantage that the probe region is defined by the user and can be readily sequenced.

For oligonucleotide probes, sequence specificity is critical to the success of the hybridization experiments. As oligonucleotide probes are defined by the researcher, the specificity of such sequences can again be tested in *in silico* comparisons to the DNA sequence databases. Additionally, there are now a number of probe-analysis software packages, in particular for the design of ribosomal RNA-based probes, to identify particular phylogenetic lineages, e.g. Probe Match; http://rdp.cme.msu.edu/probematch/search.jsp, ProbeBase (http://www.microbial-ecology.de/probebase/index.html) (16) and PRIMROSE (17). These are discussed in more detail in Chapter 15.

Once the probe sequence, amplicon or DNA fragment has been chosen the researcher will need to decide upon the choice and placement of the label that will be used to enable detection of the probe following hybridization. Classically probes were labeled using radioisotopes (mostly ^{32}P or ^{33}P) which have two disadvantages: loss of probe activity for applications requiring long exposure periods; and having to take the necessary precautions to prevent exposure to the radiochemicals being used. More recently digoxigenin (DIG) labeling (Roche) (reviewed in 18) coupled with fluorescence-based detection, chemiluminescent detection and colorimetric detection has been adopted for use in many hybridization assays (see: http://www.roche-applied-science.com/fst/products.htm?/DIG for full details of DIG-based applications). Protocols for generation and application of DIG-based probes are given in the protocols section of this chapter. Other considerations for probe labeling are the choice of location to attach the label. For oligonucleotide or polynucleotide probes, end labeling is typically used with the label attached either to the 5' end of the oligonucleotide using T4 polynucleotide kinase (19) or to the 3' end of the oligonucleotide using terminal deoxynucleotidyl transferase (20). Fragment probes can similarly be labeled either terminally, for example by end labeling (21), or more commonly by random priming (22) whereby the detection label is incorporated along the entire length of a series of DNA probes generated using Klenow enzyme to produce a pool of labeled probe molecules that are all homologous to the original template DNA. For PCR products the simplest method is to incorporate the label as a labeled nucleotide during the amplification reaction. Prior to use in DNA hybridization experiments, any fragment or PCR-derived probes will require denaturation to generate single-stranded DNA probes, typically by heating the probe to 95°C for a few minutes.

8.2.2 Choice of nucleic acid template

The design of successful hybridization experiments will also be constrained by the type of nucleic acid template to which the DNA probe will be hybridized. At its simplest level this may entail genomic DNA extracted from individual bacterial isolates. Such DNA can be analyzed either by simple dot blot hybridization, where a small volume of DNA is placed onto a hybridization membrane. Resulting hybridization analysis would confirm whether or not the genomic DNA hybridizes to the probe of interest. Alternatively, if the researcher is interested in identifying the physical location of a gene, an additional agarose gel electrophoresis and Southern transfer (1) are suggested, for example following restriction endonuclease digestion of the genomic DNA. This can be used to determine the size of restriction fragments that hybridize to the probe sequence, which can help to facilitate gel-purification based cloning of a particular restriction fragment. Southern hybridization can also be used to determine whether the gene of interest (i.e. the probe sequence) is located chromosomally or on a plasmid (23). Additional fine-scale localization of a gene of interest can be conducted via combination of pulse field gel electrophoresis with DNA hybridization (24). Southern hybridization can also be used to confirm the identity of PCR amplification products, either from cultured isolates (25) or from environmental DNA (14,15) to ensure that amplified products are in fact the gene of interest, as opposed to non-specific products. Such confirmation is especially important for analysis of amplification products from environmental DNA, where there is a greater likelihood of such false positive PCR amplification results.

As an alternative to direct isolation of genomic DNA prior to hybridization analysis, a number of studies have utilized the colony hybridization approach (26) to screen collections of environmental isolates for functional genes (10,13,27–31) and to identify common environmental species (32). In colony hybridization, the hybridization membrane is laid over the bacterial colonies on a plate, the membrane is then taken off the plate and the colonies are lysed on the membrane. The resulting crude extract is then fixed to the membrane. Whilst colony hybridization is attractive for the potential for rapid processing of bacterial colonies, there are potential pitfalls, not least that such blots typically have high background hybridization signals caused by failure to remove cellular debris. Not all bacterial isolates will be readily lysed using this method, and this may lead to an underestimation of distributions of the gene of interest in environmental isolate collections. For example Hu and Wu (33) demonstrated that colony morphology could influence the results of hybridization experiments, with dry and rough colonies of some strains, e.g. *Acinetobacter, Alcaligenes* and *Arthrobacter* spp. failing to hybridize to the universal eubacterial probe Eub 338. By contrast, when colony hybridizations were performed using the same probe but with flat and smooth colonies of the same genera, hybridization signals were detected. However, modifications of the colony hybridization protocol have improved lysis efficiencies and enabled the use of oligonucleotide probes to screen culture collections or clone libraries (32,34). Jordan and Maier (35) developed a novel adaptation of the colony hybridization method to investigate the spatial distribution of bacteria on soil surfaces. Rather than performing colony hybridization on spread plates generated from diluents of suspended soil particles, they placed an agar plate onto soil surfaces and gently pressed down onto the agar plate so that the underlying soil would adhere to the agar surface. The plate was then lifted off the soil and incubated for up to 24 h. Following a conventional colony lift and cell lysis, hybridization was carried out using a 16S rRNA gene eubacterial probe to enable enumeration of colonies. This hybridization demonstrated that culturable bacteria were unevenly distributed over the surface of the soil. Whilst this finding was in itself perhaps not surprising, this method offers the possibility, via use of multiple probes, to investigate localization of specific bacterial groups in soil, and/or, by using functional gene probes, the location of specific functional groups.

Most early applications of DNA hybridization experiments focused on identification of particular culturable bacteria using species-specific probes, and/or identifying isolates carrying particular genes of functional interests. However, with the development of methodologies to extract nucleic acids directly from environmental samples (36,37), this enabled microbial ecologists to apply DNA hybridization to detect and enumerate species and functional genes in the total as opposed to culturable microbial communities in the environment. Such applications are discussed in more detail below, but it is important also to consider the quality of nucleic acids being extracted from environmental samples. One common problem encountered when extracting DNA is the co-extraction of humic acids. Humic acids have been shown to interfere with DNA hybridization reactions (38), although a more recent study (39) demonstrated that low concentrations of humic acids (<20 ng humic acid per µl DNA) did not affect quantitative membrane hybridizations using DIG-labeled probes. Nevertheless, these studies highlight the need to ensure that community DNA used for hybridization analysis has been purified to remove as much humic acid as possible. The physical integrity of community DNA in terms of the degree of shearing or length of DNA fragments is less critical than for PCR amplifications in which in the latter, sheared DNA will result in increased chimeric formation (see Chapters 1 and 2). Therefore the choice of nucleic acid isolation method for DNA hybridization experiments is less important, and the main criterion is to remove as much of the humic acids from the sample as possible. A few reports have applied DNA hybridization to detect the expression of mRNA transcripts to provide an indication of gene expression in environmental samples. In one of the earliest studies, Tsai *et al.* (40) demonstrated the successful extraction of mRNA from soils seeded with bacteria carrying naphthalene (*nah*) degradation and mercury (*mer*) resistance genes, with subsequent detection of *nah* and *mer* sequences using slot blot and northern hybridization. Whilst their study showed that direct detection of mRNA transcripts was possible, there have been few subsequent reports of successful application of this approach to detect gene expression *in situ* by hybridization without prior seeding of bacteria. Nevertheless, Jeffrey *et al.* (41) did show how hybridization-based detection of mRNA sequences could be used to investigate gene expression. They investigated expression of the mercuric reductase-encoding *merA* gene as an indicator of mercury resistance, and combined their analysis with $^{203}Hg^{0}$-based volatalization experiments to investigate MerA-mediated Hg^{2+} reduction. In these experiments they were able to demonstrate using mRNA hybridization that the bacterial community showed acclimation following addition of Hg^{2+} ions into freshwater, with an increase in mRNA transcript numbers following addition. mRNA transcripts were also shown to increase in abundance following the additional supplementation with nutrients. Increase in transcript numbers correlated to increases in the levels of $^{203}Hg^{0}$, suggesting that transcripts were being converted to functional MerA enzymes. This elegant study showed the potential for DNA-hybridization-based analysis of mRNA in the environment, and with future developments in microarray analysis, widespread analysis of gene expression in the environment may prove feasible.

8.3 Hybridization applications in microbial ecology

Although DNA hybridization methodologies have been available for nearly 30 years, there are surprisingly few reviews discussing the application of *in vitro* DNA hybridization in microbial ecology. This is in sharp contrast to reviews describing fluorescent *in situ* hybridization (FISH)-based applications (see Chapter 9 and references therein). This subsequent section therefore reviews the application of *in vitro*-based nucleic acid hybridization approaches to study both cultured bacteria and also the total bacterial community following extraction of DNA. This section concludes with a brief review of the ongoing development of microarray-based approaches, which promises to play a major role in future studies in microbial ecology. The reader is also referred to early reviews of DNA-probing-based methods by Holben and Tiedje (42), Hazen and Jimenez (43) and Sayler and Layton (44).

8.3.1 Hybridization analysis of cultured bacteria

Early applications of DNA hybridization to microbial ecology focused on screening collections of environmental bacterial isolates for particular functional genes of interest. Such research has investigated the prevalence of heavy metal resistance genes and genes encoding biodegradation functions in bacteria from polluted habitats. The focus on metal resistance genes in particular benefited from the genetic and sequence characterization of a number of metal resistance determinants that enabled the generation of specific probes. Early studies focused especially on bacterial mercury resistance with a suite of probes developed to detect *mer* operons from both Gram-negative and Gram-positive bacteria (10,13,25,45,46). Similarly, DNA probes were developed to study distributions of a range of other heavy metal resistances including cadmium, zinc and cobalt (the czc-based probes) (29,47) nickel (48), copper (49,50) and chromate (47) both in heavy-metal-resistant bacteria and within the culturable bacterial community from a range of environmental sources including heavy-metal-polluted soils (29,47,48,50) and also piggeries (49). Subsequently, there has been considerable interest in characterization of bacterial isolates carrying biodegradation capabilities. As more sequence information has become available this has permitted the generation of an ever-increasing suite of probes including genes involved in degradation of polyaromatic hydrocarbons (*nah* genes and *ndoB*) (51,52), 2,4-dichlorophenoxyacetic acid (2,4-D) (*tfdA*) (53), dicyanide (54), and carbofuran (*mcd* gene) (55). Other functional genes for which gene probes have been applied to study gene distributions and diversity in culturable bacteria include include those involved in nitrogen cycling (denitrification) (56,57), antibiotic resistance (aminoglycoside acetyltransferases) (58) and screening for the presence of class I and class II integrase genes (59) in integrons which are responsible for formation of multidrug antibiotic resistance clusters in bacteria. The application of ribosomal RNA-derived probes to identify bacterial isolates from environmental sources is surprisingly uncommon, with most studies focusing on either FISH-based approaches (Chapter 9) or on taking a culture-independent approach via PCR generation of 16S rDNA clone libraries (Chapter 2).

8.3.2 Hybridization analysis of total community DNA

Following the development of methodologies to isolate DNA directly from environmental samples that would be representative of the total as opposed to the culturable community, there has been considerable interest in applying hybridization-based methodologies to investigate the distribution and diversity of bacterial species and genes encoding environmentally important functions. One fundamental research question is 'how many bacterial species are present within any given environment ?' and in the groundbreaking study by Torsvik (60) the principles of DNA hybridization were applied to provide estimates for bacterial species numbers in soils. Torsvik investigated heterogeneity between DNA molecules that were extracted from soil, subjected to thermal denaturation and then allowed to reassociate to form double-stranded hybrid molecules. The degree of reassociation will depend upon the extent to which single-stranded molecules will anneal (hybridize) to their counterparts. In a very simple community, reassociation will be commonplace as single-stranded sequences hybridize to complementary sequences. By contrast, in a complex community, the likelihood of sequences hybridizing to their complementary sequences will be reduced. Thus in such experiments by measuring the reassociation of DNA hybrids by spectrophotometric analysis to detect double-stranded molecules over time, the extent to which reassociation of the community DNA occurs can be determined, and used to provide estimates of overall species diversity. Using this method, Torsvik estimated that there were as many as 4000 different genomes within 1 g of soil, and this estimate is now widely reported in research that describes the complexity of environmental bacterial communities. More recent estimates have included genome numbers of 4900 and 5700 genomes per gram of soil (61).

Subsequently, variations on this approach have been developed to investigate similarities between bacterial communities from different environmental samples. In such applications, community DNA is extracted from each sample and then denatured and cross-hybridized following randomly primed-based labeling of the restriction-digested community DNA (62). The extent to which the community DNAs cross-hybridize provides an estimate of the similarities of the communities. When combined with additional denaturation–reassociation analysis, this can generate rapid estimates of community complexity. Whole community hybridization approaches were first applied to the analysis of marine bacterioplanktonic communities (62), and subsequently have been increasingly applied to the study of microbial communities in soils (63–65). For example Griffiths *et al.* (65) estimated that a series of four agricultural soils showed similarities ranging from 35 to 75%.

8.3.2.1 16S rRNA-based oligonucleotide probe analysis of bacterial communities

As discussed earlier, 16S rRNA oligonucleotide probes are most commonly used in FISH-based applications, and are reviewed by Amann and Ludwig (66). Nevertheless a number of studies have utilized the same series of probes to investigate relative abundances of the domains Bacteria, Archaea and Eukarya, or to detect and enumerate particular species of interest. Domain-specific probes were optimized by Zheng *et al.* (67) and tested both on defined mixtures of pure cultures and of rRNA samples obtained from anaerobic reactors. Similarly, domain-specific probes were used to investigate relative abundances of the three domains of life in petroleum- and oil-impacted salt marsh sediments via analysis of DNA extracted from the sediments (68). In addition to application of the domain-specific probes, hybridization probes have been applied to detect and quantify sulfate-reducing bacteria to the genus and group level (69), or to investigate vertical distributions of Chlorobiaceae and δ-proteobacteria in the water column and sediments of saline lakes (70).

8.3.2.2 Screening environmental community DNA with functional gene probes

The first report of the detection of specific functional genes from environmental samples was the detection of genes from *Bradyrhizobium japonicum* isolates that had been seeded into soil samples (71). In this study, both naturally occurring sequences (*rbcL*) and *nptII* genes that had been engineered to create a recombinant organism were successfully detected both by dot blot and Southern hybridization, with detection levels reported as low as 10^4 cells g^{-1}. Subsequently, as with the analysis of cultured isolates, the application of DNA probes to detect functional genes in community DNA isolated from environmental samples without prior seeding has focused on the detection and analysis of genes conferring metal resistance or biodegradation functions. Early reports included hybridization-based detection of *mer* sequences in water (9), and additionally, following the use of an additional PCR amplification stage to increase sensitivities, the detection of biphenyl degradation genes from sediments (15). More recently, the application of multiple probes to detect genes encoding a variety of biodegradative reactions in DNA extracted from contaminated soils has been pioneered (72–74). Such an approach aims to provide a characterization of hydrocarbon-polluted sites and an indication of the genetic potential within indigenous bacteria for possible natural attenuation or possible bioremediation of the polluted soils at a field-scale level. Probes used in these studies included those to detect genes encoding degradation of both aliphatic (*alkB*) and aromatic hydrocarbons (*nahH, nahA, todC1C2, tomA* and *xylA*) (72). Subsequently these probes were used to monitor changes in gene abundances (and by proxy of degradative bacteria) following addition of labeled aromatic hydrocarbon substrates (72) to enable degradation to be monitored, and additionally following nutrient additions (biostimulation) (73). Such studies represent a paradigm for the combinatorial analysis of microbial communities within complex environments by molecular and analytical methodologies. Similarly, investigation of degradation of the herbicide atrazine via gene-probe-based methods using *atzA, atrA* and *trzD*, to detect diverse atrazine degraders, in combination with MPN enumeration of atrazine degraders and determination of atrazine degradation rates (via addition of ^{14}C-labeled atrazine) has demonstrated that degradation genes related to those from particular bacterial genera (*Rhodococcus* and *Pseudomonas*) were more prevalant during atrazine degradation in agricultural soils than those related to the atrazine degradation genes from *Pseudomonas* (75). In addition to detection of biodegradation genes, the application of gene probes to detect and enumerate functional genes in environmental samples has included detection of genes involved in nitrogen cycling; denitrification in forest soils (76) and ammonia oxidation in lake water (77), and to investigate the diversity of plasmid replication genes representing broad host range plasmids in bacteria present in manure-treated and unamended soils (78).

8.3.3 Reverse sample genome probing (RSGP) and microarray analysis of microbial communities

Conventionally, in most hybridization experiments the target DNA (i.e. the DNA sequence that is being screened for the presence of a particular gene of interest) is immobilized on a membrane, whilst the DNA probe is supplied in solution. However, this limits the number of probes that can be screened in any single hybridization experiment and typically necessitates the requirement for stripping of the initial hybridized probe from the membrane with subsequent probing with a second probe. In RSGP (79) the opposite approach is taken, whereby the probe sequence is attached to the membrane and the community DNA being screened is labeled and supplied in solution. RSGP involves the isolation of genomic DNA from pure bacterial cultures isolated from the environmental sample of interest. Cross-hybridization analysis, as has been described above for community DNA (see 62–64), can then be used to identify bacterial genomes that show limited similarity (e.g. <70% cross-hybridization). DNA from these genomes are then denatured and applied onto a hybridization membrane, to which community DNA from the same environment that has been labeled by random priming is then hybridized. In this manner, the relative proportions of particular phylogenetic groups as indicated by hybridization signals can be determined. Dominant bacteria which have been cultured can then be analyzed further, e.g. initially by 16S rDNA sequencing, to provide presumptive identification. The application of RSGP is reviewed by Green and Voorduow (80). One attractive future development of RSGP may see the use of arrayed metagenomic clones (see Chapter 11) as probes in RSGP hybridization experiments to identify environmentally dominant sequences in the library, which can then be prioritized for sequencing. Another interesting variation on the RSGP approach has been developed by Pollard (81) to provide estimates of *in situ* bacterial growth rates of particular bacteria. In this study, genomic DNA was isolated from *Zoogloea ramigera* which is a common component of activated sludge communities. The genomic DNA was denatured and attached to a hybridization membrane, together with genomic DNA from a series of control bacteria. Subsequent methyl-[^{3}H]thymidine addition to activated sludge system resulted in the generation of bacteria containing ^{3}H-labeled DNA. Following extraction of total community DNA, hybridization reactions were then performed enabling the determination of the quantity of community DNA that had been labeled. By extraction of community DNA over time, estimates of bacterial growth rates were determined. This elegant approach could potentially be combined with the use of either a universal eubacterial probe, or alternatively an array of genus-specific probes to determine growth rates of different bacteria within a mixed community.

RSGP analysis represents a forerunner of arguably one of the most exciting recent developments in microbial ecology, namely the application of microarray analysis. Microarrays utilize essentially the same system as for RSGP, wherein it is the probe sequence that is immobilized on a solid support, either on a hybridization membrane or increasingly glass slides, as are routinely used in transcriptomic analysis of sequenced bacterial genomes. Whilst transcriptome- and also genomic-based hybridization analyses of microarrays from sequenced bacterial genomes are now relatively commonplace, the extension of such approaches to analyze total community DNA (or indeed RNA) is at present limited. Clearly there are obvious challenges concerning the complexity of environmental microbial communities, not least given the estimates of >4000 distinct genomes within a gram of soil. One of the major challenges to be overcome will relate to the design of the probes that will be used for screening of environmental DNA, to generate probes that have high specificity to the genes of interest. Such design considerations will also require careful empirical determination of hybridization and washing conditions that will constrain the hybridization stringency. Nevertheless, despite these far from insignificant difficulties, there have been a number of recent reports describing array-based analysis of total community DNA. One of the earliest studies focused on the application of 16S rRNA oligonucleotide probes which was used to study nitrifying bacteria and which demonstrated that some of the problems discussed above can successfully be overcome (82). More recently, Cho and Tiedje (83) provided a proof of concept demonstration for the application of a series of fragment probes (up to ~1 kb) in size derived from four species of *Pseudomonas*, which was then tested with the four bacterial species. Their study again suggested that there was considerable potential for future development of microarrays for analysis of complex communities. Two recent studies have reported the development of arrays to screen for functional genes. In the first study, a simple macroarray was developed containing a series of *nifH* sequences to investigate abundances of particular components of diazotrophs in marine waters (84). Although the macroarray was successfully used to detect *nifH* sequences in community DNA, detection required an initial PCR enrichment of *nifH* sequences; hence data would be qualitative rather than quantitative, highlighting that microarray analysis to detect at least some functional groups will require further developments in signal enhancement. Such potential problems are discussed in the review by Chandler (85). In the second and perhaps most exciting development, Rhee et al. (86) have reported the construction of a biodegradation array consisting of >1000 oligonucleotide probes that has been successfully tested with both bacterial cultures, microcosm enrichments and bulk community DNA. This microarray has allowed detection of a wide range of biodegradation genes, and has demonstrated that specific detection of multiple targets from environmental community DNA is achievable. The future development of other ecosystem and process-oriented microarrays offers enormous opportunities for the rapid investigation of microbial gene distribution and diversity in environmental samples. With the potential for future miniaturization and lab-on-a-chip-based approaches (87), these are extremely exciting times for microbial ecologists with future developments likely to include investigations of microbial community structure and function at the milli- or even micro-meter scale, that should revolutionize our understanding of microbial ecology.

References

1. Southern EM (1975) Detection of specific sequences among DNA fragments separated by gel electrophoresis. *J Mol Biol* **98**: 503–517.
2. Anderson MLM (1999) *Nucleic Acid Hybridization*. BIOS Scientific, Oxford, 240pp.
3. Sambrook J, Russell DW (2000) *Molecular Cloning: A Laboratory Manual*, 3rd edn. Cold Spring Harbor Laboratory Press, Cold Spring Harbor, NY.
4. Ludwig W, Dorn S, Springer N, Kirchhof G, Schleifer KH (1994) PCR-based preparation of 23S rRNA-targeted group-specific polynucleotide probes. *Appl Environ Microbiol* **60**: 3236–3244.
5. Heuer H, Hartung K, Wieland G, Kramer I, Smalla K (1999) Polynucleotide probes that target a hypervariable region of 16S rRNA genes to identify bacterial isolates corresponding to bands of community fingerprints. *Appl Environ Microbiol* **65**: 1045–1049.
6. Stahl DA, Amann R (1991) Development and application of nucleic acid probes. In: Stackebrandt E, Goodfellow M (eds), *Nucleic Acid Techniques in Bacterial Systematics*, pp. 205–242. John Wiley & Son, Ltd, Chichester.
7. Caren MP, Cattell HF, Tella RP, Webb PG, Bass JK (2004) Polynucleotide array fabrication. US Patent 20040203138.
8. Wetmur JG (1991) DNA probes: applications of the principles of nucleic acid hybridization. *Crit Rev Biochem Mol Biol* **26**: 227–259.
9. Barkay T, Liebert C, Gillman M (1989) Hybridization of DNA probes with whole community genome for detection of genes that encode microbial

responses to pollutants: *mer* genes and Hg^{2+} resistance. *Appl Environ Microbiol* **55**: 1574–1577.
10. Barkay T, Fouts DH, Olson BH (1985) Preparation of a DNA gene probe for detection of mercury resistance genes in Gram negative bacterial communities. *Appl Environ Microbiol* **49**: 686–692.
11. Couturier M, Bex F, Berquist PL, Maas WK (1988) Identification and classification of bacterial plasmid replicons. *Microbiol Rev* **52**: 375–395.
12. Osborn AM, Pickup RW, Saunders JR (2000) Development and application of molecular tools in the study of IncN-related plasmids from lakewater sediments. *FEMS Microbiol Lett* **186**: 203–208.
13. Olson BH, Lester JN, Cayless SM, Ford S (1989) Distribution of mercury resistance determinants in bacterial communities of river sediments. *Wat Res* **23**: 1209–1217.
14. Bej AK, DiCesare JL, Haff L, Atlas RM (1991) Detection of *Escherichia coli* and *Shigella* spp in water by using the polymerase chain reaction and gene probes for *uid*. *Appl Environ Microbiol* **57**: 1013–1017.
15. Erb RW, Wagner-Döbler I (1993) Detection of polychlorinated biphenyl degradation genes in polluted sediments by direct DNA extraction and polymerase chain reaction. *Appl Environ Microbiol* **59**: 4065–4073.
16. Loy A, Horn M, Wagner M (2003) ProbeBase: an online resource for rRNA-targeted oligonucleotide probes. *Nucleic Acids Res* **31**: 514–516.
17. Ashelford KE, Weightman AJ, Fry JC (2002) PRIMROSE: a computer program for generating, and estimating the phylogenetic range of 16S rRNA oligonucleotide probes and primers in conjunction with the RDP-II database. *Nucleic Acids Res* **30**: 3481–3489.
18. Höltke HJ, Ankenbauer W, Muhlegger K, Rein R, Sagner G, Seibl R, Walter T (1995) The digoxigenin (DIG) system for non-radioactive labelling and detection of nucleic acids-an overview. *Cell Mol Biol* **41**: 883–905.
19. Kumar A, Tchen P, Roullet F, Cohen J (1988) Nonradioactive labeling of synthetic oligonucleotide probes with terminal deoxynucleotidyl transferase. *Anal Biochem* **169**: 376–382.
20. Sgaramella V, Khorana HG (1972) Total synthesis of the structural gene for an alanine transfer RNA from yeast: enzymatic joining of the chemically synthesized polynucleotides to form the DNA duplex representing nucleotide sequence 1–20. *J Mol Biol* **72**: 427–444.
21. Joyce CM, Grindley ND (1983) Construction of a plasmid that overproduces the large proteolytic fragment (Klenow fragment) of DNA polymerase I of *Escherichia coli*. *Proc Natl Acad Sci USA* **80**: 1830–1834.
22. Feinberg AP, Vogelstein B (1983) A technique for radiolabeling DNA restriction endonuclease fragments to high specific activity. *Anal Biochem* **132**: 6–13.
23. Negoro S, Nakamura S, Okada H (1984) DNA-DNA hybridization analysis of nylon oligomer-degradative plasmid pOAD2: identification of the DNA region analogous to the nylon oligomer degradation gene. *J Bacteriol* **158**: 419–424.
24. Dahl KH, Lundblad EW, Røkenes TP, Olsvik Ø, Sundsfjord A (2000) Genetic linkage of the *vanB2* gene cluster to Tn*5382* in vancomycin-resistant enterococci and characterization of two novel insertion sequences. *Microbiology* **146**: 1469–1479.
25. Osborn AM, Bruce KD, Strike P, Ritchie DA (1993) Polymerase chain reaction-restriction fragment length polymorphism analysis shows divergence among *mer* determinants from gram-negative soil bacteria indistinguishable by DNA-DNA hybridization. *Appl Environ Microbiol* **59**: 4024–4030.
26. Grunstein M, Hogness DS (1975) Colony hybridization: a method for the isolation of cloned DNAs that contain a specific gene. *Proc Natl Acad Sci USA* **72**: 3961–3965.

27. Palva AM (1983) *ompA* gene in the detection of *Escherichia coli* and other Enterobacteriaceae by nucleic acid sandwich hybridization. *J Clin Microbiol* **18**: 92–100.
28. Sayler GS, Shields MS, Tedford ET, Breen A, Hooper SW, Sirotkin KM, Davis JW (1985) Application of DNA-DNA colony hybridization to the detection of catabolic genotypes in environmental samples. *Appl Environ Microbiol* **49**: 1295–1303.
29. Diels L, Mergeay M (1990) DNA probe-mediated detection of resistant bacteria from soils polluted by heavy metals. *Appl Environ Microbiol* **56**: 1485–1491.
30. Zühlsdorf MT, Wiedemann B (1992) Tn*21*-specific structures in gram-negative bacteria from clinical isolates. *Antimicrob Agents Chemother* **36**: 1915–1921.
31. Andersen SR, Sandaa RA (1994) Distribution of tetracycline resistance determinants among gram-negative bacteria isolated from polluted and unpolluted marine sediments. *Appl Environ Microbiol* **60**: 908–912.
32. Braun-Howland EB, Vescio PA, Nierzwicki-Bauer SA (1993) Use of a simplified cell blot technique and 16S rRNA-directed probes for identification of common environmental isolates. *Appl Environ Microbiol* **59**: 3219–3224.
33. Hu TL, Wu SC (2000) Comparison of colony lift with direct spotting methods of blot preparation on the effect of colony hybridization in the detection of environmental organisms. *J Microbiol Immunol Infect* **33**: 123–126.
34. Nizetic D, Drmanac R, Lehrach H (1991) An improved bacterial colony lysis procedure enables direct DNA hybridization using short (10, 11 bases) oligonucleotides to cosmids. *Nucleic Acids Res* **19**: 182.
35. Jordan FL, Maier RM (1999) Development of an agar lift-DNA/DNA hybridization technique for use in visualization of the spatial distribution of Eubacteria on soil surfaces. *J Microbiol Meth* **38**: 107–117.
36. Torsvik VL (1980) Isolation of bacterial DNA from soil. *Soil Biol Biochem* **12**: 15–21.
37. Ogram A, Sayler GS, Barkay T (1987) The extraction and purification of microbial DNA from sediments. *J Microb Methods* **7**: 57–66.
38. Tebbe CC, Vahjen W (1993) Interference of humic acids and DNA extracted directly from soil in detection and transformation of recombinant DNA from bacteria and a yeast. *Appl Environ Microbiol* **59**: 2657–2665.
39. Jordan FL, Maier RM (1999) Development of an agar lift–DNA/DNA hybridization technique for use in visualization of the spatial distribution of Eubacteria on soil surfaces. *J Microbiol Meth* **38**: 107–117.
40. Tsai YL, Park MJ, Olson BH (1991) Rapid method for direct extraction of mRNA from seeded soils. *Appl Environ Microbiol* **57**: 765–768.
41. Jeffrey WH, Nazaret S, Barkay T (1996) Detection of the *merA* gene and its expression in the environment. *Microb Ecol* **32**: 293–303.
42. Holben WE, Tiedje JM (1988) Applications of nucleic acid hybridization in microbial ecology. *Ecology* **69**: 561–568.
43. Hazen TC, Jimenez L (1988) Enumeration and identification of bacteria from environmental samples using nucleic acid probes. *Microbiol Sci* **5**: 340–343.
44. Sayler GS, Layton AC (1990) Environmental application of nucleic acid hybridization. *Annu Rev Microbiol* **44**: 625–648.
45. Gilbert MP, Summers AO (1988) The distribution and divergence of DNA sequences related to the Tn*21* and Tn*501 mer* operons. *Plasmid* **20**: 127–136.
46. Rochelle PA, Wetherbee MK, Olson BH (1991) DNA sequences encoding narrow- and broad-spectrum mercury resistance. *Appl Environ Microbiol* **57**: 1581–1589.
47. Dressler C, Kues U, Niehs DH, Friedrich B (1991) Determinants encoding resistance to several heavy metals in newly isolated copper-resistant bacteria. *Appl Environ Microbiol* **57**: 3079–3085.
48. Stoppel RD, Schlegel HG (1995) Nickel-resistant bacteria from anthropogenically

nickel-polluted and naturally nickel-percolated ecosystems. *Appl Environ Microbiol* **61**: 2276–2285.
49. Williams JR, Morgan AG, Rouch DA, Brown NL, Lee BTO (1993) Copper resistant enteric bacteria from United Kingdom and Australian piggeries. *Appl Environ Microbiol* **59**: 2531–2537.
50. Trajanovska S, Britz ML, Bhave M (1997) Detection of heavy metal ion resistance genes in Gram-positive and Gram-negative bacteria isolated from a lead-contaminated site. *Biodegradation* **8**: 113–124.
51. Ahn Y, Sanseverino J, Sayler GS (1999) Analyses of polycyclic aromatic hydrocarbon-degrading bacteria isolated from contaminated soils. *Biodegradation* **10**:149–157.
52. Hamann C, Hegemann J, Hildebrandt A (1999) Detection of polycyclic aromatic hydrocarbon degradation genes in different soil bacteria by polymerase chain reaction and DNA hybridization. *FEMS Microb Lett* **173**: 255–263.
53. Ka JO, Holben WE, Tiedje JM (1994) Genetic and phenotypic diversity of 2,4-dichlorophenoxyacetic acid (2,4-D)-degrading bacteria isolated from 2,4-D-treated field soils. *Appl Environ Microbiol* **60**: 1106–1115.
54. Teaumroong N, Schwarzer C, Auer B, Haselwandter K (1997) A non-radioactive DNA probe for detecting dicyandiamide-degrading soil bacteria. *Biol Fertil Soils* **25**: 159–162.
55. Parekh NR, Hartmann A, Fournier JC (1996) PCR detection of the *mcd* gene and evidence of sequence homology between the degradative genes and plasmids from diverse carbofuran-degrading bacteria. *Soil Biol Biochem* **28**: 1797–1804.
56. Von Berg KHL, Bothe H (1992) The distribution of denitrifying bacteria in soils monitored by DNA probing. *FEMS Microbiol Ecol* **86**: 331–340.
57. Cheneby D, Hartmann A, Henault C, Topp E, Germon JC (1998) Diversity of denitrifying microflora and ability to reduce N_2O in two soils. *Biol Fertil Soils* **28**: 19–26.
58. Lee YJ, Han HS, Seong CN, Lee HY, Jung JS (1998) Distribution of genes coding for aminoglycoside acetyltransferases in gentamicin resistant bacteria isolated from aquatic environment. *J Microbiol* **36**: 249–255.
59. Goldstein C, Lee MD, Sanchez S, Hudson C, Phillips B, Register B, Grady M, Liebert C, Summers AO, White DG, Maurer JJ (2001) Incidence of class 1 and 2 integrases in clinical and commensal bacteria from livestock, companion animals, and exotics. *Antimicrob Agent Chemother* **45**: 723–726.
60. Torsvik V, Goksoyr J, Daae FL (1990) High diversity in DNA of soil bacteria. *Appl Environ Microbiol* **56**: 782–787.
61. Chatzinotas A, Sandaa RA, Schonhuber W, Amann R, Daae FL, Torsvik V, Zeyer J, Hahn D (1998) Analysis of broad-scale differences in microbial community composition of two pristine forest soils. *Syst Appl Microbiol* **21**: 579–587.
62. Lee SH, Fuhrman JA (1991) Spatial and temporal variation of natural bacterioplankton assemblages studied by total genomic DNA cross-hybridization. *Limnol Oceanog* **36**: 1277–1287.
63. Ritz K, Griffiths BS (1994) Potential application of a community hybridization technique for assessing changes in the population structure of soil microbial communities. *Soil Biol Biochem* **26**: 963–971.
64. Griffiths BS, Ritz K, Glover LA (1996) Broad-scale approaches to the determination of soil microbial community structure: application of the community DNA hybridization technique. *Microb Ecol* **31**: 269–280.
65. Griffiths BS, Diaz Ravina M, Ritz K, McNicol JW, Ebblewhite N, Baath E (1997) Community DNA hybridization and %G+C profiles of microbial communities from heavy metal polluted soils. *FEMS Microbiol Ecol* **24**: 103–112.
66. Amann R, Ludwig W (2000) Ribosomal RNA-targeted nucleic acid probes for studies in microbial ecology. *FEMS Microbiol Rev* **24**: 555–565.

67. Zheng D, Alm EW, Stahl DA, Raskin L (1996) Characterization of universal small-subunit rRNA hybridization probes for quantitative molecular microbial ecology studies. *Appl Environ Microbiol* **62**: 4504–4513.
68. Bachoon DS, Hodson RE, Araujo R (2001) Microbial community assessment in oil-impacted salt marsh sediment microcosms by traditional and nucleic acid-based indices. *J Microbiol Meth* **46**: 37–49.
69. Devereux R, Kane MD, Winfrey J, Stahl DA (1992) Genus- and group-specific hybridization probes for determinative and environmental studies of sulfate-reducing bacteria. *Syst Appl Microbiol* **15**: 601–609.
70. Koizumi Y, Kojima H, Fukui M (2004) Dominant microbial composition and its vertical distribution in saline meromictic Lake Kaiike (Japan) as revealed by quantitative oligonucleotide probe membrane hybridization. *Appl Environ Microbiol* **70**: 4930–4940.
71. Holben WE, Jansson JK, Chelm BK, Tiedje JM (1988) DNA probe method for the detection of specific microorganisms in the soil bacterial community. *Appl Environ Microbiol* **54**: 703–711.
72. Stapleton RD, Sayler GS (1998) Assessment of the microbiological potential for the natural attenuation of petroleum hydrocarbons in a shallow aquifer system. *Microb Ecol* **36**: 349–361.
73. Stapleton RD, Ripp S, Jimenez L, Cheol-Koh S, Fleming JT, Gregory IR, Sayler GS (1998) Nucleic acid analytical approaches in bioremediation: site assessment and characterization. *J Microbiol Meth* **32**: 165–178.
74. Stapleton RD, Savage DC, Sayler GS, Stacey G (1998) Biodegradation of aromatic hydrocarbons in an extremely acidic environment. *Appl Environ Microbiol* **64**: 4180–4184.
75. Ostrofsky EB, Robinson JB, Traina SJ, Tuovinen OH (2002) Analysis of atrazine-degrading microbial communities in soils using most-probable-number enumeration, DNA hybridization, and inhibitors. *Soil Biol Biochem* **34**: 1449–1459.
76. Mergel A, Schmitz O, Mallmann T, Bothe H (2001) Relative abundance of denitrifying and dinitrogen-fixing bacteria in layers of a forest soil. *FEMS Microbiol Ecol* **36**: 33–42.
77. Hastings RC, Saunders JR, Hall GH, Pickup RW, McCarthy AJ (1998) Application of molecular biological techniques to a seasonal study of ammonia oxidation in a eutrophic freshwater lake. *Appl Environ Microbiol* **64**: 3674–3682.
78. Gotz A, Smalla K (1997) Manure enhances plasmid mobilization and survival of *Pseudomonas putida* introduced into field soil. *Appl Environ Microbiol* **63**: 1980–1986.
79. Voordouw G, Voordouw JK, Kark-Hoffschweizer RR, Fedorak PM, Westlake DWS (1991) Reverse sample genome probing, a new technique for identification of bacteria in environmental samples by DNA hybridization, and its application to the identification of sulphate-reducing bacteria in oil-field samples. *Appl Environ Microbiol* **57**: 3070–3078.
80. Greene EA, Voordouw G (2003) Analysis of environmental microbial communities by reverse sample genome probing. *J Microbiol Meth* **53**: 211–219.
81. Pollard PC (1998) Estimating the growth rate of a bacterial species in a complex mixture by hybridization of genomic DNA. *Microb Ecol* **36**: 111–120.
82. Guschin DY, Mobarry BK, Proudnikov D, Stahl DA, Rittmann BE, Mirzabekov AD (1997) Oligonucleotide microchips as genosensors for determinative and environmental studies in microbiology. *Appl Environ Microbiol* **63**: 2397–6402.
83. Cho JC, Tiedje JM (2001) Bacterial species determination from DNA-DNA hybridization by using genome fragments and DNA microarrays. *Appl Environ Microbiol* **67**: 3677–3682.
84. Jenkins BD, Steward GF, Short SM, Ward BB, Zehr JP (2004) Fingerprinting

diazotroph communities in the Chesapeake Bay by using a DNA microarray. *Appl Environ Microbiol* **70**: 1767–1776.

85. Chandler DP (2002) Advances towards integrated biodetection systems for environmental molecular microbiology. *Curr Issues Mol Biol* **4**: 19–32.
86. Rhee S-K, Liu X, Wu L, Chong SC, Wan X, Zhou J (2004) Detection of genes involved in biodegradation and biotransformation in microbial communities by using 50-mer oligonucleotide microarrays. *Appl Environ Microbiol* **70**: 4303–4317.
87. Liu W-T, Zhu L (2005) Environmental microbiology-on-a-chip and its future impacts. *Trends Biotechnol* **23**: 174-179.

Protocols

The following sections outline three applications of nucleic acid hybridization:

8.1 Rapid extraction of total genomic DNA from culturable isolates for dot blot hybridization.

8.2 Southern hybridization of PCR products on agarose gels to confirm identity of amplification products.

8.3 Dot blot hybridization of total nucleic acids from environmental sources.

Each of the methods utilizes digoxigenin (DIG)-labeled probes. The reader is referred to the applications guide produced by Roche for full details on the use of DIG-labeled probes (http://www.roche-applied-science.com/fst/products.htm?/DIG).

Protocol 8.1: Rapid extraction of total genomic DNA from culturable isolates for dot blot hybridization

This protocol first describes a rapid method for total genomic DNA isolation that can be used as an alternative to colony blot hybridization to prepare genomic DNA from large numbers of environmental isolates (or clones) to screen for the presence of a particular gene of interest, in this case the *bphC* gene encoding a C-2,3-O biphenyl dioxygenase in *Burkholderia* sp. LB400 (15). During the method cells are lysed and genomic DNA visualized by agarose gel electrophoresis prior to loading the samples onto a hybridization membrane using a dot (or slot) blot apparatus (typically 96 dots to a membrane). This method has the advantage that the genomic DNA is physically visualized on the gel. By contrast, in colony blot hybridization, a negative signal could be due either to the absence of a particular gene in an isolate or to failure to lyse the isolate during the colony blot lysis stage. This protocol then describes the preparation of a digoxigenin (DIG)-labeled probe that is used to screen the genomic DNA samples by DNA hybridization, with subsequent chemiluminescent detection of the DIG probe.

MATERIALS

Reagents

Reagents for DNA isolation

Cell lysis buffer: TE (pH 8.0) containing 5% (v/v) antifoam emulsion (Sigma) and 0.1 mg ml^{-1} xylene cyanol

1 M NaOH saturated with SDS (sodium dodecyl sulfate)

Acid phenol/chloroform/isoamylalcohol (25:24:1); note: use unbuffered phenol

Ethidium bromide

0.2 N NaOH

Hybridization membrane (Roche)

Reagents for DNA probe preparation

PCR DIG probe synthesis kit (Roche)

Two oligonucleotide primers:

PS4U: 5′-CTCCAGCCATACTCGACCTC-3′

PS4D: 5′-ATCGCCGTTCAGCAGGGCGA-3′

Taq DNA polymerase

Sterile distilled water

PCR purification kit (Qiagen or similar)

Hybridization membrane (Roche)

Reagents for DNA hybridization

20 × SSC: 3 M NaCl, 0.3 M sodium citrate (pH 7.0)

2 × SSC: 1:10 dilution of 20 × SSC

Maleic acid buffer: 0.1 M maleic acid, 0.15 M NaCl and adjust pH to 7.5 using NaOH

Blocking reagent stock solution: Dissolve Blocking Reagent (Roche) in maleic acid buffer (see above) to a final concentration of 10% (w/v) with stirring and heating on a stir plate. The blocking reagent must be heated while it dissolves in the maleic acid buffer. Boiling will cause the reagent to coagulate so care should be taken to avoid this. The solution will be turbid.

Hybridization solution: 5 × SSC, 0.1% N-lauroylsarcosine, 0.02% SDS, and 1% Blocking Reagent (from stock solution—see below)

Washing solutions:	4 ×:	4 × SSC, 1% SDS
	2 ×:	2 × SSC, 0.1% SDS
	0.01 ×:	0.5 × SSC, 0.1% SDS

Reagents for chemiluminescent detection

Maleic acid buffer: See above

Blocking solution: Dilute blocking reagent stock solution (see above) 1 in 10 with maleic acid buffer

DIG antibody: Anti-digoxigenin-AP-Fab fragments, conjugated to alkaline phosphatase (Roche)

Detection buffer: 100 mM Tris-HCl, 100 mM NaCl (pH 9.5)

Washing buffer (WB): Add 0.3% (w/v) Tween 20 to maleic acid buffer

CSPD substrate (Roche)

Note: Blocking reagent, DIG antibody and CSPD substrate are components of the DIG luminescent detection kit (Roche)

Equipment

Microcentrifuge

Agarose gel electrophoresis apparatus and power source

UV transilluminator and gel documentation system/Polaroid camera

Dot blot manifold apparatus

Vacuum pump

Oven (120°C)

PCR thermal cycler

Densitometry equipment (or scanner and gel documentation software, e.g. Biorad).

Hybridization oven and bottles (Hybaid, Techne) or similar

68°C and 95°C waterbaths

Gilson micropipettes

Heated bag sealer

X-ray cassette

Additional materials

1.5 ml Eppendorf tubes

Glass universal bottles (or similar)

A4 plastic page protectors (or use Saran Wrap)

X-ray film and developing chemicals

METHODS

DNA isolation

1. Set up 0.5 ml liquid cultures (nutrient broth) of each environmental isolate, and grow overnight at 30°C with shaking. Include cultures of positive and negative controls, i.e. bacterial strains containing the DNA probe sequence, or that lack the sequence of interest.
2. Harvest cells by centrifugation at 13 000 rpm for 3 min in Eppendorf tubes, and discard supernatant.
3. Resuspend the cell pellet in 100 µl of cell lysis buffer.
4. To lyse the cells add 20 µl of 1 M NaOH saturated with SDS, and invert tubes for 1 min.
5. Vortex the cells for an additional 1 min.
6. Load 20 µl of the crude cell lysate on to a 0.8% agarose gel. Run gel at 100 V for 1 h.
7. Stain gel in dilute ethidium bromide solution (1 µg ml^{-1}) and visualize gel under UV light (302 nm). (Note: ethidium bromide is a powerful mutagen, and is moderately toxic. Gloves should be worn at all times when working with ethidium bromide stained gels, and any solutions containing this dye.) Photograph the gel under UV light at 302 nm.
8. Whilst the gel is running, add 80 µl of acid phenol/chloroform/isoamylalcohol to the remaining crude cell extracts. Invert the tubes for 1 min, and then centrifuge at 13 000 rpm for 5 min.
9. The upper aqueous layer (containing purified genomic DNA) can be used directly for dot blot hybridization. By comparison of the band intensities of the genomic DNA extracts, equivalent amounts of DNA from each isolate can be estimated.
10. Aliquot 10 µl of 0.2 NaOH into a series of Eppendorf tubes, and add the genomic DNA (between 5–15 µl depending on band intensity) to the tubes. Mix briefly by pipetting, and then transfer the tubes containing the denatured DNA immediately to ice.

11. Transfer the entire denatured sample (up to 20 µl) onto the hybridization membrane by using a dot blot manifold and leave for 30 min.
12. Apply suction for ~30 s using a vacuum pump.
13. Disassemble the dot blot manifold and then bake the membrane at 120°C for 30 min. The membrane is now ready to be used for DNA hybridization.

Preparation of digoxigenin-labeled DNA probe by PCR

This method describes the use of the PCR DIG probe synthesis kit (Roche) to generate a probe for the *Burkholderia* sp. LB400 *bphC* gene (15). Employing this method, digoxigenin-11-dUTP (DIG-11-dUTP) is incorporated by Taq DNA polymerase during the PCR.

1. Set up a 50 µl PCR using the PCR DIG probe synthesis kit containing the following: 5 µl of 10 × DIG-PCR buffer (Roche), 5 µl DIG-labeled dNTPs (Roche), 20 pmol of each of the forward and reverse primers, 5 U of Taq DNA polymerase, ~100 ng template DNA (*Burkholderia* sp. LB400 genomic DNA), made up to 30 µl with sterile distilled water.
2. Carry out the PCR using the following cycle: initial denaturation at 94°C for 2 min, followed by 30 cycles of 94°C for 1 min, 60°C for 1 min, 72°C for 2 min, and a final extension of 72°C for 10 min. Run 3 µl of the PCR product on a 1% agarose gel to confirm amplification.
3. Purify the remainder of the probe using a Qiagen PCR purification kit, and elute the purified PCR product in a final volume of 50 µl. Check 3 µl of the purified PCR product on a 1% agarose gel.
4. The yield of the DIG-labeled probe can be estimated by comparison to the labeled control DNA supplied in the probe synthesis kit. Briefly: set up a a dilution series for both the labeled probe and the control DNA (undiluted to 10^{-4}), and spot the samples onto a hybridization membrane (two rows of spots with ~1 cm between spots).

5. Place the membrane in an oven at 120°C for 30 min to fix the DNA to the membrane.
6. Carry out chemiluminescent detection of the labeled probe and control DNA as detailed below (see Chemiluminescent detection).
7. The yield of the probe can be determined either approximately, or by densitometry by comparison to the hybridization signals of the control DNA.

DNA hybridization (including washing steps)

1. Set the temperature of the hybridization oven to 68°C.
2. Lay the hybridization membrane (DNA side up) in a shallow tray in 50–100 ml of 2 × SSC. (Note: handle membranes with non-powdered gloves and/or blunt-ended forceps.) Roll up the membrane (DNA side facing inside the roll) and place the membrane into a hybridization bottle. Add ~5 ml of 2 × SSC, and rotate the bottle until the membrane unwinds and is in contact with the glass, ensuring that there are no air bubbles between the membrane and the glass. Multiple membranes can be placed side by side in the hybridization bottle. Pour away the 2 × SSC from the bottle.
3. Add 10 ml of hybridization buffer (pre-heated at 68°C) to the hybridization bottle, and place the bottle in the hybridization oven and leave to rotate at medium speed for 2 h. This stage is prehybridization.
4. Denature the double-stranded DNA probe by boiling 10 µl of the probe (100 ng) in an Eppendorf tube in a waterbath (>95°C for 10 min). Then immediately transfer the Eppendorf tube containing the probe onto ice to prevent renaturation of the two DNA strands.
5. Pour the hybridization solution from the hybridization bottle into a glass universal bottle. Pipette the denatured probe into the decanted prehybridization solution and mix well (this is now the hybridization solution).
6. Pour the hybridization solution back into the hybridization bottle and set up in the oven as

before. Leave the membrane to hybridize overnight at 68°C.

Following hybridization, a series of washing steps using washing solutions containing decreasing salt concentrations are performed to sequentially remove unbound probe from the membrane (non-specific binding), and from where the DNA probe may have bound weakly to non-homologous DNA sequences (e.g. showing less than ~70% identity to the probe).

7. Discard the hybridization solution from the Hybaid bottle. Note: for DIG-labeled probes, the probe can be retained, if frozen immediately, and then reused for future hybridization experiments.
8. Add 50 ml of pre-warmed (68°C) 4 × washing solution into the hybridization bottle. Rotate in the oven at 68°C for 5 min. Pour off the washing solution. Repeat the process with fresh pre-warmed 4 × washing solution.
9. Repeat the process in step 8, using pre-warmed 2 × washing solution at 68°C, but with rotatation for 15 min (two washes).
10. Repeat the washing stage by addition of 50 ml of pre-warmed 0.1 × washing solution at 68°C (this wash is carried out only once). Rotate for 15 min. Pour off the washing solution.

Chemiluminescent detection

DIG-labeled probes can be detected by a variety of techniques: fluorescence-based detection, chemiluminescent detection and colorimetric detection. The following method describes chemiluminescent detection, whereby the anti-DIG antibodies conjugated to alkaline phosphatase bind to the digoxigenin incorporated into the probe. A chemiluminescent alkaline phosphatase substrate (CSPD) is then added, which, following removal of the phosphate group, emits a luminescent signal that can be detected using X-ray film.

1. Take the hybridization membrane out of the hybridization bottle and wash the membrane briefly in ~100 ml of maleic acid buffer in a plastic tray.

2. Add 20 ml of blocking solution. Transfer the membrane into a clean hybridization bottle and add 20 ml of blocking solution. Incubate the membrane in the oven (with rotation) for 30 min at room temperature.
3. Dilute the DIG antibody 1:10 000 in blocking solution in a glass universal, i.e. 2 μl of antibody in 20 ml blocking solution.
4. Discard the blocking solution from the hybridization bottle containing the membrane and pour in the antibody solution (from step 3) and incubate the membrane in the diluted antibody solution in the oven (with rotation) for 30 min. at room temperature.
5. Pour off the antibody solution from the membrane. Add 50 ml of washing buffer (WB) and return the bottle containing the membrane to the oven for 15 min at room temperature. Repeat this step.
6. Discard the washing buffer (WB) and transfer the membrane to a plastic tray containing ~100 ml of detection buffer for 2 min at room temperature. Discard the detection buffer and place the drained membrane (DNA side up) between the sheets of an A4 plastic page protector (pre-cut so that it opens easily).
7. Dilute the CSPD substrate 1:100 in detection buffer in an Eppendorf tube, i.e. add 10 μl of CSPD substrate to 990 μl detection buffer.
8. Apply the substrate to the membrane: lift the top sheet of the plastic page protector and pipette the diluted substrate (1 ml) carefully over the membrane, pipetting the drops evenly over the membrane. Lower the top sheet of plastic carefully and press out any air bubbles by wiping outward from centre. Incubate at room temperature for 5 min.
9. Prepare a fresh page protector, and transfer the membrane into this. Seal the membrane within the dry page protector using a heated bag sealer.
10. Incubate the membrane enclosed in the page protector for 15 min at 37°C. After the incubation time a steady state reaction is reached between the enzyme and substrate.

Multiple exposures to X-ray film can then be performed during the next 24–48 h.

11. Spots on the resulting autoradiographs can then be scored for hybridization signals, and additionally the relative strengths of hybridization signals can be determined via densitometry to identify isolates carrying genes either closely related to the probe sequence, or genes that are more divergent. Check also that the probe has only bound to the positive control and not to any negative controls (distilled water) and/or DNA controls that lack the probe sequence.

Protocol 8.2: Southern hybridization of PCR products on agarose gels to confirm identity of amplification products

This protocol describes a Southern transfer (1) based screen to confirm the identity of PCR amplification products. PCR is routinely used in microbial ecology to detect the presence of particular species or genes in environmental samples. However, the visualization of a PCR product by agarose gel electrophoresis does not confirm that the PCR product is actually the gene of interest. This is especially true when studying functional genes for which amplification products may vary in size. Hence additional analysis should be undertaken to confirm the identity of the amplified products. This can be done via DNA sequencing, or alternatively, to screen larger numbers of PCR products, Southern hybridization methods can be employed. Southern hybridization as opposed to dot blot hybridization of the PCR product is preferable, as the former will confirm whether (or not) the probe will hybridize to the amplification product of the expected size. For amplification of functional genes from environmental samples the generation of multiple PCR products is not uncommon. The following method thus describes post-PCR analysis to screen a series of amplification products of the *bphC* gene (see above) from either individual isolates or from environmental DNA.

MATERIALS

Reagents

Type II DNA loading buffer: 0.25% bromophenol blue, 0.25% xylene cyanol, 15% Ficoll Type 400

Ethidium bromide

Sterile distilled water

Denaturing solution: 0.5 N NaOH, 1.5 M NaCl

Neutralization solution: 0.5 M Tris-HCl, 3 M NaCl (pH 7.5)

20 × SSC and 2 × SSC: See Protocol 8.1.

Equipment

Gel electrophoresis apparatus and power supply

Shaking table

Oven (120°C)

Additional materials

1 kb DNA ladder (Invitrogen) or similar

Plastic tray

Hybridization membrane (Roche)

Forceps

Scissors

3mm Whatmann paper

Absorbent paper towels

Cling film

Pencil

METHODS

1. Load 10 µl of PCR products (using 2 µl of Type II DNA loading buffer) onto a 1% agarose gel. Include also 10 µl of the PCR negative control, a positive control (e.g. restriction digest of plasmid containing cloned probe sequence, and a DNA marker (e.g. 1 kb ladder, Invitrogen). Run gel at 80 V for 1–2 h to achieve good separation of DNA fragments.
2. Stain gel in dilute ethidium bromide solution (1 µg ml^{-1}) and visualize gel under UV light

(302 nm). (Note: ethidium bromide is a powerful mutagen, and is moderately toxic. Gloves should be worn at all times when working with ethidium bromide-stained gels, and any solutions containing this dye.) Place a ruler by the side of the gel and then photograph gel.

3. Rinse the gel with distilled water (dH_2O) in a shallow plastic tray using enough water to submerge the gels.
4. Denature the DNA by submerging the gel in denaturing solution for 15 min at room temperature in the plastic tray, place on a shaking table. Discard the denaturing solution and repeat the process.
5. Rinse the gel again with dH_2O as before.
6. Neutralize the DNA: Submerge the gel in neutralization solution for 15 min at room temperature (on the shaking table as before). Discard the neutralization solution and repeat the process. Whilst the gel is soaking in neutralization solution continue with Stage 7.
7. Prepare the nylon membrane: cut a piece of nylon membrane to the same size of the gel (the gel can be trimmed of excess material). Lay the membrane in a tray containing 2 × SSC. Use enough 2 × SSC to saturate the membrane. (Note: nylon membranes should be handled by the edges with non-powdered gloves and/or blunt-ended forceps.)
8. After the neutralization washes are completed, set up a Southern transfer apparatus using capillary transfer as follows. Place a support (e.g. a thick glass plate) over a plastic tray and place on top of this (in order):
   (i) two pieces of 3 mm Whatman filter paper wetted with 20 × SSC. Cut the filter paper slightly wider than the gel and long enough to hang over the glass and into the tray with an additional 5 cm in length;
   (ii) the gel (face up);
   (iii) the hybridization membrane;
   (iv) five pieces of 3 mm Whatman filter paper (cut to the same size as the gel);

(v) a wad of absorbent paper towels ~5 cm high (cut to the same size as the gel);

(vi) a second smaller glass plate with a heavy weight (~0.5 kg) positioned on top.

9. Half fill the plastic tray with 20 × SSC buffer. Make sure that the Whatman paper hangs down into the buffer and that it is saturated in all areas. Cover any exposed areas of buffer with cling film. Leave the Southern transfer apparatus overnight.

10. The next day, remove the weights, top glass plate and paper towel wad. Carefully remove the pieces of Whatman paper and then the nylon membrane. Mark (using a pencil) the side of the membrane that was against the gel; this face of the membrane should face inwards (i.e. into solution) when the membrane is transferred to a hybridization tube. Additionally cut off a small triangle from the top right corner of the membrane to allow future orientation.

11. Visualize the gel under UV light after blotting, to confirm that all of the DNA has been transferred to the membrane.

12. Bake the hybridization membrane DNA side up between two pieces of Whatman paper in an oven at 120°C for 30 min. Baked membranes can be stored for future hybridization at room temperature.

13. Hybridization, washing and detection stages can then be carried out as described for Protocol 8.1. The resulting autoradiographs can then be compared to the photograph of the agarose gel (enlarged to the equivalent size as the autoradiograph). Check that the probe has bound to the positive control but not to the DNA ladder, and check also that there is no hybridization signal in the negative control. This would indicate contamination in the original PCR. PCR products that hybridize to the probe (in this case the *bphC* gene) are confirmed as *bphC* gene-related sequences. Other PCR products generated may be non-specific amplicons or alternatively more divergent sequences, and can be characterized further by DNA sequencing.

Protocol 8.3: Dot blot hybridization of total nucleic acids from environmental sources

This method can be used to detect the presence of specific genes of interest in an environmental sample; also to compare relative gene abundances in different environmental samples using densitometry.

MATERIALS

Reagents, equipment and materials are as stated above.

METHODS

1. Total genomic DNA can be isolated from environmental sources as detailed in Chapter 1.
2. Prepare a dilution series (undiluted to 10^{-4}) in Eppendorf tubes of each environmental DNA sample (2 μl of genomic DNA and 18 μl sterile distilled water). Prepare also a series of tubes containing a positive control (genomic DNA, cloned construct or PCR product containing the probe), negative controls (DNA samples as above but lacking the probe sequence) and also a reagent negative (20 μl sterile distilled water).
3. Add 20 μl of 0.2 N NaOH to each of the above samples to denature the DNA, and mix by pipetting sample up and down; then transfer tubes immediately onto ice.
4. Load 20 μl (in duplicate) of the denatured DNA onto the hybridization membrane using a dot blot manifold, and leave for 30 min. Apply a vacuum across the manifold for ~30 s.
5. Place the membrane in an oven at 120°C for 30 min to fix the DNA to the membrane.

6. Hybridization, washing and detection stages can then be carried out as described for Protocol 8.1. Spots on the resulting autoradiographs can then be scored for hybridization signals, and the relative strengths of hybridization signals can be determined via densitometry (of the duplicate spots) to determine relative gene abundances in the environmental samples. Check also that the probe has only bound to the positive control and not to any negative controls (distilled water) and/or DNA controls that lack the probe sequence.

Fluorescence *in situ* hybridization for the detection of prokaryotes

9

Holger Daims, Kilian Stoecker and Michael Wagner

9.1 Introduction

Determining the structure and dynamics of microbial communities is a core component of microbial ecology. Traditionally, microbiologists used time-consuming cultivation-based methods to detect prokaryotes in the environment and to estimate their abundance. However, today it is common knowledge that the vast majority of prokaryotes cannot be cultured in the laboratory by applying standard methods (1). Furthermore, cultivation approaches underestimate the actual environmental abundance even of culturable bacteria since efficient dispersal of biofilms is not always possible and many bacteria that are in principle culturable can occur in a so-called 'viable but not culturable' state (2). Therefore, for most ecosystems the majority of numerically and functionally important microbial community members are not available as pure cultures. This situation might improve with the more widespread implementation of innovative cultivation techniques (3–7) but detailed censuses of complex microbial communities require a fundamentally different approach.

Since the mid 1980s 5S and (later) 16S rRNA gene sequence retrieval from complex communities (see Chapter 2) has allowed microbial ecologists to detect and phylogenetically assign microorganisms without the need of a preceding cultivation step (8). The widespread use of this rRNA approach has led to the identification of a surprisingly high number of previously unknown prokaryotes and to the discovery of >40 novel bacterial phyla since 1987 (9–11). Today, almost 100 000 16S rRNA sequences have been deposited in public databases (12), enormously improving our perception of prokaryotic diversity.

Microbial ecologists not only want to determine the phylogeny of environmental microorganisms but are also interested in the morphology, localization, abundance and activity of these organisms. These goals can be tackled by applying fluorescence *in situ* hybridization with nucleic acid probes that target signature regions of rRNA molecules (13–15). In this chapter, we describe the basis of FISH, provide protocols for its application in environmental microbiology and briefly discuss how FISH can be combined with other techniques in order to determine the activity and function of the detected organisms.

9.2 Fundamentals of rRNA FISH

rRNA molecules are ideal target molecules for detection of prokaryotes because they are ubiquitously distributed, contain conserved and variable sequence regions and are naturally amplified within microbial cells as integral parts of the ribosome. In its most widespread format, FISH uses fluorescently labeled oligonucleotide probes for detection of 16S rRNA within prokaryotic cells. The degree of conservation of the probe target sequence determines the phylogenetic depth of the group of organisms targeted by the probe. For example, species, genus, family, order, and domain-specific probes can be designed (14). Generally, the high degree of conservation of rRNA molecules makes them unsuitable to differentiate between strains of a given species. During the last 15 years, hundreds of rRNA-targeted oligonucleotide probes for *in situ* detection of various entities of prokaryotes (from the domain to the species level) have been published and can be accessed online together with annotated information by referring to probeBase [http://www.microbial-ecology.net/probebase; (16)]. Using these probes and the recommended hybridization conditions, the respective target organisms can be specifically detected in appropriately fixed environmental samples within a few hours after sampling by epifluorescence or confocal laser scanning microscopy (see *Figure 9.1*). In addition, software tools are available to design new rRNA-targeted oligonucleotide probes (17–19) for FISH. It is recommended always to use more than one probe for the detection of a particular target organism in order to increase its reliability. Multiple probe hybridization is possible because probes labeled with different fluorescent dyes can be applied simultaneously in the same hybridization experiment (20). *Table 9.1* shows characteristics of the most frequently applied fluorochromes for FISH.

Positive FISH signals from a prokaryotic cell (preferably conferred by more than one specific oligonucleotide probe) in an environmental sample are used to identify the organism and to show its presence in the system. However, for several reasons the absence of FISH signals with specific probes does not necessarily mean that the target organism is not present in the environment being studied.

(i) With a requirement for 10^3–10^4 target cells per ml of sample, the detection limit of FISH is relatively high. It is possible to lower the detection limit in some cases by pre-filtration of aquatic samples containing

Table 9.1 Excitation and emission wavelengths of fluorescent dyes suitable for labeling rRNA-targeted oligonucleotide probes

Fluorochrome	Excitation (nm)	Emission (nm)	Fluorescence colour
Oregon Green 488	490, 493	514, 520	Green
5(6)-Carboxyfluorescein-N-hydroxysuccinimide ester (FLUOS)	492	518	Green
Cy3	514	566	Orange-red
Tetramethyl rhodamine isothiocyanate (TRITC)	550	573	Red
Cy5	649	666	Near infrared

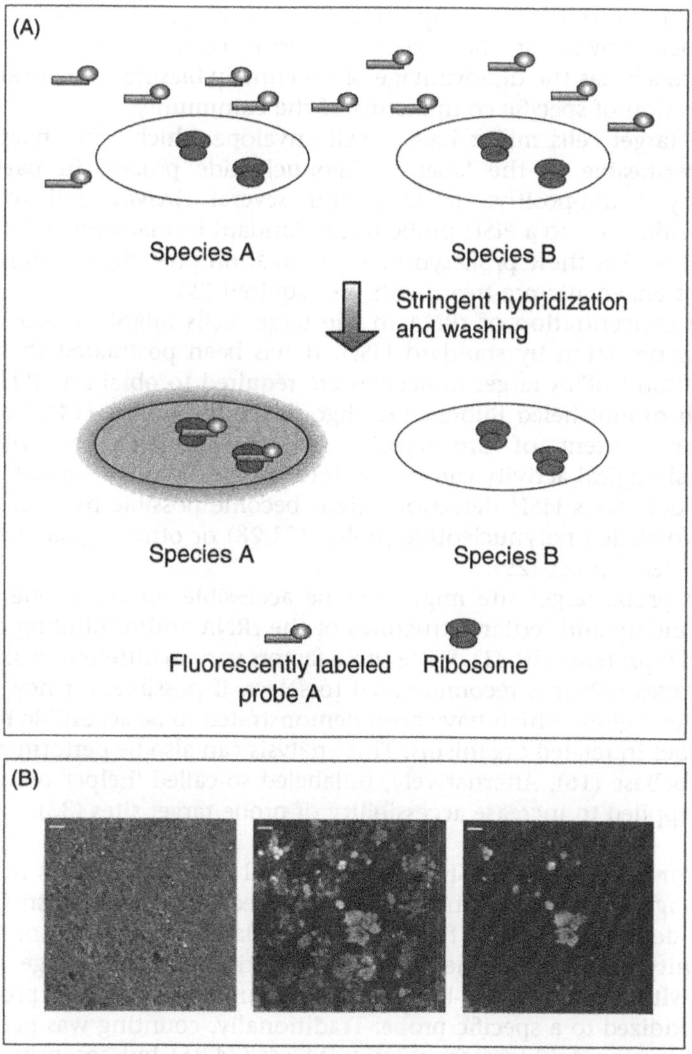

Figure 9.1

(A) Principle of FISH with rRNA-targeted oligonucleotide probes. A fluorescently labeled probe (probe A) is applied to specifically detect cells of its target species A in an environmental sample. When the correct hybridization and washing conditions are applied the probe does not bind to ribosomes of species B, which occurs in the same habitat. (B) Application of FISH to specifically detect *Nitrospira*-like bacteria in a complex biofilm sample. Left panel: DIC micrograph of the biofilm. Central panel: FISH of the biofilm with the EUB338 probe mix, which detects most known Bacteria (71). Right panel: FISH of the biofilm with probe S-G-Ntspa-0662-a-A-18, which detects the genus *Nitrospira* (41). Bar = 10 µm.

planktonic cells (21) or by including a short pre-enrichment step to induce growth of the target organisms (22), although the latter approach has the disadvantage of potentially biasing numerical representation of specific components of the community.

(ii) The target cells might have a cell envelope which, after fixation, is impermeable to the labeled oligonucleotide probes. In particular, many Gram-positive bacteria and several *Archaea* fail to show hybridization to a FISH probe if the standard formaldehyde fixation is applied. For these prokaryotes, other fixation procedures and in some cases enzymatic pre-treatments are required (23–27).

(iii) The concentration of rRNA in the target cells might be too low for their detection by standard FISH. It has been postulated that a few thousand rRNA target molecules are required to obtain a FISH signal with monolabeled fluorescent oligonucleotide probes (14). The ribosome content of prokaryotic cells in dormancy or with low physiological activity can be significantly below this threshold value. In such cases FISH detection might become possible by using either multilabeled polynucleotide probes (21,28) or other signal amplification techniques (29).

(iv) The probe target site might not be accessible for the probe due to secondary and tertiary structures of the rRNA and/or binding of ribosomal proteins (30,31). Since these factors vary in different prokaryotic lineages (32) it is recommended to select, if possible, for new probes target regions which have been demonstrated to be accessible for FISH probes in related organisms. This analysis can also be performed using probeBase (16). Alternatively, unlabeled so-called 'helper probes' can be applied to increase accessibility of probe target sites (33).

A major advantage of FISH compared to all other techniques in microbial ecology is that the abundance of the detected microorganisms can be directly determined. Most frequently the relative abundance of specific community members is measured by counting the percentage of cells stained with a general DNA-binding dye or a universal bacterial probe that also hybridized to a specific probe. Traditionally, counting was performed manually using epifluorescence microscopy (34,35) but recently a semi-automated counting procedure has been reported (36). In samples containing biofilms or aggregated bacteria, counting of individual cells is imprecise and very tedious. Therefore, confocal laser scanning microscopy and digital image analysis has been used to semi-automatically measure the biovolume of specifically stained cells (37,38). Biovolume measurements can be used to determine both the relative (39) and absolute abundance (40) of specific components of a microbial community. Besides abundance, the localization of FISH-stained microorganisms can be observed microscopically (41,42). Co-localizations of bacterial populations in biofilms are signposts for interactions and thus provide valuable indicators to identify metabolic networks between community members.

FISH has been successfully applied in many environmental samples (43). Nevertheless, it is important to keep in mind that FISH is not suited for every kind of sample. In particular, if samples contain abiotic material or cells with a high intrinsic fluorescence (for example cyanobacterial microbial mats), fluorescence-based detection becomes difficult. Similar problems arise if samples contain components to which oligonucleotide probes adsorb. Furthermore, if bacteria grow within material which is not translucent (for example soil particles), quantitative FISH analyses are, if at all, only possible after thin-sectioning of the samples. For such samples it is recommended to apply other techniques, such as quantitative dot blot hybridization (44), quantitative real-time PCR (see Chapter 6) and oligonucleotide microarrays (45), which also allow determining the structure of complex microbial communities.

9.3 Structure–function analyses: combination of FISH with other techniques

To understand the relationships between microorganisms and their environment, data on the structure and dynamics of microbial communities must be combined with information regarding the activity and ecophysiology of its members.

FISH using rRNA-targeted oligonucleotide probes has also been used to estimate the activity of bacteria in the environment (46). For this purpose the FISH signal intensity per detected cell, which reflects the cellular ribosome concentration, is measured by digital image analysis. A strong FISH signal is interpreted as an indication of physiological activity because for several bacteria a relatively tight correlation between ribosome content and physiological activity has been observed in pure culture experiments (47). However, many factors influence the stability of prokaryotic rRNA in environmental bacteria (48,49), challenging the assumption that all prokaryotic cells which are detectable by standard FISH are active. For example, several slow-growing chemolithoautotrophic bacteria retain high rRNA levels even after prolonged periods of inactivity (50–52). Several recent studies indicate that precursor rRNA sequences, which flank the mature rRNA sequence on the primary transcript of the *rrn* operons, might be better targets than the rRNA itself for inferring the activity of microorganisms via FISH (51,53,54). However, the low degree of conservation of precursor rRNA sequences complicates probe design for broader phylogenetic target groups.

In recent years various creative combinations of rRNA FISH with other methods have been developed in order to monitor metabolic activity and ecophysiology of the detected organisms. For example, incorporation of the halogenated thymidine analogue bromodeoxyuridine into the DNA of replicating cells can be detected *in situ* by immunofluorescence. This activity staining can directly be combined with FISH-based identification of the cells (55). More specific information on functional parameters of probe-stained prokaryotes in structured environments such as biofilms can be obtained by combining FISH with microelectrode measurements (56). However, single cell resolution cannot be achieved using this combination of methods. In a few case studies, modified FISH protocols have been successfully applied to *in situ* detection of prokaryote mRNA (28,57,58). However, due to methodological problems associated with the requirement for using additional signal amplification strategies and the absence of genome sequence information for most uncultured prokaryotes, these techniques have not yet found widespread application in microbial ecology.

The combined use of rRNA FISH with microautoradiography (59,60) allows microbial ecologists to investigate the substrate uptake patterns of prokaryotes in environmental samples under different conditions at single cell resolution (41,61,62). For this purpose, the environmental sample is incubated with radiolabeled substrates under conditions of interest. After FISH staining and microautoradiography, substrate incorporation of probe-stained cells is detected by silver-grain formation in the emulsion located on top of the cells and thereby provides direct information on activity and ecophysiology of the detected organism within its habitat. Together with other emerging technologies such as stable isotope probing (see Chapter 7) and the isotope array (63), FISH microautoradiography will provide hitherto unachievable insights into structure–function relationships of complex microbial communities.

References

1. Schleifer K-H (2004) Microbial diversity: facts, problems and prospects. *Syst Appl Microbiol* **27**: 3–9.
2. Roszak DB, Colwell RR (1987) Survival strategies of bacteria in the natural environment. *Microbiol Rev* **51**: 365–379.
3. Keller M, Zengler K (2004) Tapping into microbial diversity. *Nat Rev Microbiol* **2**: 141–150.
4. Rappé MS, Connon SA, Vergin KL, Giovannoni SJ (2002) Cultivation of the ubiquitous SAR11 marine bacterioplankton clade. *Nature* **418**: 630–633.
5. Zengler K, Toledo G, Rappé M, Elkins J, Mathur EJ, Short JM, Keller M (2002) Cultivating the uncultured. *Proc Natl Acad Sci USA* **99**: 15681–15686.
6. Kaeberlein T, Lewis K, Epstein SS (2002) Isolating "uncultivable" microorganisms in pure culture in a simulated natural environment. *Science* **296**: 1127–1129.
7. Huber R, Huber H, Stetter KO (2000) Towards the ecology of hyperthermophiles: biotopes, new isolation strategies and novel metabolic properties. *FEMS Microbiol Rev* **24**: 615–623.
8. Olsen GJ, Lane DJ, Giovannoni SJ, Pace NR, Stahl DA (1986) Microbial ecology and evolution: a ribosomal RNA approach. *Annu Rev Microbiol* **40**: 337–365.
9. Hugenholtz P, Goebel BM, Pace NR (1998) Impact of culture-independent studies on the emerging phylogenetic view of bacterial diversity. *J Bacteriol* **180**: 4765–4774.
10. Cho J-C, Vergin KL, Morris RM, Giovannoni SJ (2004) *Lentisphaera araneosa* gen nov, sp nov, a transparent exopolymer producing marine bacterium, the description of a novel bacterial phylum, Lentisphaerae. *Environ Microbiol* **6**: 611–621.
11. Fieseler L, Horn M, Wagner M, Hentschel U (2004) Discovery of a novel candidate phylum "Poribacteria" in marine sponges. *Appl Environ Microbiol* **70**: 3724–3732.
12. Cole JR, Chai B, Marsh TL *et al.* (2003) The Ribosomal Database Project (RDP-II): previewing a new autoaligner that allows regular updates and the new prokaryotic taxonomy. *Nucl Acids Res* **31**: 442–443.
13. DeLong EF, Wickham GS, Pace NR (1989) Phylogenetic stains: ribosomal RNA based probes for the identification of single cells. *Science* **243**: 1360–1363.
14. Amann RI, Ludwig W, Schleifer K-H (1995) Phylogenetic identification and in situ detection of individual microbial cells without cultivation. *Microbiol Rev* **59**: 143–169.

15. Wagner M, Horn M, Daims H (2003) Fluorescence in situ hybridization for the identification of prokaryotes. *Curr Opinion Microbiol* **6**: 302–309.
16. Loy A, Horn M, Wagner M (2003) probeBase: an online resource for rRNA-targeted oligonucleotide probes. *Nucl Acids Res* **31**: 514–516.
17. Ludwig W, Strunk O, Westram R *et al.* (2004) ARB: a software environment for sequence data. *Nucl Acids Res* **32**: 1363–1371.
18. Pozhitkov AE, Tautz D (2002) An algorithm and program for finding sequence specific oligonucleotide probes for species identification. *BMC Bioinformatics* **3**: 9.
19. Ashelford KE, Weightman AJ, Fry JC (2002) PRIMROSE: a computer program for generating and estimating the phylogenetic range of 16S rRNA oligonucleotide probes and primers in conjunction with the RDP-II database. *Nucl Acids Res* **30**: 3481–3489.
20. Amann R, Snaidr J, Wagner M, Ludwig W, Schleifer KH (1996) In situ visualization of high genetic diversity in a natural microbial community. *J Bacteriol* **178**: 3496–3500.
21. Pernthaler A, Preston CM, Pernthaler J, DeLong EF, Amann R (2002) Comparison of fluorescently labelled oligonucleotide and polynucleotide probes for the detection of pelagic marine bacteria and archaea. *Appl Environ Microbiol* **68**: 661–667.
22. Fang Q, Brockmann S, Botzenhart K, Wiedenmann A (2003) Improved detection of Salmonella spp in foods by fluorescent in situ hybridization with 23S rRNA probes: a comparison with conventional culture methods. *J Food Prot* **66**: 723–731.
23. Roller C, Wagner M, Amann R, Ludwig W, Schleifer K-H (1994) In situ probing of gram-positive bacteria with high DNA G + C content using 23S rRNA-targeted oligonucleotides. *Microbiology* **140**: 2849–2858.
24. Meier H, Amann R, Ludwig W, Schleifer K-H (1999) Specific oligonucleotide probes for in situ detection of a major group of gram-positive bacteria with low DNA G+C content. *Syst Appl Microbiol* **22**: 186–196.
25. Burggraf S, Mayer T, Amann R, Schadhauser S, Woese CR, Stetter KO (1994) Identifying members of the domain Archaea with rRNA-targeted oligonucleotide probes. *Appl Environ Microbiol* **60**: 3112–3119.
26. de los Reyes FL, Ritter W, Raskin L (1997) Group-specific small-subunit rRNA hybridization probes to characterize filamentous foaming in activated sludge systems. *Appl Environ Microbiol* **63**: 1107–1117.
27. Davenport RJ, Curtis TP, Goodfellow M, Stainsby FM, Bingley M (2000) Quantitative use of fluorescent in situ hybridization to examine relationships between mycolic acid-containing actinomycetes and foaming in activated sludge plants. *Appl Environ Microbiol* **66**: 1158–1166.
28. Wagner M, Schmid M, Juretschko S, Trebesius KH, Bubert A, Goebel W, Schleifer K-H (1998) In situ detection of a virulence factor mRNA and 16S rRNA in Listeria monocytogenes. *FEMS Microbiol Lett* **160**: 159–168.
29. Pernthaler A, Pernthaler J, Amann R (2002) Fluorescence in situ hybridization and catalyzed reporter deposition for the identification of marine bacteria. *Appl Environ Microbiol* **68**: 3094–3101.
30. Fuchs BM, Wallner G, Beisker W, Schwippl I, Ludwig W, Amann R (1998) Flow cytometric analysis of the in situ accessibility of *Escherichia coli* 16S rRNA for fluorescently labelled oligonucleotide probes. *Appl Environ Microbiol* **64**: 4973–4982.
31. Behrens S, Fuchs BM, Mueller F, Amann R (2003) Is the in situ accessibility of the 16S rRNA of *Escherichia coli* for Cy3-labeled oligonucleotide probes predicted by a three-dimensional structure model of the 30S ribosomal subunit? *Appl Environ Microbiol* **69**: 4935–4941.

32. Behrens S, Ruhland C, Inacio J, Huber H, Fonseca A, Spencer-Martins I, Fuchs BM, Amann R (2003) In situ accessibility of small-subunit rRNA of members of the domains Bacteria, Archaea, Eucarya to Cy3-labeled oligonucleotide probes. *Appl Environ Microbiol* **69**: 1748–1758.
33. Fuchs BM, Glöckner FO, Wulf J, Amann R (2000) Unlabelled helper oligonucleotides increase the in situ accessibility to 16S rRNA of fluorescently labelled oligonucleotide probes. *Appl Environ Microbiol* **66**: 3603–3607.
34. Wagner M, Amann R, Lemmer H, Schleifer K-H (1993) Probing activated sludge with oligonucleotides specific for Proteobacteria: inadequacy of culture-dependent methods for describing microbial community structure. *Appl Environ Microbiol* **59**: 1520–1525.
35. Wagner M, Erhart R, Manz W, Amann R, Lemmer H, Wedl D, Schleifer K-H (1994) Development of an rRNA-targeted oligonucleotide probe specific for the genus Acinetobacter and its application for in situ monitoring in activated sludge. *Appl Environ Microbiol* **60**: 792–800.
36. Pernthaler J, Pernthaler A, Amann R (2003) Automated enumeration of groups of marine picoplankton after fluorescence in situ hybridization. *Appl Environ Microbiol* **69**: 2631–2637.
37. Bouchez T, Patureau D, Dabert P, Juretschko S, Doré J, Delgenès P, Moletta R, Wagner M (2000) Ecological study of a bioaugmentation failure. *Environ Microbiol* **2**: 179–190.
38. Schmid M, Twachtmann U, Klein M, Strous M, Juretschko S, Jetten M, Metzger JW, Schleifer K-H, Wagner M (2000) Molecular evidence for a genus-level diversity of bacteria capable of catalyzing anaerobic ammonium oxidation. *Syst Appl Microbiol* **23**: 93–106.
39. Juretschko S, Loy A, Lehner A, Wagner M (2002) The microbial community composition of a nitrifying-denitrifying activated sludge from an industrial sewage treatment plant analyzed by the full-cycle rRNA approach. *Syst Appl Microbiol* **25**: 84–99.
40. Daims H, Ramsing NB, Schleifer K-H, Wagner M (2001) Cultivation-independent, semiautomatic determination of absolute bacterial cell numbers in environmental samples by fluorescence in situ hybridization. *Appl Environ Microbiol* **67**: 5810–5818.
41. Daims H, Nielsen JL, Nielsen PH, Schleifer KH, Wagner M (2001) In situ characterization of Nitrospira-like nitrite-oxidizing bacteria active in wastewater treatment plants. *Appl Environ Microbiol* **67**: 5273–5284.
42. Juretschko S, Timmermann G, Schmid M, Schleifer K-H, Pommering-Röser A, Koops H-P, Wagner M (1998) Combined molecular and conventional analyses of nitrifying bacterium diversity in activated sludge: *Nitrosococcus mobilis* and Nitrospira-like bacteria as dominant populations. *Appl Environ Microbiol* **64**: 3042–3051.
43. Amann R, Fuchs BM, Behrens S (2001) The identification of microorganisms by fluorescence in situ hybridisation. *Curr Opin Biotechnol* **12**: 231–236.
44. Raskin L, Poulsen LK, Noguera DR, Rittman BE, Stahl DA (1994) Quantification of methanogenic groups in anaerobic biological reactors by oligonucleotide probe hybridization. *Appl Environ Microbiol* **60**: 1241–1248.
45. Loy A, Lehner A, Lee N, Adamczyk J, Meier H, Ernst J, Schleifer K-H, Wagner M (2002) Oligonucleotide microarray for 16S rRNA gene-based detection of all recognized lineages of sulfate-reducing prokaryotes in the environment. *Appl Environ Microbiol* **68**: 5064–5081.
46. Poulsen LK, Licht TR, Rang C, Krogfelt KA, Molin S (1995) Physiological state of *Escherichia coli* BJ4 growing in the large intestines of streptomycin-treated mice. *J Bacteriol* **177**: 5840–5845.
47. Schaechter M, Maaløe O, Kjeldgaard NO (1958) Dependency on medium and

temperature of cell size and chemical composition during balanced growth of *Salmonella typhimurium*. *J Gen Microbiol* **19**: 592–606.

48. Binder BJ, Liu YC (1998) Growth rate regulation of rRNA content of a marine *Synechococcus* (Cyanobacterium) strain. *Appl Environ Microbiol* **64**: 3346–3351.
49. Oda Y, Slagman S-J, Meijer WG, Forney LJ, Gottschal JC (2000) Influence of growth rate and starvation on fluorescent in situ hybridization of *Rhodopseudomonas palustris*. *FEMS Microbiol Ecol* **32**: 205–213.
50. Wagner M, Rath G, Amann R, Koops H-P, Schleifer K-H (1995) In situ identification of ammonia-oxidizing bacteria. *Syst Appl Microbiol* **18**: 251–264.
51. Schmid M, Schmitz-Esser S, Jetten M, Wagner M (2001) 16S-23S rDNA intergenic spacer and 23S rDNA of anaerobic ammonium-oxidizing bacteria: implications for phylogeny and in situ detection. *Environ Microbiol* **3**: 450–459.
52. Morgenroth E, Obermayer A, Arnold E, Brühl A, Wagner M, Wilderer PA (2000) Effect of long-term idle periods on the performance of sequencing batch reactors. *Wat Sci Tech* **41**: 105–113.
53. Oerther DB, Pernthaler J, Schramm A, Amann R, Raskin L (2000) Monitoring precursor 16S rRNAs of *Acinetobacter* spp in activated sludge wastewater treatment systems. *Appl Environ Microbiol* **66**: 2154–2165.
54. Cangelosi GA, Brabant WH (1997) Depletion of pre-16S rRNA in starved *Escherichia coli* cells. *J Bacteriol* **179**: 4457–4563.
55. Pernthaler A, Pernthaler J, Schattenhofer M, Amann R (2002) Identification of DNA-synthesizing bacterial cells in coastal North Sea plankton. *Appl Environ Microbiol* **68**: 5728–5736.
56. Schramm A, de Beer D, van den Heuvel JC, Ottengraf S, Amann R (1999) Microscale distribution of populations and activities of *Nitrosospira* and *Nitrospira* spp along a macroscale gradient in a nitrifying bioreactor: quantification by in situ hybridization and the use of microsensors. *Appl Environ Microbiol* **65**: 3690–3696.
57. Hodson RE, Dustman WA, Garg RP, Moran MA (1995) In situ PCR for visualization of microscale distribution of specific genes and gene products in prokaryotic communities. *Appl Environ Microbiol* **61**: 4074–4082.
58. Tolker-Nielsen T, Holmstrom K, Molin S (1997) Visualization of specific gene expression in individual *Salmonella typhimurium* cells by in situ PCR. *Appl Environ Microbiol* **63**: 4196–4203.
59. Lee N, Nielsen PH, Andreasen KH, Juretschko S, Nielsen JL, Schleifer K-H, Wagner, M (1999) Combination of fluorescent in situ hybridization and microautoradiography—a new tool for structure–function analyses in microbial ecology. *Appl Environ Microbiol* **65**: 1289–1297.
60. Ouverney CC, Fuhrman JA (1999) Combined microautoradiography-16S rRNA probe technique for determination of radioisotope uptake by specific microbial cell types in situ. *Appl Environ Microbiol* **65**: 1746–1752.
61. Nielsen JL, Christensen D, Kloppenborg M, Nielsen PH (2003) Quantification of cell-specific substrate uptake by probe-defined bacteria under in situ conditions by microautoradiography and fluorescence in situ hybridization. *Environ Microbiol* **5**: 202–211.
62. Kindaichi T, Ito T, Okabe S (2004) Ecophysiological interaction between nitrifying bacteria and heterotrophic bacteria in autotrophic nitrifying biofilms as determined by microautoradiography-fluorescence in situ hybridization. *Appl Environ Microbiol* **70**: 1641–1650.
63. Adamczyk J, Hesselsoe M, Iversen N, Horn M, Lehner A, Nielsen PH, Schloter M, Roslev P, Wagner M (2003) The isotope array, a new tool that employs substrate-mediated labelling of rRNA for determination of microbial community structure and function. *Appl Environ Microbiol* **69**: 6875–6887.

64. Weber S, Stubner S, Conrad R (2001) Bacterial populations colonizing and degrading rice straw in anoxic paddy soil. *Appl Environ Microbiol* **67**: 1318–1327.
65. Dedysh SN, Derakshani M, Liesack W (2001) Detection and enumeration of methanotrophs in acidic Sphagnum peat by 16S rRNA fluorescence in situ hybridization, including the use of newly developed oligonucleotide probes for Methylocella palustris. *Appl Environ Microbiol* **67**: 4850–4857.
66. Gieseke A, Bjerrum L, Wagner M, Amann R (2003) Structure and activity of multiple nitrifying bacterial populations co-existing in a biofilm. *Environ Microbiol* **5**: 355–369.
67. Moter A, Leist G, Rudolph R, Schrank K, Choi BK, Wagner M, Göbel UB (1998) Fluorescence in situ hybridization shows spatial distribution of as yet uncultured treponemes in biopsies from digital dermatitis lesions. *Microbiology* **144**: 2459–2467.
68. Thimm T, Tebbe CC (2003) Protocol for rapid fluorescence in situ hybridization of bacteria in cryosections of microarthropods. *Appl Environ Microbiol* **69**: 2875–2878.
69. Lathe R (1985) Synthetic oligonucleotide probes deduced from amino acid sequence data Theoretical and practical considerations. *J Mol Biol* **183**: 1–12.
70. Glöckner FO, Amann R, Alfreider A, Pernthaler J, Psenner R, Trebesius K-H, Schleifer K-H (1996) An in situ hybridization protocol for detection and identification of planktonic bacteria. *Syst Appl Microbiol* **19**: 403–406.
71. Daims H, Brühl A, Amann R, Schleifer K-H, Wagner M (1999) The domain-specific probe EUB338 is insufficient for the detection of all Bacteria: development and evaluation of a more comprehensive probe set. *Syst Appl Microbiol* **22**: 434–444.
72. Schramm A, Fuchs BM, Nielsen JL, Tonolla M, Stahl DA (2002) Fluorescence in situ hybridization of 16S rRNA gene clones (Clone-FISH) for probe validation and screening of clone libraries. *Environ Microbiol* **4**: 713–720.
73. Ouverney CC, Armitage GC, Relman DA (2003) Single-cell enumeration of an uncultivated TM7 subgroup in the human subgingival crevice. *Appl Environ Microbiol* **69**: 6294–6298.

Protocol 9.1

FISH with rRNA-targeted probes consists of the following steps (in this order):

1. Fixation and dehydration of the samples.
2. Hybridization of the samples with the probes and washing.
3. Microscopic evaluation.

The following protocol describes in detail how FISH should be performed to obtain optimal and reproducible results. Additional protocols are provided that explain how microbial populations in biofilms and flocs can be quantified using FISH, how new rRNA-targeted probes are designed, and how the optimal hybridization stringency for a new probe is determined.

MATERIALS

Reagents

Reagents for cell fixation:

Phosphate buffer [20:80 (v/v) mixture of 200 mM NaH_2PO_4 and 200 mM Na_2HPO_4, pH of the buffer mixture should be 7.2–7.4]

3 × phosphate-buffered saline (PBS) [390 mM NaCl, 15% (v/v) phosphate buffer, pH of the buffer should be 7.2–7.4]

1 × PBS

4% paraformaldehyde (PFA) solution. For 50 ml PFA solution, heat 33 ml distilled water to 65°C; add 2 g PFA while stirring; add NaOH until the paraformaldehyde is dissolved; add 16.6 ml 3 × PBS; leave solution to cool to room temperature; adjust pH to 7.2–7.4; filter the solution using syringes and 0.2 µm filters. Store the solution at –20°C. Caution: PFA is toxic. Wear gloves and a dust mask.

96% (v/v) ethanol

Ice

Reagents for coating microscope slides:

Ethanolic KOH [10% (v/v) KOH in 96% (v/v) ethanol]

Gelatine solution [0.1% (w/v) gelatine, 0.01% (w/v) chromium potassium sulfate; heat to 60–70°C to dissolve gelatine]

Acidic ethanol [1% (v/v) HCl in 70% (v/v) ethanol]

0.01% poly-L-lysine

Reagents for *in situ* hybridization (including dehydration of fixed samples):

50%, 80% and 96% (v/v) ethanol

Double-distilled water (ddH_2O)

5 M NaCl

Tris buffer (1 M Tris/HCl, pH 8.0)

Formamide (toxic, wear gloves and work in a fume hood especially when handling warm hybridization buffer that contains formamide). Note: Purchase the highest quality of formamide (molecular biology grade). Lower grade formamide can be contaminated with cations that reduce hybridization stringency.

10% (w/v) sodium dodecyl sulfate (SDS)

0.5 M ethylenediaminetetraacetate (EDTA), pH 8.0

Hybridization buffer [180 µl 5 M NaCl, 20 µl Tris buffer, formamide and ddH_2O (both to be varied depending on stringency; see *Table 9.2*), 1 µl 10% SDS]

Washing buffer [1 ml Tris buffer, 5 M NaCl and 0.5 M EDTA (see *Table 9.3*), top up with ddH_2O to 50 ml]

Fluorescently labeled rRNA-targeted oligonucleotide probes (probe concentration should be 30 ng/µl of probes labeled with Cy3 and Cy5 and 50 ng/µl of probes labeled with FLUOS). Probe solutions must be stored in the dark at –20°C.

Unlabeled competitor oligonucleotides (same concentration as corresponding probe)

Ice-cold ddH_2O

Antifadent (e.g. Citifluor AF1, Citifluor Ltd. London, UK)

Hand-warm 0.5–1% (w/v) agarose

SYBR Green I (FMC Bioproducts, Rockland, USA)

Table 9.2 Composition of hybridization buffers used to achieve different hybridization stringency

	0%	5%	10%	15%	20%	25%	30%	35%	40%	45%	50%
5 M NaCl	180	180	180	180	180	180	180	180	180	180	180
1 M Tris-HCl	20	20	20	20	20	20	20	20	20	20	20
ddH_2O	799	749	699	649	599	549	499	449	399	349	299
FA	0	50	100	150	200	250	300	350	400	450	500
10% SDS	1	1	1	1	1	1	1	1	1	1	1

Volumes are in µl. The final volume is always 1 ml. The top row indicates the formamide concentrations in the corresponding buffers. FA = formamide; SDS = sodium dodecyl sulfate.

Equipment

1.5 ml plastic tubes

50 ml screw-cap tubes

0.2 µm filters and syringes

Laboratory balance

Gloves, dust mask

Centrifuge for 1.5 ml tubes; centrifuge for larger tubes is needed for fixation of larger sample volumes

Fridge (4°C) and freezer (–20°C)

Magnetic stirrer with heater (60–70°C)

Drying oven (46°C)

Microscope slides (Teflon-coated slides partitioned into 6–10 fields are especially useful for FISH) and suitable cover slips

Glass or plastic containers and a slide holder for immersing microscope slides

Tissue paper

Rack to hold 50 ml screw-cap tubes in a horizontal position

Water bath (48°C)

Tweezers

Fume hood

Oil-free compressed air

Epifluorescence or confocal laser scanning microscope, filters and lasers that match the excitation and emission wavelengths of the fluorochromes (see *Table 9.1*)

Image analysis software capable of detecting objects in images and of measuring their area (in pixels) and brightness

Current database of 16S and 23S rRNA sequences

Software for probe design and probe evaluation (sequence mismatch detection), for example ARB (17).

METHODS

Fixation of cells

1. Samples should be fixed as soon as possible after sampling and should not be frozen prior to fixation. Freezing and subsequent thawing may destroy microbial cells.
2. If possible the microorganisms in the sample should be transferred to 1 × PBS prior to fixation. Bacterial cultures or enrichments are harvested by centrifugation and resuspended in 1 × PBS before they are fixed to remove media compounds. Depending on the concentration of prokaryotes, samples from aquatic environments are either fixed directly (e.g. activated sludge) or microbial cells are concentrated prior to fixation. Biofilm samples, which are well suited to FISH analysis, can also be obtained by placing slides or cover slips directly into an environment. After their removal the attached biofilm is fixed by dipping these glass supports into the appropriate fixation solution. For peat and soil samples it is recommended to extract microbial cells prior to fixation and FISH (64,65).
3. Decide whether to fix the biomass with PFA or with ethanol. Cell walls of Gram-negative prokaryotes are strengthened by fixation with the cross-linking agent PFA. This step prevents cell lysis during hybridization and storage. Fixation with PFA renders the cell walls of many Gram-positive cells impermeable to oligonucleotide probes. Samples should be fixed with ethanol instead of PFA for FISH detection of Gram-positive cells. Note: The storage life of Gram-negative cells in ethanol-fixed samples is significantly

shorter than after PFA fixation. For most environmental samples it is recommended to prepare PFA- as well as ethanol-fixed aliquots.

A. Fixation of Gram-negative cells

A.1. Mix sample with ice-cold 4% PFA solution (3 vol. PFA solution, 1 vol. sample).

A.2. Incubate the mixture at 4°C (do not freeze) for 3–12 h. Longer fixation times or higher fixation temperatures may render the cell envelopes of Gram-negative cells less permeable to oligonucleotide probes.

A.3. Centrifuge fixed sample (5 min, 4°C, 15 000 *g*) and replace supernatant with ice-cold 1 × PBS. Repeat this step two or three times in order to remove residual PFA.

A.4. Resuspend the sample in 1 vol. ice-cold 1 × PBS, and then add 1 vol. ice-cold 96% (v/v) ethanol.

A.5. Store the sample at –20°C. Samples fixed according to this protocol can be stored for several months to years.

B. Fixation of Gram-positive cells

B.1. Mix 1 vol. sample with 1 vol. ice-cold 96% (v/v) ethanol and store the sample at –20°C.

B.2. For some Gram-positive cells, additional enzymatic pre-treatment might be necessary to ensure sufficient permeabilization of the cell envelope (24,27).

C. Cryosectioning of the samples

For subsequent FISH detection of microbial cells, some samples need to be sectioned prior to the hybridization and washing procedure. Different protocols and sectioning techniques are used for this purpose (66–68).

Coating of microscope slides

1. Slides can be coated with either gelatine or poly-L-lysine in order to improve the adhesion of sample material to the glass surface. Biomass often detaches from uncoated slides during the dehydration, hybridization and washing procedures. Coated slides should be stored dry and dust-free.

A. Coating with gelatine

A.1. Clean the slides in ethanolic KOH for 1 h.

A.2. Air dry the slides.

A.3. Warm gelatine solution to 60–70°C.

A.4. Dip the slides for a few seconds into the gelatine solution.

A.5. Air dry the slides for ≥3 h in upright position.

B. Coating with poly-L-lysine

B.1. Clean the slides in acidic ethanol for 5 min.

B.2. Dip the slides for 5 min at room temperature into 0.01% poly-L-lysine.

B.3. Dry the slides for 1 h at 46°C in upright position.

Dehydration of fixed samples

1. Apply 5–30 µl of PFA- or ethanol-fixed sample material onto a microscope slide (or onto one field of a Teflon-coated slide).
2. Dry for ~15 min at 46°C or longer at room temperature.
3. Dip slide for 3 min each into 50%, 80% and 96% (v/v) ethanol. The dehydrating effect of the ethanol concentration series disintegrates cytoplasmic membranes which thus become permeable to oligonucleotide probes.
4. Dry the slides for a couple of minutes at 46°C.

***In situ* hybridization**

Prepare 1 ml of fresh hybridization buffer (see *Table 9.2*). The correct hybridization stringency, which ensures that a probe is specific, is achieved by adding formamide to the hybridization buffer. Formamide interferes with the hydrogen bonds that stabilize nucleic acid duplexes. Instead of adding formamide, high hybridization stringency can also be achieved by lowering the salt concentration in the buffer or by increasing the hybridization temperature. The salt concentration is not modified here because the hybridization reaction proceeds with slower kinetics when the concentration of cations is low. Short hybridization times are possible with this protocol because all hybridization buffers contain a concentration of Na^+ cations that supports the quick formation of DNA–RNA heteroduplexes. Using temperature to adjust the stringency would make it difficult to simultaneously hybridize several samples to different probes, requiring different hybridization conditions, as separate ovens are required for every hybridization experiment. The pH of the buffer is adjusted by addition of Tris-HCl. SDS, which reduces the surface tension of the hybridization buffer, allows the buffer to spread more evenly over the biomass on the microscope slides and to penetrate thicker samples and microbial aggregates more easily. Note: The total Na^+ concentration in the washing buffer is calculated based on the formula published by Lathe (69) assuming that 1% formamide is equivalent to an increase of the hybridization temperature of 0.5°C:

$$[Na^+] = 10^{\frac{-0.5\,[\%\,FA_{Hyb}] - 0.76}{16.61}}$$

When calculating the amount of 5 M NaCl required for the wash buffers (see *Table 9.3*) it must be considered that the addition of EDTA to the washing buffers of higher stringency adds a further 0.01 M $[Na^+]$.

1. Thaw the oligonucleotide probe solutions. Thawed probes should be kept on ice and should be protected from light. Note: Use small volumes of oligonucleotide probe solutions because fluorescently labeled oligonucleotides are damaged if they are repeatedly frozen and thawed.

Table 9.3 Composition of washing buffers used to achieve different washing stringency

	0%	5%	10%	15%	20%	25%	30%	35%	40%	45%	50%
5 M NaCl	9.00	6.30	4.50	3.18	2.15	1.49	1.02	0.70	0.46	0.30	0.18
1 M Tris-HCl	1	1	1	1	1	1	1	1	1	1	1
0.5 M EDTA	0	0	0	0	0.5	0.5	0.5	0.5	0.5	0.5	0.5
ddH_2O	to 50 ml	to 50 ml	to 50 ml	to 50 ml	to 50 ml	to 50 ml	to 50 ml	to 50 ml	to 50 ml	to 50 ml	to 50 ml

Volumes are in ml. The final volume is always 50 ml. The top row indicates the formamide concentration in the corresponding hybridization buffer.

2. Add 1 µl of each probe to 10 µl of hybridization buffer, mix well and apply the mixture onto the dehydrated sample on a microscope slide (or onto the fields of a Teflon-coated slide which contains sample).
3. Put a piece of tissue paper into a 50 ml screw-top plastic tube and pour the remaining hybridization buffer onto the tissue paper.
4. Immediately place the slide horizontally into the tube and close the tube. Place the tube in a horizontal position onto a rack and incubate it in an oven at 46°C for 1–5 h (an incubation of 90 min is sufficient in most cases). The tightly sealed plastic tube functions as a moisture chamber preventing the evaporation of hybridization solution from the slide. In particular, the evaporation of formamide can cause non-specific probe binding to non-target cells.
5. Prepare 50 ml of washing buffer (see *Table 9.3*) in a 50 ml tube and pre-heat it to 48°C in a water bath. The NaCl concentration in the washing buffer matches the formamide concentration in the hybridization buffer. A highly stringent hybridization (high formamide concentration in the hybridization buffer) requires a highly stringent washing step (low Na^+ concentration in the washing buffer). Formamide is not added to the washing buffer for stringency adjustment in order to reduce the amount of toxic liquid waste. If the hybridization buffer contains ≥20% or more formamide, 0.5 M EDTA must be added to the washing buffer. EDTA captures trace amounts of bivalent cations, which frequently occur as contaminants of chemical reagents and reduce the stringency of the washing step because they stabilize the DNA–RNA hybrids. The washing step is performed at 48°C, i.e. slightly more stringent than the hybridization. The pH of the washing buffer is adjusted by addition of Tris-HCl.
6. Remove the screw-cap tube with the slide from the hybridization oven, immediately wash away the hybridization buffer with a small volume of pre-warmed washing buffer, and transfer the slide into the remaining

washing buffer (caution: use tweezers and work in a fume hood because formamide is toxic). This should be done quickly to minimize cooling of the hybridization buffer on the slide. Cooling reduces the stringency and causes non-specific binding of probes to non-target organisms.

7. Put the tube containing the washing buffer and the slide back into the water bath and incubate for 10–15 min at 48°C.
8. Take the slide out of the tube and dip it for 2–3 s into ice-cold ddH_2O in order to remove residual washing buffer from the slide. This is necessary because fluorescent salt crystals hamper the microscopic observation of FISH results. However, the low ionic strength of ddH_2O destabilizes nucleic acid duplexes and therefore the water must be cold in order to minimize dissociation of probes from the ribosomes of their target organisms.
9. Air dry the slide as quickly as possible (the use of compressed air is recommended). Fast drying also reduces probe dissociation.
10. Dried slides can be stored in the dark at –20°C for several weeks without significant loss of probe-conferred fluorescence signal.
11. It should be noted that even with partitioned Teflon-coated slides, only hybridizations with probes requiring the same formamide concentration can be performed on the same slide at the same time since each slide is hybridized in one moisture chamber and is later immersed in one specific washing buffer. However, a sample can be sequentially hybridized to probes which require different formamide concentrations. Perform the most stringent hybridization and washing steps first and then the others in order of decreasing stringency.
12. Autofluorescence of cells and nonspecific adhesion of probes or fluorochromes to cell envelopes must be distinguished from specific probe binding. This is accomplished by doing separate FISH experiments with a nonsense probe (not targeting prokaryotic rRNA) labeled with the same fluorescent dye as the other probes. This probe should not detect

any bacterial cells, and fluorescence observed after FISH with that probe must either be caused by non-specific adhesion or be due to autofluorescence. Note that different fluorescent dyes might differ in non-specific binding to cell components. Therefore, control hybridizations with a nonsense probe must be performed with all dyes used in subsequent experiments.

Microscopy

1. Apply two drops of antifadent close to the left and right end of a slide after FISH and washing have been completed (frozen slides should be warmed to room temperature prior to this step).
2. Put a microscope cover slip on top and wait until the antifadent has spread over the whole slide. Caution: too much antifadent can blur the microscope image.
3. Observe the sample under an epifluorescence microscope or confocal laser scanning microscope equipped with suitable filters or lasers, respectively (excitation and emission wavelengths of fluorochromes often used for FISH are listed in *Table 9.1*).
4. Slides embedded in antifadent can be stored at 4°C (do not freeze) for several days before the probe-conferred fluorescence begins to decline. Alternatively, the antifadent can be removed with ddH_2O and the dried slides can be stored at –20°C for longer periods.

Quantification of microorganisms by FISH

1. Cell numbers of planktonic microorganisms can be determined by direct counting of FISH-stained cells under an epifluorescence microscope. For this purpose, defined volumes of liquid samples are filtered and FISH is performed on the filters according to Glöckner *et al.* (70). The cells are then counted directly on the filter. Microorganisms that are embedded in dense aggregates, flocs and biofilms cannot accurately be counted by this approach. Instead, the following protocol can be used to quantify their relative biovolume in an environmental sample.

2. Apply 10–15 μl of fixed sample onto a microscope slide (or onto a field of a Teflon-coated slide). Dry the sample and repeat this step two or three times until a thick layer of biomass is obtained on the slide. This procedure prevents single cells from selectively accumulating on the slide surface while the sample is being dried. Accumulation of cells on the slide surface can cause significant quantification bias because the proportion of these cells in the total microbial community is underestimated when the focus plane of the microscope is above the slide surface, and is overestimated when the focus plane is exactly on the slide surface. Gelatine- or poly-L-lysine-coated slides should be used in order to prevent detachment of such thick layers of biomass from the slide during dehydration and FISH. In addition, the slides with the biomass should be covered with a thin layer of agarose to keep the biomass in place.
3. Dip the slide horizontally into hand-warm 0.5–1% agarose for a few seconds and place it horizontally on a cold surface such as a Petri dish lid that has been placed on ice.
4. When the agarose has solidified, carefully remove excess agarose from both sides of the slide (but not from fields containing biomass).
5. Dehydrate the sample in an ethanol concentration series as described above.
6. Dry the slide for ~20 min at 46°C. Dried agarose forms a thin 'film', which functions like a glue, retains the biomass on the slides and does not usually interfere with subsequent FISH and microscopy, although thicker or denser agarose layers can increase background fluorescence.
7. Perform FISH with a specific probe (referred to as 'probe 1' in the following text) and a domain-specific probe such as the EUB338 probe mix, which is specific for Bacteria [(71); referred to as 'probe 2']. Probes 1 and 2 must be labeled with different fluorochromes. Ensure that the hybridization conditions match the stringency requirements for probe 1.

8. Instead of probe 2 a nucleic acid stain such as SYBR Green I can be used to stain the total microbial biomass. SYBR Green I stains DNA and RNA. Pure DNA stains such as DAPI are not suitable for this quantification protocol because the area of the same cells is smaller in images of the DAPI stain than in images of probe-conferred fluorescence. Samples are stained with SYBR Green I after FISH with probe 1 and washing. Apply 20 μl of a 10 000-fold dilution of SYBR Green I onto each sample and incubate for 10 min in the dark at room temperature. Wash the slides briefly with distilled water, then air dry slides (do not use compressed air to avoid possible inhalation of the nucleic acid stain).
9. Acquire digital images of the detected organisms by using a confocal laser scanning microscope. Randomly select 20–30 different microscope fields and record at each position separate optical sections (image pairs) of the cells stained by probe 1 and of the biomass stained by probe 2. The pinhole of the microscope should be adjusted to generate optical sections with a thickness of 1 μm. In every image pair, the images of cells that are detected by both probes 1 and 2 must be congruent. Detector settings of the confocal microscope can be varied to optimize this condition. If the images of such cells are not congruent, the relative area and biovolume of the populations stained by probe 1 cannot be determined accurately (see below).
10. Measure the area of the probe-stained cells in each image pair. Various freeware and commercial image analysis programs offer functions to determine the area (in pixels) of objects in images.
11. Calculate for each image pair the percentage of the area of the organisms stained by probe 1 relative to the area of the biomass stained by probe 2.
12. Average the relative areas obtained for the 20–30 image pairs. Provided that the image pairs were recorded at random positions, the average relative area of the cells stained by probe 1 is also an estimate of their relative biovolume in the sample. An extended

version of this protocol allows determination of absolute cell densities for biofilm organisms (40).

Probe design

1. Properly designed rRNA-targeted oligonucleotide probes detect a clearly defined group of phylogenetically related organisms (the target group may also consist of only one known organism). Check whether an existing probe already targets the organisms you wish to detect *in situ*. The probeBase database (16) offers comprehensive lists of probes and their target organisms.
2. If you are designing a new probe, collect a set of high-quality, full-length rRNA sequences of probe target organisms whose phylogenetic affiliation has clearly been determined. You will also need a large rRNA sequence database that must contain as many full-length sequences of probe non-target organisms as possible.
3. Use specialized computer software [e.g. ARB (17)] to search the target rRNA sequences for sites where they differ from all non-target sequences. These sites are potential binding positions for specific oligonucleotide probes. Their typical length is between 15 and 25 nucleotides.
4. Identify the exact mismatch positions in the potential probe binding regions where the base sequences of target and non-target organisms differ. The mismatch positions affect the specificity of a probe. Mismatches in the centre of a probe binding site destabilize duplexes formed by probes and non-target rRNA more effectively than mismatches at the borders of the binding region.
5. Identify the type of the base mismatches. Uracil (in the probe binding region on the rRNA) forms base pairs with thymine but also with guanine (both on the DNA probe). Therefore, the contribution of G:U mismatches to probe specificity is limited. G:A and G:G mismatches are also less effective than A:A, A:C, C:C, C:U and U:U mismatches.

6. Probes often have only one or two weak mismatches to rRNA regions of some non-target organisms. Non-specific probe binding to rRNA in such organisms can be suppressed by applying highly stringent hybridization conditions and by using unlabeled competitor oligonucleotides. Competitors perfectly match the rRNA sequence of particular non-target organisms at the probe binding site. When used in FISH experiments in a 1:1 molar ratio to the probe, they suppress non-specific probe binding because they bind with higher affinity than the probe to the rRNA of the non-target organisms. If competitors are required they should be used together with the respective probes in all FISH experiments including those performed for probe evaluation.

Probe evaluation

1. Test a new probe in FISH experiments with an environmental sample or a pure culture containing probe-target organisms in order to check whether the probe binds to the ribosomes of the target cells. These hybridizations should be performed under non-stringent conditions by using a hybridization buffer without formamide and the corresponding washing buffer. If no probe-conferred fluorescence is observed, the ribosome content of the target cells is too low for FISH; or the cell envelope of the target organisms is impermeable; or the binding site of the new probe is blocked by rRNA secondary and tertiary structure or ribosomal proteins.
2. The optimal hybridization conditions for new probes are determined in a series of FISH experiments with increasing formamide concentrations in the hybridization buffers and correspondingly decreasing NaCl concentrations in the washing buffers. If possible, obtain pure cultures of a probe target and a non-target organism that has only one base mismatch to the probe sequence. Environmental samples containing the respective organisms can be used if pure cultures are not available. Alternatively, rRNA genes of probe target and non-target

organisms cloned into plasmid vectors can be expressed in *Escherichia coli* cells. The *E. coli* cells are then used in FISH experiments to evaluate probes targeting the heterologously expressed rRNA (72,73). Fix the cells appropriately.

3. Prepare hybridization buffers with formamide concentrations ranging from 0 to 70% (in steps of 5 or 10%) and the corresponding washing buffers.
4. Apply 5–10 µl of fixed target and non-target cells onto the appropriate number of separate microscope slides (or onto different fields of Teflon-coated slides).
5. Dehydrate the cells and perform FISH with the new probe and the different hybridization and washing buffers. The length of hybridization and washing steps must be the same for all slides. Also use the same probe solution for all hybridizations.
6. Observe the cells by epifluorescence microscopy. Note the formamide concentration where the non-target cells no longer fluoresce (a low background autofluorescence of cells is normal and must not be misinterpreted as probe-conferred fluorescence). This is the minimal formamide concentration for specific FISH with the new probe.
7. Note the formamide concentration where the fluorescence of the probe target cells starts to decrease. A formamide concentration of 5% below this value is considered to be optimal for hybridization with the respective probe since hybridization to the target organisms is not affected but probe binding to most non-target organisms with more than one or two mismatches to the probe is most likely suppressed. Keep in mind that unknown non-target organisms with novel sequence patterns at the probe binding site may occur in environmental samples. Therefore the highest possible formamide concentration for the new probe, which is determined in this step, should be used for FISH analysis of environmental samples.
8. In order to evaluate new probes more accurately one can measure, by digital image

analysis, the probe-conferred fluorescence of cells after FISH with different formamide concentrations. For this purpose, record images that contain ≥100 probe target and suitable non-target cells, respectively, per formamide concentration by using an epifluorescence microscope with a CCD camera or a confocal laser scanning microscope (if a confocal laser scanning microscope is used, each optical slice should be as thick as 2 μm to capture all fluorescence emitted by a single cell). It is important that all pictures are acquired with the same instrument settings (e.g. exposure time and detector sensitivity).

9. Use suitable image analysis software to quantify the brightness of the single cells in all recorded images. Determine the average cell brightness for each formamide concentration and plot these values against the formamide concentrations. *Figure 9.2* shows three probe dissociation curves that were obtained by this approach. One can

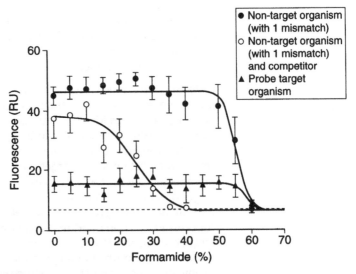

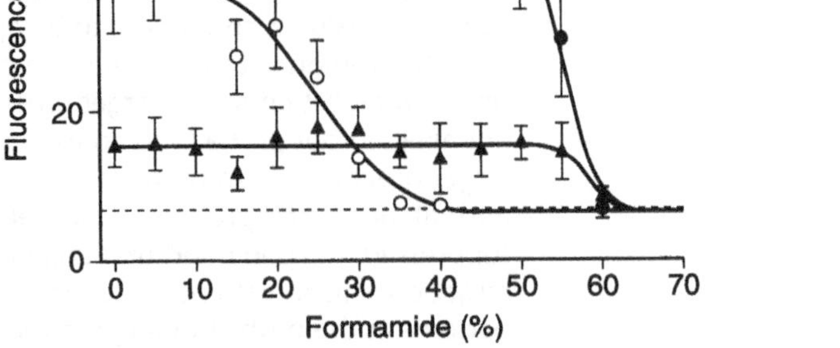

Figure 9.2

Dissociation curves determined for an oligonucleotide probe after FISH to a probe target organism and to a non-target organism that has only one base mismatch to the probe sequence. Specific detection of the probe target organism is only possible when an unlabeled competitor oligonucleotide is used in addition to the probe. The minimum and optimum formamide concentrations for the probe are 40 and 55%, respectively. Figure modified from (41).

infer from these curves that the probe detected its target organism with 0–55% formamide in the hybridization buffer. However, the probe also detected a non-target organism with only one mismatch at the probe binding site over the same range of formamide concentrations. The fluorescence emitted by the non-target cells after FISH was even higher than the fluorescence emitted by the probe target cells, probably because the ribosome content of the non-target cells was higher than the ribosome content of the probe target cells. Applying formamide concentrations >55% would not improve the situation because the probe would then also not bind to the ribosomes of the target organism. The solution was to use an unlabeled competitor oligonucleotide, with a complementary sequence to the rRNA of the non-target organism. With this competitor in the hybridization buffer the probe did not detect the non-target organism >40% formamide. Provided that the competitor is used, 40% is the minimum and 55% is the optimum formamide concentration for the probe.

Lessons from the genomes: microbial ecology and genomics

10

Andrew S. Whiteley, Mike Manefield, Sarah L. Turner and Mark J. Bailey

10.1 Introduction

Microbial ecology is the study of interactions between microbes and their biotic and abiotic environments. The responses of microbes to their environment and the impacts they have upon it are encoded in the genomes of individuals. For this simple reason the young, yet established, field of microbial ecology and the rapidly expanding field of genomics are entwined. This chapter considers what genomics has offered microbial ecology to date and what it might offer in the future.

Microbes are constantly adapting to their environments, and their responses can be divided into two distinct categories. To one category belongs the immediate responses already encoded on the genome, occurring through phenotypic changes mediated by the complex regulatory systems governing gene expression. This type of physiological response is rapid yet relatively limited in its versatility. The other type of response is known as adaptive evolution, which involves changes in the genetic composition of an organism, as selected for by the environment. This type of genetic response constitutes a relatively slow, longer-term response to environmental change but appears to be infinitely versatile.

The distinction between genetic and physiological responses is reflected in the neat conceptual and practical partitioning of genomics into structural and functional aspects. 'Structural genomics' is the analysis of genome composition and architecture, which requires the sequencing of an organism's entire genetic material. For our purposes, the ecologically significant characteristics of genomes become apparent by comparing the genome structure of organisms with different roles in different ecosystems. Structural genomics corresponds to an analysis of genetic adaptation to environmental change. 'Functional genomics' is the analysis of global gene expression from genomes. It seeks to characterize the regulatory systems and their effect upon gene expression underlying physiological responses to environmental change.

To reflect the partition inherent to the study of genomics, this chapter consists of a section on what lessons microbial ecologists can learn from structural genomics followed by a section on what can be learnt from functional genomics. The chapter will conclude with a discussion on the opportunities offered to the field of microbial ecology by genomic-based technologies, including applications in the natural environment.

10.2 Structural genomics

When Fleishman *et al.* (1) published the first microbial genome, that of *Haemophilus influenza*, they stated that 'a prerequisite to understanding the complete biology of an organism is the determination of its entire genome sequence'. This is one prerequisite but genome sequences do not provide a complete biological understanding *per se.* This shortfall arises because the complete ecology of any bacterium has not been fully described. Whilst we may have a good understanding of the genetics and physiology of individual bacterial strains under laboratory conditions, the functional role of substantial proportions of every bacterial genome remains unknown and awaits further investigation.

The *H. influenza* Rd genome was reported to contain 1743 open reading frames (ORFs), of which 736 (42%) had 'no role assignment'. Of these hypothetical ORFs, 347 matched other hypothetical proteins in the databases and 389 were 'ORFans' unique to the *H. influenza* genome, with no matches to any other sequences in the databases (1). There were few surprises from the genome; most of the predicted core metabolic functions necessary for free-living bacteria were identified, except for those already predicted to be absent from earlier physiological studies (e.g. three components of the tricarboxylic acid cycle: citrate synthase, isocitrate dehydrogenase and aconitase). Bioinformatic interrogation of the genome sequence also allowed functional grouping of genes, e.g. (potential) virulence genes and previously unidentified genes, under the regulation of well-described transcriptional regulators.

The only ecological information presented (1) highlighted that *H. influenza* Rd is a non-pathogenic commensal resident of the human upper respiratory mucosa. Type b strains, which cause meningitis, contained genes for pathogenicity that were largely absent from the sequenced genome. Preliminary comparisons with the available data for the *Escherichia coli* K12 genome sequence indicated that many of the ORFs unique to one or other organism (i.e. that might encode for physiological differences) belonged to the 'hypothetical' class. Since function could not be inferred for these ORFs, neither could the likely ecological significance of their presence or absence. The authors emphasized that comparative genomics would become more powerful as more genomes became available.

10.3 Comparative genomics

Complete genome sequences are available for 172 Bacteria and 19 Archaea, with 267 other prokaryotic genome projects in progress (as of September 2004; for updates see:
http://www.ncbi.nlm.nih.gov/genomes/MICROBES/Complete.html). The key aim of comparative genomic studies is to better understand fundamental aspects of bacterial genome organization, in respect of gene content, order, regulation and evolution. This is achieved by undertaking intra-species/genus

genome comparisons and is logically extended to comparisons of bacterial species that are distantly related but occupy a similar ecological niche, or alternatively to closely related bacterial species that occupy divergent ecological niches. The former comparisons might include intracellular pathogens (*Mycobacteria, Chlamydia, Rikettsia*) and/or symbionts (*Buchnera*) of eukaryotes, to identify shared genome traits. The latter might include members of the Rhizobiaceae, which include the facultative intracellular pathogens *Brucella* spp. and plant/soil associated rhizobial species (2). However, the majority of the genomes completed to date are of pathogenenic species, and so, the diversity and representation of environmental ecotypes for comparison is somewhat restricted.

10.3.1 Small genomes

The genomes of two *Buchnera aphidicola* spp., isolated from different host aphids, *Schizaphis graminum* (*Buchnera Sg*) and *Acrythosiphon pisum* (*Buchnera Ap*), offer the simplest genomes for comparison. As obligate intracellular symbionts that are maternally (vertically) transmitted between aphid generations, their ecology is relatively simple: they only ever experience a single largely invariant environment. The symbiosis is thought to have originated ~150 million years ago (MYA) and the two host species to have diverged ~50 MYA (3). Ecologically, *Buchnera* provide the aphid host with metabolic pathways for the synthesis of essential amino acids, absent from phloem sap upon which the aphids feed. The two buchneral genomes are relatively small (~641 kbp) and share 526 of their 564 (*Buchnera Ap*) and 545 (*Buchnera Sg*) genes, all of which occur in the same order (synteny) in each genome (3). Comparisons show that the majority of differences are due to point mutations that generate pseudogenes and gene losses, with *Buchnera Ap* containing 13 and *Buchnera Sg* 38 pseudogenes respectively. Some of these recent gene inactivations are consistent with known ecological differences in that five of the *Buchnera Sg* pseudogenes are components of the cysteine biosynthetic pathway. This correlates with the higher levels of cysteine in the phloem of grasses, compared with pea, the main food plants of *S. graminum* and *A. pisum* respectively. As such, a combination of random mutation and the absence/reduction of selective pressures to retain the functional pathway have probably contributed to the inactivation of these genes (4).

Pseudogenes represent relatively recent gene inactivation processes since, over time, functionally redundant DNA sequences will be deleted from the genome (4). Pairwise comparisons indicate that there are 14 gene sequences (excluding pseudogenes) present exclusively in *Buchnera Ap* and four in *Buchnera Sg*. All of these genes have similarity to hypothetical proteins in *E. coli*, their closest free-living relative. All 18 were therefore likely to have been present in the last common ancestor of the *Buchnera* lineage, implying that differential loss or maintenance reflects host/niche adaptation. The ecological relevance of these differences is, as yet, unclear. Unfortunately, the *Buchnera* system is not amenable to genetic manipulation and so genes cannot be mutated to assess their effects on host survival and fitness and thus investigate their ecological relevance.

Despite their small genome sizes, ~14% of the open reading frames

('ORFs') of the *Buchnera* genomes code for 'hypothetical' proteins of unknown function. Doolittle (5) noted that on average, 'at least 25% of the genes (in each bacterial genome) remain 'hypothetical''. This reaches a maximum in the archaeon, *Aeropyrum pernix*, within which 57% of the ORFs could not be assigned a function at the time of publication (6). The absence of functional gene assignments is perhaps most surprising for the laboratory workhorses *Escherichia coli* K12 (7) and *Bacillus subtilis* (8), the genetics of which have been studied intensively for >50 years. Such studies have relied on phenotypic differences, identifiable in laboratory assays, and have probably overlooked genes that enable *E. coli* to survive in the mammalian gut or *B. subtilis* in soil. For example, Yu *et al.* (9) deleted up to 313 kbp of the *E. coli* genome without affecting the growth rate of the mutant in laboratory culture. This is a clear indication of substantial levels of functional redundancy and/or large numbers of genes which are only environmentally regulated, and represents a significant problem when analyzing an organism's genome in the context of laboratory cultivation. Such considerations could mask the impact of directed gene mutation analyses *in vitro*, could miss environmentally regulated genes, and might explain why obligate intracellular bacteria can survive with significantly (~10-fold) smaller genomes.

10.3.2 Large genomes

The natural environment is highly complex, due to spatial and temporal heterogeneity (physical, chemical and biological). In contrast to the intracellular bacteria, all the larger genomes sequenced to date are from soil-associated bacteria. The largest examples include *Mesorhizobium loti* (7.04 Mbp, NC_002678) (10), *Bradyrhizobium japonicum* (9.1 Mbp, NC_004463) (11), *Streptomyces coelicolor* (8.67 Mbp, NC_003888) (12), *Pirellula sp.* (7.14 Mbp, NC_5027) (13) and *Pseudomonas aeruginosa* (6.26 Mbp, NC_002516) (14). The latter organism is described as an opportunistic pathogen of humans, but should be more accurately classified as a ubiquitous environmental bacterium. Other large genomes include the plant-associated bacteria *Sinorhizobium meliloti* (6.8 Mbp, NC_003047) (15), *Agrobacterium tumefaciens* (5.6 Mbp, NC_003304) (16), *Bacillus anthracis* (5.23 Mbp, NC_003397) and *Bacillus cereus* (5.43 MBp, NC_003909) (17) and *Xanthomonas* spp. (5.1 and 5.2 Mbp, NC_003902 and NC_003919 respectively) (18). The large genome size of *P. aeruginosa* correlated with high numbers of regulators enable it to sense and respond to environmental fluctuations (14); the same is possibly true for the genome of the *S. meliloti*, containing large numbers of ORFs with homology to membrane-associated transporter proteins (15).

10.3.3 The horizontal gene pool (HGP)

Comparison of benign and pathogenic strains (*E. coli* K12 vs. H:O157) (19) or benign and closely related pathogenic species (*B. subtilis* vs. *B. cereus* vs. *B. anthracis*) (17,20) have identified genes involved in pathogenicity. Such genes are often associated with pathogenicity islands and/or mobile genetic elements (e.g. phage, plasmids, transposons), responsible for gene transfers between strains and species (21). Horizontally transferred genes should, in theory, display 'patchy' distributions across evolutionary lineages, within species and populations, and may be readily identified by %GC contents that differ markedly from adjacent chromosomal sequences. Such distributions have been described for pathogenicity traits in *E. coli* (22), *Vibrio chloerae* (23) Streptococci (24,25), *Helicobacter pylori* (26) and *Shewanella* spp. (27). Berg and Kurland (4) suggest that acquisition of pathogenicity determinants represents adaptation by niche expansion. Similar evidence exists for non-pathogenic bacteria isolated from different environments, where ecologically relevant genes are found on 'Ecoislands' and plasmids (28,29).

The preceding sections using comparative genomics seem to indicate that gene loss is a function of mutation and reduced selection, whilst the horizontal acquisition of advantageous genes enables niche expansion. Wernegreen *et al.* (30) suggested that *Buchnera* spp. have reduced numbers of transposons and other repeat sequences. Such repeat sequences are hotspots for recombination, enabling gene acquisition, gene losses, gene duplications and genome rearrangements (31). These data might predict that organisms occupying more heterogeneous environments (variable in space and time) will have a greater genetic potential. This genetic potential will be manifest both quantitatively (more DNA) and qualitatively (greater potential for recombination) and that a large proportion of genomic information will be associated with the HGP. Horizontal gene transfers provide a means for rapid adaptation to new or changing environments (4,32,33). As such, Levin and Bergstrom (32) discuss the population genetic and ecological factors that might influence whether genes encoding such particular phenotypic traits will persist as part of the HGP or become incorporated into the chromosome as a more permanent genomic component.

10.3.4 Insights into phylogeny

Bacterial genomes are dynamic, undergoing gene losses and gene acquisition by horizontal transfer or gene duplication. This dynamism can result in different (non-congruent) tree topologies when phylogenies are constructed from several different genes for a defined group of organisms (34). One means of constructing more robust phylogenies is to combine data for several genes (35), reducing the contribution of aberrant or weak evolutionary signals. An alternative approach is to use whole genome gene content and/or gene order as a measure of phylogenetic relatedness (36). Reassuringly, these approaches produce phylogenies that largely agree with established single gene phylogenies, such as 16S ribosomal RNA, suggesting that, whilst horizontal gene transfers contribute significantly to short-term bacterial adaptation, these are not sufficient to blur the evolutionary relationships of bacterial lineages. However, as a note of caution for this type of analysis, Gogarten *et al.* (37) should be consulted for discussion of why this interpretation may be misleading.

Comparative genomics is providing interesting data about the likely processes that underpin bacterial evolution and patterns of adaptation, as a basis for understanding bacteria in an ecological context. As a simple example, one key observation is that the horizontal gene pool is an important component of bacterial adaptation and that the amount of horizontally transferred DNA increases with genome size (38). However, one of the major challenges is that genomic databases are biased towards pathogenic bacteria, a challenge which needs to be met in terms of representation of 'ecologically' relevant bacteria within genome sequencing programs. Despite this pitfall, methodologies that rely upon the functional analyses of genomic data (e.g. microarrays and proteomics) are being applied to environmental bacteria and are the subject of the next section.

10.4 Functional genomics in microbial ecology

Whilst structural and comparative genomics provide useful information pertaining to how an organism has adapted and evolved over time to survive in its given environment, it does not, in itself, describe how a microbe reacts to environmental change through immediate physiological responses. However, functional genomics (the genome-wide analysis of gene expression) provides a powerful means to describe how microorganisms will respond to their environment. Global analyses of mRNA transcript, protein and metabolite production constitute the three tiers of functional genomics, each with their own unique limitations and advantages. The methodological detail associated with transcriptomics, proteomics and metabolomics are beyond the scope of this review and are discussed elsewhere (39–41). Of greater interest here are the uses to which functional genomics can be put by the microbial ecologist.

10.4.1 Characterizing microbial responses to environmental change

To date, the characterization of genome-wide gene expression profiles has found greatest utility in analyzing the response of a microbe to changes in environmental parameters under controlled laboratory conditions. The history of cataloging changes in global protein expression of an organism under various experimental conditions is already long and illustrious (42). For example, global changes in protein expression of *E. coli* K-12 in response to common environmental challenges such as starvation for carbon, nitrogen, phosphate and individual amino acids, oxidative stress, heat shock (42 and 50°C), shifts to and from anaerobic conditions and treatment with naladixic acid, quinone, 2,4-dinitrophenol and cadmium chloride, have been characterized (43).

The analysis of any given proteome and the ability to isolate proteins of interest via two-dimensional polyacrylamide gel electrophoresis is limited to subsets of an organism's total protein content (based primarily on solubility, pI and size). Consequently this restricts the utility of such analysis for the description of genome-wide gene expression. Morevover, the inherent difficulties associated with the isolation of membrane fractions (44) may necessitate the separate analysis of the membrane subproteome and cytosolic proteomes (45). By contrast, the application of microarrays enables the genome-wide assessment of gene expression in terms of mRNA transcript production (transcriptome analysis). For example, Wei *et al.* (46) have used high-density microarrays of 4290 open reading frames from *E. coli* K-12 to profile the global transcriptional response of this organism in minimal and rich media. Increasingly, transcriptome- and proteome-based analyses are now being used in combination to investigate global gene expression in various organisms under various conditions (47,48). Such studies offer the potential for substantially improving our understanding of how microbes perceive and react to their surrounding environment.

Isolating the entire transcriptome of an organism is a simpler task than isolating an entire proteome, but there are a number of limitations to which transcriptomics is bound. The primary caveat is that the analyses of transcriptomes as a means of describing genome-wide gene expression ignores post-transcriptional and post-translational levels of control in gene expression. A secondary issue that limits the application of transcriptional analyses is the requirement for the production of the DNA microarrays themselves. Until recently, microarrays have been relatively difficult to fabricate but array developments by companies such as Affymetrix (http://www.affymetrix.com/) and Nimblegen (http://www.nimblegen.com) facilitate microarray production 'off the shelf' for 'model' organisms, or for those which are being sequenced independently by other researchers.

10.4.2 Relating specific genes to specific functions

One of the key challenges facing biologists is to ascribe a function to the large numbers of so-called hypothetical genes in genome sequences. Indeed, one of the most significant findings of genome sequencing projects has been to reveal how many such genes exist, with their individual roles in defining metabolic, physiological, structural and morphological diversity still awaiting discovery. A primary objective in microbial ecology is to relate specific genes to particular ecological functions. Innovative approaches to discover environmentally responsive genes include *in vivo* expression technology (49,50), or hypothesis-driven directed mutagenesis which generates mutants with altered global gene expression. Functional genomic analyses can be used to characterize the effects of such mutations on gene expression and thereby relate the cellular role of a gene product to an ecologically relevant gene.

Thompson *et al.* (51) compared the transcriptome and the proteome of a putative ferric uptake regulator (*fur*) mutant of *Shewanella oneidensis* with those of a wild type strain to confirm the role of *fur* in the regulation of siderophore-mediated iron assimilation. The results additionally implicated *fur* in energy metabolism, thereby suggesting a regulatory framework underlying the ability of *S. oneidensis* to generate energy via the reduction of insoluble ferric iron. This framework underpins the proliferation of this organism under oxygen-limited, iron-abundant conditions. Another example where functional genomics has offered a powerful means of examining mutant strains in microbial ecology is the characterization of genes regulated by intercellular signals in bacteria. DeLisa *et al.* (52) employed an *E. coli* DNA microarray to determine which genes display altered expression levels during quorum sensing by investigating gene expression in a *luxS* mutant that was unable to synthesize the signalling pheromone AI-2. In experiments involving the exogenous addition of the AI-2 signal to growth medium they found that >5% of *E. coli* ORFs were responsive to the presence of the signal, including genes involved in cell division, morphogenesis and cell surface architecture. A further variation on the use of functional genomics in the analysis of mutants involves the characterization of mutations derived from evolutionary processes. Ferea *et al.* (53) cultured *Saccharomyces cerevisiae* in glucose-limited chemostats over 250 generations and compared genome-wide transcription of the resulting strains with that of the parent. In the context of natural selection, DNA microarrays revealed that all successful populations had alterations in their central metabolism to allow the complete oxidation of glucose (53).

Metabolome analysis has also been exploited in the characterization of mutants (39). Like proteins, metabolites have direct functional roles in cellular activities. Metabolomics, however, is complicated by the fact that in most cases there is no direct relationship between genes and metabolites, in contrast to the obvious link between mRNA transcripts and proteins. Nevertheless, the utility of metabolomics in revealing the phenotypes of apparent silent mutations in *S. cerevisiae* has been demonstrated (54). Whilst this investigation did not in itself ask questions pertinent to microbial ecology, there is no reason why the approach cannot be extended to the characterization of phenotypes encoded by genes exhibiting habitat-dependent expression in microorganisms of ecological relevance.

10.4.3 Expression of genes from components of a genome

Gene transfer between components of a genome and between genomes is a phenomenon of great interest to the microbial ecologist. Autonomous and mobile genetic elements such as plasmids and phages, constituting distinct components of an organism's genome, encode traits enabling the proliferation of occupied hosts in certain environments and can thus underpin microbial population dynamics (4). The constitutive expression and the induction or repression of environmentally responsive genes encoded on mobile genetic elements can be assessed using the tools of functional genomics (55). Whilst this application awaits broad exploitation, it is likely, in time, to answer questions regarding the selective advantages of traits encoded on mobile genetic elements (MGE) and provide insight into the mechanisms dictating their spread through microbial communities and the effect of MGE transmission on the rise and fall of host populations.

Microorganisms typically have a limited ability to move significant distances within their given environment and hence evade environmental challenges. Environmental perturbations will usually elicit physiological responses by the microbe that may be mediated via environmental control of gene regulation in the cell. Consequently, investigation of the regulatory mechanisms controlling such physiological responses will contribute substantially to our understanding of microbial ecology. It can, however, be argued that the response of pure cultures of bacteria under controlled laboratory conditions to perturbations in specific environmental parameters may not represent the response of any given organism to the same perturbation *in situ*. At the heart of such an argument is the recognition that microbes do not perceive individual environmental parameters but are at the mercy of the interplay between all such parameters, including the confounding effects of cohabiting members in mixed microbial communities. This is not to say that *in vitro* transcriptomic, proteomic and metabolomic analyses have no part to play in microbial ecology, but it is important to recognize that it is not feasible to conduct comprehensive multifactorial functional genomic studies of microbial responses to all environmental parameters *in vitro*. The question then arises 'can the tools of functional genomics be used to deconstruct microbial interactions of a biotic or abiotic nature *in situ*?'

10.5 The application of genomic tools *in situ*

10.5.1 Organism diversity

Significant developments in microbial ecology can be unequivocally linked to the adoption and use of 16S rRNA as a phylogenetic marker (see Chapter 2). 16S rRNA sequencing has revealed that the diversity of organisms in the natural environment is extensive, with prokaryotes occupying the major portion of the 'tree of life' (56). Molecular ecologists are interested in the microbial diversity in an environmental subsample, in terms of the numbers of species, their richness and evenness.

The need to define microbial community compositions *in situ* has led to the array-based probing technologies being the first genomic tools to be applied to microbial ecology *in situ*. Guschin *et al.* (57) used a specific suite of 16S rDNA probes directed to key nitrifying genera immobilized in a simple array format. Successful detection of specific groups was demonstrated using reverse hybridization techniques using fluorescently labeled rRNA or *in vitro*-transcribed targets generated from cloned rDNA. A similar strategy was subsequently used for PCR-amplified rDNA (58), or *in vitro*-transcribed rRNA targets with chaperone-detector probes, for successful biotin-label-based detection of *Geobacter* and *Desulfovibrio* groups. These data revealed that fragmented rRNA hybridized more readily than intact rRNA and a minimum detection limit of 10^6 cells was determined. Furthermore, these data indicated that soil extracts spiked into reactions inhibited the efficiency of rRNA hybridization, a major consideration for future environmental studies (58).

More recently, an array system termed the SRP-PhyloChip has been designed and applied to examine all known members of the sulfate-reducing prokaryotes, using 132 16S rRNA targeted 18-mer oligonucleotides (59). The work represented one of the first comprehensive attempts at array-based 16S analysis in the environment, with numbers of oligonucleotides approaching those that would be desirable in future analyses. Although a portion of the development work was devoted to clinical samples, the chip was also tested with DNA samples from a hypersaline microbial mat albeit following PCR-based amplification of 16S rDNA prior to subsequent labeling and array analysis. The authors elegantly demonstrated the requirements of scale up for array technology in terms of numbers of probes used, the requirements to design and refine a probe set, and the potential complexity of analysis required to obtain meaningful signals from environmental samples. However, despite these technological issues, the chip accurately detected sulfate-reducing organisms in environmental samples that could be subsequently confirmed by array-independent technologies such as PCR amplification, sequencing and gene-specific analyses (59).

Technologically, the preceding work highlighted several limitations that still need to be overcome to allow successful application of the technology in the environment. Key points for optimization within array formats include the quality of the probe design, the type and quality of nucleic acids that are hybridized with the array (purity and size of the DNA, RNA, or cDNA, and the strength of the fluorescent labels used), and finally the strategies for, and accuracy of, signal quantification. Perhaps the most significant factor is to determine the optimal hybridization stringency conditions required for accurate analyses. For arrays, if hybridization is to be effective the probe suite needs to be empirically analyzed under different stringency conditions (e.g. hybridization temperature and salt concentrations) with reference targets of known divergence from the probe sequence. Establishment of hybridization conditions which encompass high stringency for exact match probes but also allow hybridization at moderate divergence (e.g. 20% sequence divergence) will be the most applicable to the analysis of environmental nucleic acids. This is due to the expectation of divergent targets within a given gene family in the environment and the need for this diversity to be accounted for. Specifically, using the dissociation temperature (T_d), calculated by melting profiles (60) of the probe is a good basis for assessing hybridization conditions and stringency in multi-probe formats. Secondarily, Liu *et al.* (60) suggest the use of nested probes of increasing phylogenetic resolution, a view reinforced by the SRP PhyloChip analyses (59). Further, the incorporation of redundant probes to critically examine stringency during hybridizations could be utilized as a form of internal standardization (48,49).

10.5.2 Functional genes

Although a knowledge of the organisms present in any given environment is a prerequisite for many applications of microbial ecology, it is also fundamental to establish the function(s) of the constituents of the communities. Microbial functionality regulates the majority of biogeochemical cycles in the biosphere, serves as a reservoir for potential bioprospecting and provides potential solutions through 'green' microbial biotechnologies. An understanding of the diversity and amount (gene dosage) of functional genes present and, in the future, the degree to which they are expressed at the mRNA level are central components to the understanding of microbial functionality. To examine the diversity of genes in the environment, two approaches have been taken utilizing genomic technology. The first takes the form of shotgun cloning and sequencing, the so-called 'metagenomics' approach. The second approach is clearly an analogue of rRNA analysis in the environment, in that array formats have been used to look at the presence and/or expression of specific pathways.

10.5.3 Metagenomics

Metagenomics (see also Chapter 11) is the application of structural genomics to the environment, where gene libraries are constructed from the shotgun cloning of environmentally recovered DNA followed by sequence analyses (61). The technique provides a comprehensive way of isolating environmentally relevant genes, and as such is heavily utilized as a tool in the field of novel gene/pathway discovery. However, within an ecological context metagenomics suffers from two pitfalls. First, it may be difficult to assign a taxonomic framework to the genes observed unless a 16S rRNA gene or robust phylogenetic marker is located within a sequenced fragment, or sequencing is restricted to taxonomically identified fragments (61). Second, the characterization of 'unknown' genes is hampered in that a function for any given gene can only be assigned on the basis of similarity to known functional genes. This problem is exacerbated as most of the genomic datasets are from eukaryotes or clinically important bacteria, as opposed to prokaryotes from the environment. Despite these points, several major successes have been afforded by environmental metagenomic approaches.

The discovery that heterotrophic processes in the open ocean were contributed to by heterotrophs which contained light-harvesting systems for energy generation (proteorhodopsin) was a direct result of a metagenomics approach (62,63). This significant finding was elucidated through sequencing of oceanic-derived genomic libraries constructed in bacterial artificial chromosome ('BAC') libraries. Subsequently, proteorhodopsin homologues linked with a 16S rRNA gene representative of the SAR86 cluster in oceanic waters were observed. Homologies with the well-described bacterial pigments could be identified and the active nature of the pathway in the environments subsequently demonstrated (64). Significantly, the metagenomic approach also led to an insight into the ecology of the organisms containing proteorhodopsin, through the elucidation of spectral variants occupying distinct geographical locations. Similar approaches using BAC library technology applied to soil have indicated the potential of metagenomics for accessing genes from uncultured soil organisms, such as the members of the *Acidobacterium* phyla, and the expression of their metabolites in a recombinant host (65). Applications of metagenomics are beginning to reveal the gene contents and differences between members of the Crenarcheota (66,67), an environmentally important group containing non-thermophilic members with widespread distributions in marine, freshwater and soil habitats.

On a larger scale, metagenomics has been employed to describe the genomic content of entire ecosystems. In the first of these studies, Tyson *et al.* (68) assembled genome sequences of the dominant prokaryotes (*Leptospira* and *Ferroplasma*) together with three other partial genomes using shotgun sequencing of a small insert plasmid library derived from plasmid DNA isolated from an acid mine drainage biofilm. In the second study, Venter *et al.* (69) have applied similar methods to study the bacterial community present in surface waters from the Sargasso Sea. The authors generated a phenomenal 1.045 billion base pairs of non-redundant sequence that they assembled into near-complete or partial genome sequences from an estimated 1800 bacterial species. In particular, large sequence scaffolds were generated that showed strong similarities to *Burkholderia, Shewanella,* SAR86 and *Prochlorococcus* genomes, but in addition sequences were derived from an estimated 148 novel phylotypes. These ecosystem level datasets, when combined with new projects, such as the Sorcerer II expedition (http://www.sorcerer2expedition.org) aiming to generate environmental genomic datasets from both marine and terrestrial ecosystems following a circumnavigation of the world, represent an outstanding opportunity to expand our knowledge and understanding of the microbial diversity in the natural environment.

10.5.4 Functional gene arrays

Metagenomic approaches reveal a large number of functional genes of interest in the environment. However, in order to assess the distribution and functioning of these genes, analyses other than the labour-intensive sequencing-based strategies are required, ideal candidates being those of the array technologies. The first attempt at a relatively comprehensive array-based detection of the genes present in environmental samples concentrated on the genes involved in nitrogen and methane cycling. Wu *et al.* (70) used immobilized probes constructed by PCR amplification of 89 *nirS, nirK, amoA* and *amoP* sequences on a glass slide array. Hybridization with labeled targets from cultures and environmental samples provided some interesting insights into gene detection in the environment and, predictably, highlighted some of the pitfalls previously discussed. Under the hybridization conditions employed (65°C), targets with ≥80–85% identity could be efficiently detected (rising to 90% at 75°C). Furthermore, they estimated that target abundances of 1 ng of pure genomic DNA and 25 ng of environmental DNA were required for *nirS* detection. These levels of detection, in principle, seem hopeful when considering potential applications in ecosystems such as soils and water. In terms of signal quantification, linear signal intensities were observed for pure genomic DNA in the range of 1 to 100 ng, but the signal intensity was found to vary, depending upon the degree of target divergence, as would be theoretically predicted. The authors highlighted that a key requirement for future quantification of both 16S and functional gene arrays is the ability to differentiate signal intensities which arise from variations in target abundance from those that arise simply due to sequence divergence. As suggested, since signal intensity at lower stringencies of hybridization was relatively unaffected by sequence divergence (if <20%; although the effect became more pronounced at higher stringencies), a sequential hybridization strategy could be employed to differentiate abundance/divergence signal variations.

Whilst signal intensity criteria are important issues to resolve in environmental applications of arrays, the issue of standardization and sample-to-sample comparison are also key areas for development. Single stringency hybridizations with exact match probes and targets have been employed to demonstrate the use of internal standardization as a tool for more robust quantification in array technologies (71). Internal array standardization allows multiple samples, with differing target concentrations, to be compared on a rational basis by correcting for uneven array printing and hybridization variation. Moreover, internal standardization affords the ability to quantify mRNA levels in the environment from multiple samples, a criterion currently limited to comparison of two environmental conditions or bacterial strains (e.g. wild type vs. mutant) using current competitive dual-color hybridizations.

Despite such challenges, microbial ecologists seem set to embrace array-based methods for investigating gene distribution, diversity and expression. Recent applications of array approaches have included development of small microarrays (68 probes) targeted at methane monooxygenase (*pmoA*) and ammonia monooyxgenase (*amoA*) genes for studying methanotrophs in landfills (72), and a macroarray (40 probes) consisting of PCR-amplified environmental *nifH* genes used to investigate spatial variability in marine diazotroph communities (73). Most recently, Rhee *et al.* (74) reported the development, testing and application of a 1662 probe 50-mer microarray based on genes involved in biodegradation and metal resistance, and used this to show that degradation genes from members of the *β-proteobacteria* were dominant in napthalene-amended microcosms.

10.5.5 Diversity, function and process

A primary concern for microbial ecologists is that the tools of genomics are problematic when analyzing organisms of interest *in situ* amidst a background of non-target microbes (e.g. within a complex soil community). Moreover, it could be said that the analyses tend to lack a context, in terms of directly linking the functional pathways/organisms with environmental processes, e.g. carbon cycling. In this respect we anticipate the development of complementary approaches to relate the genomic analyses of an organism to the process it performs within a mixed community *in situ*.

One way of using genomic technologies in an environmental context is the design and implementation of experimental techniques which can separate out the true signal of a target organism/group in mixed microbial communities from all of the organisms present within the sample. At present, cells (75) or extracted rRNA (76) for standard or genomic analyses can be recovered from environments with reference to a phylogenetic framework by oligonucleotide probe capture. One interesting facet of the rRNA hybridization capture method is the ability to link the phylogenetic analysis with a process the organism may be performing (76) using stable isotope detection within their nucleic acids (e.g. ^{13}C or ^{15}N). Quantification of isotopic signatures within recovered nucleic acids will, in principle, provide a very powerful way to determine the nature and rates of processes and link this with genomic detection strategies. Other applications include pulsing stable-isotope-labeled substrates into communities and recovering 'heavy' DNA containing the genomes of the organisms that were involved in processing the substrate (77) (see also Chapter 6), and the recovery of RNA to access rRNA or the transcriptome (78,79). More recently, Adamczyk *et al.* (80) have combined radioactive isotopes of carbon, in the form of ^{14}C-bicarbonate, with a microarray analysis to investigate ammonia-oxidizing bacteria in activated sludge. This elegant work first detected the groups present by microarray hybridization and subsequently measured radioactive incorporation directly on the arrays, to give a combined phylogeny and function analysis.

The additional combination of the aforementioned approaches with the future application of isotope-labeled protein strategies such as ICAT (isotope-coded affinity tag) (81) offers enormous potential for investigation of microbial community function at the level of the genome, metagenome, transcriptome and proteome.

There are several techniques that do not directly link molecular signatures to biogeochemistry through isotopic tracer studies, but which operationally define subgroups with microbial communities. Such methods include fluorescent *in situ* hybridization (e.g. 82) (see Chapter 9) and/or activity measures (83–85) coupled to flow cytometric cell sorting for the recovery of specific bacterial populations. We predict that these tools will become a central form of sample collection for *in situ* studies prior to genomic applications in the coming years. Techniques that allow directed genomic analyses will target the community in an effective way, could link genomics to microbially driven processes and substantially increase the success of genomic analyses *in situ*.

10.6 Lessons learned and future perspectives

> *'There will come a time when you believe everything is finished. That will be the beginning.'*
>
> *Louis L'Amour*

The discipline of microbial ecology is firmly rooted in understanding the ecology of microbes in the 'real world'. This encompasses microbial evolution, adaptation, community structure and function, together with their effects upon biogeochemical cycles at scales which span millimeters, kilometers, continents and oceans. Genomics is a tool which aids us in this endeavor. As we have stated in the preceding sections, certain obstacles need to be overcome prior to the widespread use of genomics in microbial ecology. These obstacles are by no means insurmountable, and we would formalize these as five 'lessons' learned for future consideration.

(i) *Progress will be directly proportional to database size.* The current size of databases with fully annotated genomes and functional analyses for environmental organisms limits genomic application for microbial ecologists. A key requirement is the analysis of multiple representatives of the key environmental groups or lineages thus far identified. Specific tasks involved with this are: assessing culture requirements, isolation of pure cultures, directed sequencing programs and effective functional analyses (both in silico and biochemically) of these isolates.

(ii) *The horizontal gene pool.* The importance of the HGP to the evolution of environmental microbes and their pathways cannot be understated. As such, intensive sequencing/functional analyses for mediators of the HGP (plasmids, transposons etc.) should be undertaken for a range of habitats and processes, and their results given equal significance to those placed upon chromosomal sequences.

(iii) *Ex situ* vs. *in situ*. Genomic *ex situ* studies using cultured isolates can be used to determine underlying principles of bacterial survival/adaptation/functionality in the environment. However, in order to extrapolate *ex situ* experiments to observations made *in situ*, great care must be required for the *ex situ* experimental design and execution. For example, microbes continuously experience multiple stresses and significant variation in both the type and concentrations of nutrient sources. Microbes rarely, if ever, exist in a 'batch' culture scenario with optimal growth conditions. Experimental manipulations using genomic analyses should reflect the continuous and multifactoral lifestyle that microbes experience.

(iv) *Signals and noise.* The successful application of genomic approaches *in situ* requires the development of methods that provide significant resolution of signal against noise. An example would be the requirements for the detection of the 16S rRNA genes of a specific species against a background of large numbers of co-habiting organisms. A large amount of research on standardization and calibration is still required prior to the widespread application of genomic technologies in environmental samples.

(v) *Genomics and biogeochemistry.* An important role of genomics in microbial ecology will be to relate features of microbial communities detectable by genomic strategies to environmental processes. As an example, the determination of abundances of methane oxidizers and their functional transcripts should be related to methane oxidation rates. The absence of a link between environmental process measurements and genomic detection of organism and transcript abundances will ultimately lead to an increase in information at the expense of ecological understanding.

References

1. Fleischman RD *et al.* (1995) Whole genome random sequencing and assembly of *Haemophilus influenza* Rd. *Science* **269**: 496–512.
2. Paulsen IT *et al.* (2002) The *Brucella suis* genome reveals fundamental similarities between animal and plant pathogens and symbionts. *Proc Natl Acad Sci USA* **99**: 13148–13153.
3. Tamas I, Klasson L, Canback B, Naslund AK, Eriksson AS, Wernegreen JJ, Sandstrom JP, Moran N, Andersson SGE (2002) 50 million years of genomic stasis in endosymbiotic bacteria. *Science* **296**: 2376–2379.
4. Berg OG, Kurland CG (2002) Evolution of microbial genomes: sequence acquisition and loss. *Mol Biol Evol* **19**: 2265–2276.
5. Doolittle RF (2002) Microbial genomes multiply. *Nature* **416**: 697–700.
6. Kawarabayasi Y, Hino Y, Horikawa H *et al.* (1999) Complete genome sequence of an aerobic hyper-thermophilic crenarchaeon, *Aeropyrum pernix* K1. *DNA Res* **6**: 83–101.
7. Blattner FR, Plunkett III G, Bloch CA *et al.* (1997) The complete genome sequence of *Escherichia coli* K12. *Science* **277**: 1453–1462.
8. Kunst F, Ogasawara N, Moszer I *et al.* (1997) The complete genome sequence of the Gram-positive bacterium *Bacillus subtilis*. *Nature* **390**: 249–256.
9. Yu BJ, Sung BH, Koob MD, Lee CH, Lee JH Lee WS, Kim MS, Kim SC (2002) Minimisation of the *Escherichia coli* genome using Tn5-targeted Cre/*loxP* excision system. *Nature Biotechnol* **20**: 1018–1023.
10. Kaneko T, Nakamura Y, Sato S *et al.* (2000) Complete genome structure of the nitrogen-fixing symbiotic bacterium *Mesorhizobium loti*. *DNA Res* **7**: 331–338.
11. Kaneko T, Nakamura Y, Sato S *et al.* (2002) Complete genomic sequence of nitrogen fixing symbiotic bacterium *Bradyrhizobium japonicum* USDA110. *DNA Res* **9**: 189–197.
12. Bentley SD, Chater KF, Cerdeno-Tarraga AM *et al.* (2002) Complete genome sequence of the model actinomycete *Streptomyces coelicolor* A3(2). *Nature* **417**: 141–147.
13. Glockner FO, Kube M, Bauer M *et al.* (2003) Complete genome sequence of the

marine planctomycete *Pirellula* sp strain 1. *Proc Natl Acad Sci USA* **100**: 8298–8303.

14. Stover CK, Pham XQ, Erwin AL *et al.* (2000) Complete genome sequence of *Pseudomonas aeruginosa* PAO1, an opportunistic pathogen. *Nature* **406**: 959–964.
15. Galibert F, Finan TM, Long SR *et al.* (2001) The composite genome of the legume symbiont *Sinorhizobium meliloti*. *Science* **293**: 668–672.
16. Wood DW, Setubal JC, Kaul R *et al.* (2001) The genome of the natural genetic engineer *Agrobacterium tumefaciens* C58. *Science* **294**: 2317–2323.
17. Read TD, Peterson SN, Tourasse N *et al.* (2003) The genome sequence of *Bacillus anthracis* Ames and comparison to closely related bacteria. *Nature* **423**: 81–86.
18. da Silva AC, Ferro JA, Reinach FC *et al.* (2002) Comparison of the genomes of two *Xanthomonas* pathogens with differing host specificities. *Nature* **23**: 459–463.
19. Perna NT, Plunkett III G, Burland V *et al.* (2001) Genome sequence of enterohaemorrhagic *Escherichia coli* 0157:H7. *Nature* **409**: 529–533.
20. Read TD, Salzberg SL, Pop M *et al.* (2002) Comparative genome sequencing for discovery of novel polymorphisms in *Bacillus anthracis*. *Science* **296**: 2028–2033.
21. Hacker J, Kaper JB (2000) Pathogenicity islands and the evolution of microbes. *Annu Rev Microbiol* **54**: 641–679.
22. Reid SD, Herbelin H, Bumbaugh AC, Selander RK, Whittam TS (2000) Parallel evolution in pathogenic *Escherichia coli*. *Nature* **406**: 64–67.
23. Dziejman M, Balon E, Boyd D, Fraser CM, Heidelberg JF, Mekalanos JJ (2002) Comparative genomic analysis of *Vibrio cholerae*: genes that correlate with cholera endemic and pandemic disease. *Proc Natl Acad Sci USA* **99**: 1556–1561.
24. Coffey TJ, Enright MC, Daniels M, Morona JK, Hrniewicz W, Paton JC, Spratt BG (1998) Recombinational exchanges at the capsular polysaccharide biosynthetic locus lead to frequent serotype changes among natural isolates of *Streptococus pneumoniae*. *Mol Microbiol* **27**: 73–83.
25. Beres SB, Sylva GL, Barbian KD *et al.* (2002) Genome sequence of a serotype M3 strain of group A *Streptococcus*: phage-encoded toxins, the high-virulence phenotype, and clone emergence. *Proc Natl Acad Sci USA* **99**: 10078–10083.
26. Salama N, Guillemin K, McDaniel TK, Sherlock G, Tomkins S, Falkow S (2000) A whole-genome microarray reveals genetic diversity among *Helicobacter pylori* strains. *Proc Natl Acad Sci USA* **97**: 14668–14673.
27. Murray AE, Lies D, Nealson K, Zhou J, Tiedje JM (2001) DNA/DNA hybridisation to microarrays reveals gene-specific differences between closely related microbial genomes. *Proc Natl Acad Sci USA* **98**: 9853–9858.
28. Burrus V, Pavlovic G, Decaris B, Geudon G (2002) Conjugative transposons: the tip of the iceberg. *Mol Microbiol* **46**: 601–610.
29. Van der Meer JR, Sentchilo V (2003) Genomic islands and evolution of catabolic pathways in bacteria. *Curr Opin Microbiol* **14**: 248–254.
30. Wernegreen JJ, Ochman H, Jones IB, Moran NA (2000) Decoupling of genome size and sequence divergence in a symbiotic bacterium. *J Bacteriol* **182**: 3867–3869.
31. Mavingui P, Flores M, Guo X, Davilla G, Perret X, Broughton WJ, Palacios R (2002) Dynamics of genome architecture in *Rhizobium* sp NGR234. *J Bacteriol* **184**: 171–176.
32. Levin BR, Bergstrom CT (2000) Bacteria are different: observations, interpretations, speculations, and opinions about the mechanisms of adaptive evolutions in prokaryotes. *Proc Natl Acad Sci USA* **97**: 6981–6985.
33. Lilley AK, Young JPW, Bailey MJ (2000) Bacterial population genetics: do plasmids maintain diversity and adaptation? In: Thomas CM (ed.) *The Horizontal Gene Pool: Bacterial Plasmids and Gene Spread*. Harwood Academic Publishers, pp. 287–300.
34. Snel B, Bork P, Huynen MA (1999) Genome phylogeny based on gene content. *Nature Genet* **21**: 108–109.

35. Baldauf SL, Roger AJ, Wenk-Siefert I, Doolittle WF (2000) A kingdom-level phylogeny of eukaryotes based on combined protein data. *Science* **290**: 972–977.
36. Korbel JO, Snel B, Huynen MA, Bork P (2002) SHOT: a web server for the construction of genome phylogenies. *Trends Genet* **18**: 158–162.
37. Gogarten PJ, Doolittle WF, Lawrence JG (2002) Prokaryotic evolution in light of gene transfer. *Mol Biol Evol* **19**: 2226–2238.
38. Ragan MA, Charlebois RL (2002) Distributional profiles of homologous open reading frames among bacterial phyla: implications of vertical and lateral transmission. *Int J Syst Evol Microbiol* **52**: 777–787.
39. Oliver SG (2002) Functional genomics: lessons from yeast. *Phil Trans R Soc Lond* **357**: 17–23.
40. Lockhart DJ, Winzeler EA (2000) Genomics, gene expression and DNA arrays. *Nature* **405**: 827–836.
41. Harrington CA, Rosenow C, Retief J (2000) Monitoring gene expression using DNA microarrays. *Curr Opin Microbiol* **3**: 285–291.
42. O'Farrell PH (1975) High resolution two-dimensional electrophoresis of proteins. *J Biol Chem* **250**: 4007–4021.
43. Van Bogelen R, Abshire K, Pertsemilidis A, Clark RL, Neidhart F (1996) Gene-protein database of *Escherichia coli* K-12. In: Neidhart FC (ed.) *Escherichia coli and Salmonella: Cellular and Molecular Biology*, 2nd edn, pp. 2067–2117. American Society for Microbiology, Washington, DC.
44. Santoni V, Molloy M, Rabilloud T (2000) Membrane proteins and proteomics: un amour impossible. *Electrophoresis* **21**: 1054–1070.
45. Nouwens AS, Cordwell SJ, Larsen MR, Molloy MP, Gillings M, Willcox MDP, Walsh BJ (2000) Complementing genomics with proteomics: The membrane subproteome of *Pseudomonas aeruginosa* PAO1. *Electrophoresis* **21**: 3797–3809.
46. Wei Y, Lee J-M, Richmond C, Blattner FR, Rafalski JA, LaRossa RA (2001) High-density microarray-mediated gene expression profiling of *Escherichia coli*. *J Bacteriol* **183**: 545–556.
47. Yoon SH, Han MJ, Lee SY, Jeong KJ, Yoo SJ (2003) Combined transcriptome and proteome analysis of *Escherichia coli* during high cell density culture. *Biotechnol Bioengng* **81**: 753–767.
48. Mostertz J, Scharf C, Hecker M, Homuth G (2004) Transcriptome and proteome analysis of *Bacillus subtilis* gene expression in response to superoxide and peroxide stress. *Microbiology* **150**: 497–512.
49. Mahan MJ, Slauch JM, Mekalanos JJ (1993) Selection of bacterial virulence genes specifically induced in host tissue. *Science* **259**: 686–688.
50. Rainey PB, Heithoff DM, Mahan MJ (1997) Single-step conjugative cloning of bacterial gene fusions involved in microbe–host interactions. *Mol Gen Genet* **256**: 84–87.
51. Thompson DK, Beliaev AS, Giometti CS *et al.* (2002) Transcriptional and proteomic analysis of a ferric uptake regulator (Fur) mutant of *Shewanella oneidensis*: possible involvement of Fur in energy metabolism, transcriptional regulation, and oxidative stress. *Appl Environ Microbiol* **68**: 881–892.
52. DeLisa MP, Wu C-F, Wang L, Valdes JJ, Bentley WE (2001) DNA microarray-based identification of genes controlled by autoinducer 2-stimulated quorum sensing in *Escherichia coli*. *J Bacteriol* **183**: 5239–5247.
53. Ferea TL, Botstein D, Brown PO, Rosenzweig RF (1999) Systematic changes in gene expression patterns following adaptive evolution in yeast. *Proc Natl Acad Sci USA* **96**: 9721–9726.
54. Raamsdonk LM, Teusnik B, Broadhurst D *et al.* (2001) A functional genomics strategy that uses metabolome data to reveal the phenotype of silent mutations. *Nature Biotech* **19**: 45–50.
55. Erdner DL, Eisen J, Sobecky PA (2001) Whole plasmid sequencing and expres-

sion analysis of cryptic marine plasmids. Abstract in the conference proceedings of Aquatic Sciences 2001. American Society for Limnology and Oceanography, Waco.
56. Hugenholtz P, Goebel BM, Pace NR (1998) Impact of culture-independent studies on the emerging phylogenetic view of microbial diversity. *J Bacteriol* **180**: 4765–4774.
57. Guschin DY, Mobarry BK, Proudnikov D, Stahl DA, Rittman BE, Mirzabekov AD (1997) Oligonucleotide microchips as genosensors for determinative and environmental studies in microbiology. *Appl Environ Microbiol* **63**: 2397–2402.
58. Small J, Call DR, Brockman FJ, Straub TM, Chandler DP (2001) Direct detection of 16S rRNA in soil extracts by using oligonucleotides microarrays. *Appl Environ Microbiol* **67**: 4708–4716.
59. Loy A, Lehner A, Lee N, Adamczyk J, Meier H, Ernst J, Scheifer KH, Wagner M (2002) Oligonucleotide microarray for 16S rRNA gene based detection of all recognised lineages of sulphate reducing prokaryotes in the environment. *Appl Environ Microbiol* **68**: 5064–5081.
60. Liu W, Mirzabekov AD, Stahl DA (2001) Optimisation of an oligonucleotides microchip for microbial identification studies: a non-equilibrium dissociation approach. *Environ Microbiol* **3**: 619–629.
61. Stein JL, Marsh TL, Wu KY, Shizuya H, DeLong EF (1996) Characterisation of uncultivated prokaryotes: Isolation and analysis of a of a 40 kilobase pair genome fragment from a planktonic marine archaeon. *J Bacteriol* **178**: 591–599.
62. Beja O, Suzuki MT, Koonin EV *et al.* (2000) Construction and analysis of bacterial artificial chromosome libraries from a marine microbial assemblage. *Environ Microbiol* **2**: 516–529.
63. Beja O, Aravind L, Koonin EV *et al.* (2000) Bacterial rhodopsin: Evidence for a new type of phototrophy in the sea. *Science* **289**: 1902–1906.
64. Beja O, Spudich EN, Spudich JL, Leclerc M, DeLong EF (2001) Proteorhodopsin phototrophy in the Ocean. *Nature* **411**: 786–789.
65. Rondon MR, August PR, Bettermann AD *et al.* (2000) Cloning the soil metagenome: a strategy for accessing the genetic and functional diversity of uncultured microorganisms. *Appl Environ Microbiol* **66**: 2541–2547.
66. Beja O, Koonin EV, Aravind L, Taylor LT, Seitz H, Stein JL, Bensen DC, Feldman RA, Swanson RV, DeLong EF (2002) Comparative genomic analysis of archaeal genotypic variants in a single population, and in two different oceanic provinces. *Appl Environ Microbiol* **68**: 335–345.
67. Quaiser A, Ochsenreiter T, Klenk H-P, Kletzin A, Treusch AH, Meurer G, Eck J, Sensen CW, Schleper C (2002) First insight into the genome of an uncultivated crenarchaeote from soil. *Environ Microbiol* **4**: 603–611.
68. Tyson GW, Chapman J, Hugenholtz P, Allen EE, Ram RJ, Richardson PM, Solovyev VV, Rubin EM, Rokshar DS, Banfield JF (2004) Community structure and metabolism through reconstruction of microbial genomes from the environment. *Nature* **428**: 37–43.
69. Venter JC, Remington K, Heidelberg JF *et al.* (2004) Environmental genome shotgun sequencing of the Sargasso Sea. *Science* **304**: 66–74.
70. Wu L, Thompson DK, Li G, Hurt RA, Tiedje JM, Zhou J (2001) Development and evaluation of functional gene arrays for detection of selected genes in the environment. *Appl Environ Microbiol* **67**: 5780–5790.
71. Cho J-C, Tiedje JM (2002) Quantitative detection of microbial genes by using DNA microarrays. *Appl Environ Microbiol* **68**: 1425–1430.
72. Stralis-Pavese N, Sessitsch A, Weilharter A, Reichnauer T, Riesing J, Csontos J, Murrell JC, Bedrosey L (2004) Optimization of diagnostic microarray for application in analysing landfill methanotroph communities under different plant covers. *Environ Microbiol* **6**: 347–363.

73. Jenkins BD, Steward GF, Short SM, Ward BB, Zehr JP (2004) Fingerprinting diazotroph communities in the Chesapeake Bay by using a DNA microarray. *Appl Environ Microbiol* **70**: 1767–1776.
74. Rhee S-K, Liu X, Wu L, Chong SC, Wan X, Zhou J (2004) Detection of genes involved in biodegradation and biotransformation in microbial communities by using 50-mer oligonucleotide microarrays. *Appl Environ Microbiol* **70**: 4303–4317.
75. Stoffels M, Ludwig W, Schleifer KH (1999) rRNA probe-based cell fishing of bacteria. *Environ Microbiol* **1**: 259–271.
76. MacGregor BJ, Bruchert V, Fleischer S, Amann R (2002) Isolation of small-subunit rRNA for stable isotopic characterization. *Environ Microbiol* **4**: 451–464.
77. Radajewski S, Ineson P, Parekh NR, Murrell JC (2000) Stable-isotope probing as a tool in microbial ecology. *Nature* **403**: 646–649.
78. Manefield M, Whiteley AS, Ostle N, Ineson P, Bailey MJ (2002) Technical considerations for RNA based stable isotope probing: an approach to associating microbial diversity with microbial community function. *Rapid Commun Mass Spectr* **16**: 2179–2183.
79. Manefield M, Whiteley AS, Griffiths RI, Bailey MJ (2002) RNA stable isotope probing: a novel means of linking microbial community function to phylogeny. *Appl Environ Microbiol* **68**: 5367–5373.
80. Adamczyk J, Hesselsoe M, Iversen N, Horn M, Lehner A, Nielsen PH, Schloter M, Roslev P, Wagner M (2003) The isotope array, a new tool that employs substrate-mediated labelling of rRNA for determination of microbial community structure and function. *Appl Environ Microbiol* **69**: 6875–6887.
81. Gygi SP, Rist B, Gerber SA, Turecek F, Gelb MH, Aebersold R (1999) Quantitative analysis of complex protein mixtures using isotope-coded affinity tags. *Nature Biotechnol* **17**: 994–999.
82. Wallner G, Fuchs B, Spring S, Beisker W, Amann R (1997) Flow sorting of microorganisms for molecular analysis. *Appl Environ Microbiol* **63**: 4223–4231.
83. Servais P, Courties C, Lebaron P, Troussellier M (1999) Coupling bacterial activity measurements with cell sorting by flow cytometry. *Microb Ecol* **38**: 180–189.
84. Servais P, Agogue H, Courties C, Joux F, Lebaron P (2001) Are the actively respiring cells (CTC+) those responsible for bacterial production in aquatic environments? *FEMS Microbiol Ecol* **35**: 171–179.
85. Whiteley AS, Griffiths RI, Bailey MJ (2003) Analysis of the microbial functional diversity within water stressed soil communities by flow cytometric analysis and CTC+ cell sorting. *J Microbiol Methods* **54**: 257–267.

Metagenomic libraries from uncultured microorganisms

11

Doreen E. Gillespie, Michelle R. Rondon, Lynn L. Williamson and Jo Handelsman

11.1 Introduction

Most microorganisms in the environment are refractory to culturing by standard methods. In many habitats, as few as 0.1–1% of the viable cells are culturable, and community diversity estimated by molecular methods that bypass culturing far exceeds that which can be accounted for by culturing. Analyses of environments as diverse as soil, hot springs, oceans, mammalian cavities and insect tissue have revealed uncultured species richness (1–6). Many thousands of bacterial species remain to be discovered and studied.

Most of our knowledge of microbiology is derived from cultured microorganisms. A new approach in microbiology, known as metagenomics, environmental genomics, or community genomics, has been developed to access information about the biology of the uncultured majority of microorganisms (7,8). Metagenomics is the culture-independent analysis of the metagenome, which refers to the collective genomes of the organisms in an environment (9). The general approach is to extract DNA directly from an environmental sample, clone the DNA into a plasmid vector, and introduce the cloned DNA into a culturable host (*Figure 11.1*). Assessment of the phylogenetic diversity represented by the DNA in the libraries reflects the diversity of the assemblage from which the libraries were constructed (10–13). Functional and sequence-based analyses provide insight into the physiological capacities of community members (11,14,15). These analyses, conducted in parallel, can lead to an understanding of the roles of uncultured microorganisms in complex communities and to the discovery of new tools for biotechnology (9,16). Metagenomic analysis is in its infancy, but it has already yielded insights into microbial diversity and function.

11.2 Factors affecting experimental design

Many of the methods currently in use for environmental library construction have been adapted from techniques used to generate large-insert libraries from eukaryotic organisms, including cloning in bacterial artificial chromosomes (BACs) and cell lysis in agarose plugs (17,18). Library

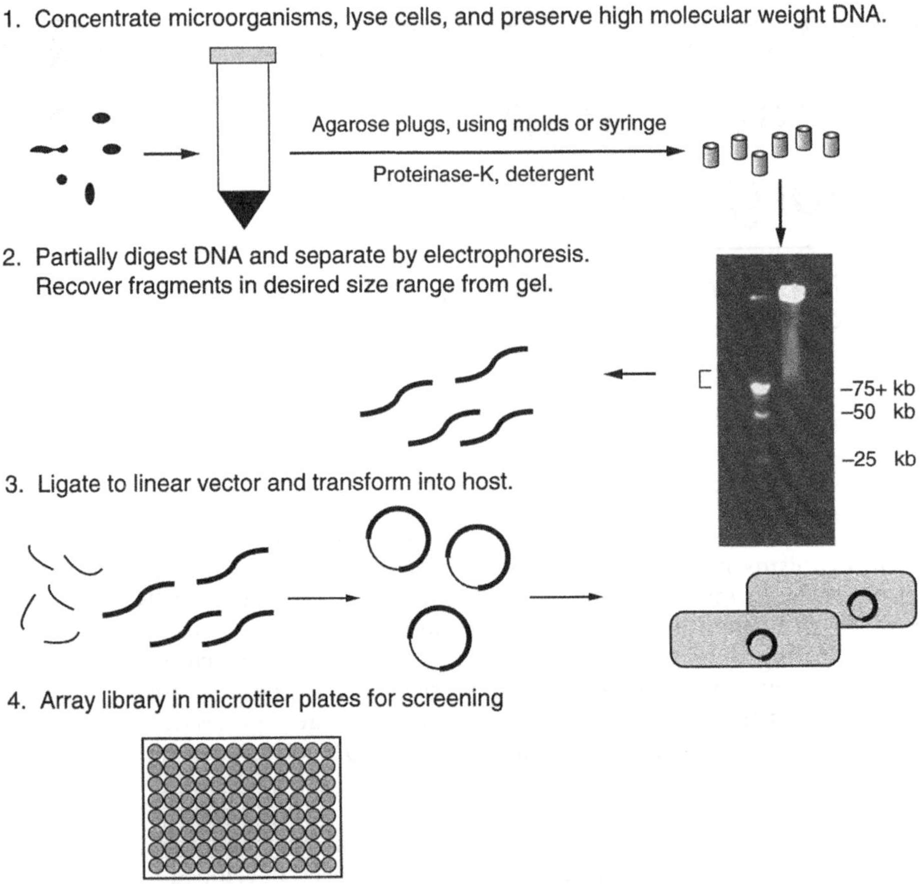

1. Concentrate microorganisms, lyse cells, and preserve high molecular weight DNA.
Agarose plugs, using molds or syringe
Proteinase-K, detergent
2. Partially digest DNA and separate by electrophoresis.
Recover fragments in desired size range from gel.
–75+ kb
–50 kb
–25 kb
3. Ligate to linear vector and transform into host.
4. Array library in microtiter plates for screening

Figure 11.1

Construction of a metagenomic library, adapted from Stein *et al.* (7).

construction using environmental DNA presents challenges that are not at issue for the construction of genomic libraries of organisms in pure culture. In this section, we address critical challenges that direct the selection of the source and method of preparation of environmental DNA, as well as the choice of cloning vector and host strain. *Table 11.1* summarizes the environmental sources, vectors and applications of libraries that have been described to date.

11.2.1 Environmental source

As indicated in *Table 11.1*, metagenomic libraries have been constructed from DNA samples from diverse environments. The first environmental library was constructed with DNA isolated from oceanic microorganisms, which were concentrated from sea water prior to extraction of DNA (7). Other microbial habitats, such as soil, can present challenges different from sea water for direct DNA isolation, because chemical contaminants and par-

Table 11.1 Metagenomic libraries from diverse environments

Environment	DNA extraction method	Vector	Average insert size (kb)	Genes and/or functions identified	References
Sea water	Agarose plug	BAC (pIndigo BAC536)	60–80 (two libraries)	Photosystem gene clusters; rhodopsin; 16S rRNA	(10,15,39,40)
Sea water	Direct	Plasmid (Lambda Zap II)	1.8–5.4	Chitinases	(43)
Sea water	Agarose plug	Fosmid (pFOS1)	35–45	*radA* genes	(12)
Sea water	Agarose plug	Fosmid (pFOS1)	35–45	16S rRNA	(7,13)
Soil	Direct	Cosmid	NR	N-acyl amino acid antibiotics; Violacein gene cluster	(41,42)
Soil	Direct	Plasmid (pBluescript SK+)	5–8	4-Hydroxybutyrate utilization; lipases; esterases; Na^+/H^+ antiporter	(14,45,46)
Soil	Direct	BAC (pBTP2)	37	Small molecule antibiotic	(47)
Soil	Direct	BAC (pBeloBAC11)	27–43 (two libraries)	Amylases; lipases; hemolytic activity; triaryl cation antibiotic; DNase; 16S rRNA	(11,50)
Soil	NR	NR	NR	Small molecule antibiotics	(51)
Enrichment cultures from horse excrement and soil	Direct	Cosmid (pWE15)	30–40 (multiple libraries)	Biotin biosynthesis operons	(44)
Marine sponge	Agarose plug	Fosmid (pFOS1)	35–45	DNA polymerase; 16S rRNA	(26,48,49)
Beetle	Direct	Cosmid (pWEB)	NR	Pederin biosynthesis gene cluster	(52)
Human oral flora	Direct	TOPO-XL	NR	Tetracycline resistance gene	(53)

NR: not reported.

ticulates interfere with DNA isolation, purification and digestion and/or with methods designed to concentrate cells (19–21) (see also Chapter 1). For this reason, it is helpful to extract and concentrate the microorganisms prior to cell lysis and DNA isolation. Although the isolation of DNA from organisms only after physical separation from soil particles may introduce a bias in the diversity of microorganisms represented in the final sample, due to the tight association of some bacteria with soil particles, this bias may be balanced by the purity and quality of the DNA, thus facilitating the construction of metagenomic libraries. Similar concerns should be considered when targeting DNA from prokaryotes found in close association with eukaryotic cells, where an additional challenge may be the separation and removal of eukaryotic cells or DNA from the prokaryotic sample.

11.2.2 DNA isolation

Metagenomic analysis may entail cloning of small or large fragments of DNA. Small fragments are adequate if single or small groups of genes are sufficient for an analysis, whereas large inserts are required to study multigenic pathways, genomic organization, or for extensive DNA sequence analysis. Whatever the size of the desired insert, the preparation of DNA is critical to the success of library construction. The use of partial digestion to prepare DNA for cloning necessitates starting with DNA at least three times longer than the desired insert size. Therefore, if inserts of 100 kb are desired, the pre-digested DNA should be ≥300 kb. DNA of this size is vulnerable to shearing by pipetting, centrifugation, freeze–thaw cycling, and other steps commonly used for extraction of DNA from environmental samples or pure cultures. Cell lysis and DNA isolation using the agarose plug method, in which cells are lysed while immobilized in agarose, is the gentlest means of obtaining high molecular weight DNA and is described in this chapter. An alternative protocol for direct isolation of DNA from environmental samples is also described for situations in which cell separation and concentration are not feasible (see also Chapter 1). Simpler techniques to lyse cells, such as bead-beating, may be used when smaller (2–5 kb) library inserts are desired.

11.2.3 Vector selection

The choice of vector is guided by the intended uses of the library and by the desired insert size. Bacterial artificial chromosomes (BACs) are most commonly used for large insert libraries, due to their ability to accommodate inserts over 300 kb in size (22,23). BAC maintenance at single copy may be desirable due to potential lethality caused by toxic gene products expressed from the insert DNA. BACs with inducible copy number, such as the pCC1BAC™ vector (Epicentre, Madison, WI, USA), permit library maintenance at low copy number, but the higher copy number resulting from induction provides easier BAC isolation and higher expression levels of gene products, thus facilitating identification of active clones in a functional screen. For small insert libraries, a high copy number vector with an inducible promoter is desired to maximize the chance of gene expression. Among alternative vectors to be considered are cosmids, that carry intermediate size inserts (38–52 kb) (24), and bacteriophage-derived P1 plasmids, that can maintain inserts up to 95 kb (25). Several commercial cosmid kits are currently available to simplify vector preparation.

11.2.4 Host cell selection

Escherichia coli strain DH10B [F$^-$ *mcr*A Δ(*mrr-hsd*RMS-*mcr*BC) Φ80*lac*ZΔM15 Δ*lac*X74 *rec*A1 *end*A1 *ara*Δ139 Δ(*ara, leu*)7697 *gal*U *gal*K λ$^-$ *rps*L (Str^R) *nup*G] has been most commonly used as the host strain for large insert BAC libraries, because it accepts and maintains large plasmids. Alternative host strains may also be considered for particular screening applications or to find genes expressed in other species, but their ability to accept and maintain large plasmids should first be assessed. Use of commercially available competent cells for electroporation can ensure high transformation rates. If functional screening for activities endogenous to the host strain is desired, a mutant lacking the activity should be used so that the activity expressed from the environmental DNA can be detected against the host background.

11.3 Summary

DNA isolated from microorganisms that have thus far proved resistant to culturing contains valuable information for a range of disciplines, from biotechnology to microbial ecology. Metagenomic libraries are becoming an essential resource in the characterization of community composition and function. Although these libraries appear to represent a significant shift away from traditional culturing, they will also yield insights that may enable successful culturing of previously uncultured microbes.

Comprehensive study of the genomes of complex communities requires large libraries, in contrast to genome analysis of an organism in pure culture. For example, to provide even single coverage of an environment containing 100 5–Mb genomes would require 5000 clones with average insert size of 100 kb. Even the simplest environment, thought to contain a single organism, may contain strain diversity that will only be accessible by a direct cloning approach (26), thus highlighting the requirement for high level coverage of the collective genomes in any environment. Thoughtful design, construction, and storage of these libraries will maximize the information obtained from them.

References

1. Schmidt TM, DeLong EF, Pace NR (1991) Analysis of a marine picoplankton community by 16S ribosomal RNA gene cloning and sequencing. *J Bacteriol* **173**: 4371–4378.
2. Stahl DA, Lane DJ, Olsen GJ, Pace NR (1985) Characterization of a Yellowstone hot spring microbial community by 5S rRNA sequences. *Appl Environ Microbiol* **49**: 1379–1384.
3. Pryde SE, Richardson AJ, Stewart CS, Flint HJ (1999) Molecular analysis of the microbial diversity present in the colonic wall, colonic lumen, and cecal lumen of a pig. *Appl Environ Microbiol* **65**: 5372–5377.
4. Giovannoni SJ, Britschgi TB, Meyer CL, Field KG (1990) Genetic diversity in Sargasso Sea bacterioplankton. *Nature* **345**: 60–63.
5. Paster BJ, Boches SK, Galvin JL, Ericson RE, Lau CN, Levanos VA, Sahasrabudhe A, Dewhirst FE (2001) Bacterial diversity in human subgingival plaque. *J Bacteriol* **183**: 3770–3783.
6. O'Neill SL, Giordano R, Colbert AME, Karr TL, Robertson HM (1992) 16S rRNA phylogenetic analysis of the bacterial endosymbionts associated with cytoplasmic incompatibility in insects. *Proc Natl Acad Sci USA* **89**: 2699–2702.
7. Stein JL, Marsh TL, Wu KY, Shizuya H, DeLong EF (1996) Characterization of uncultivated prokaryotes: isolation and analysis of a 40-kilobase-pair genome fragment from a planktonic marine archaeon. *J Bacteriol* **178**: 591–599.

8. Rondon MR, Goodman RM, Handelsman J (1999) The Earth's bounty: assessing and accessing soil microbial diversity. *Trends Biotechnol* **17**: 403–409.
9. Handelsman J, Rondon MR, Brady SF, Clardy J, Goodman RM (1998) Molecular biological access to the chemistry of unknown soil microbes: a new frontier for natural products. *Chem Biol* **5**: R245–249.
10. Beja O, Suzuki MT, Heidelberg JF, Nelson WC, Preston CM, Hamada T, Eisen JA, Fraser CM, DeLong EF (2002) Unsuspected diversity among marine aerobic anoxygenic phototrophs. *Nature* **415**: 630–633.
11. Rondon MR, August PR, Bettermann AD *et al.* (2000) Cloning the soil metagenome: a strategy for accessing the genetic and functional diversity of uncultured microorganisms. *Appl Environ Microbiol* **66**: 2541–2547.
12. Sandler SJ, Hugenholtz P, Schleper C, DeLong EF, Pace NR, Clark AJ (1999) Diversity of *radA* genes from cultured and uncultured *Archaea*: comparative analysis of putative RadA proteins and their use as a phylogenetic marker. *J Bacteriol* **181**: 907–915.
13. Vergin KL, Urbach E, Stein JL, DeLong EF, Lanoil BD, Giovannoni SJ (1998) Screening of a fosmid library of marine environmental genomic DNA fragments reveals four clones related to members of the order *Planctomycetales*. *Appl Environ Microbiol* **64**: 3075–3078.
14. Henne A, Schmitz RA, Bomeke M, Gottschalk G, Daniel R (2000) Screening of environmental DNA libraries for the presence of genes conferring lipolytic activity on *Escherichia coli*. *Appl Environ Microbiol* **66**: 3113–3116.
15. Beja O, Suzuki MT, Koonin EV *et al.* (2000) Construction and analysis of bacterial artificial chromosome libraries from a marine microbial assemblage. *Environ Microbiol* **2**: 516–529.
16. Osburne MS, Grossman TH, August PR, MacNeil IA (2000) Tapping into microbial diversity for natural products drug discovery. *ASM News* **66**: 411–417.
17. Shizuya H, Birren B, Kim UJ, Mancino V, Slepak T, Tachiri Y, Simon M (1992) Cloning and stable maintenance of 300-kilobase-pair fragments of human DNA in *Escherichia coli* using an F-factor-based vector. *Proc Natl Acad Sci USA* **89**: 8794–8797.
18. Frijters ACJ, Zhang-ZZ, Van-Damme M, Wang GL, Ronald PC, Michelmore RW (1997) Construction of a bacterial artificial chromosome library containing large EcoRI and HindIII genomic fragments of lettuce. *Theor Appl Genet* **94**: 390–399.
19. Tebbe CC, Vahjen W (1993) Interference of humic acids and DNA extracted directly from soil in detection and tranformation of recombinant DNA from bacteria and yeast. *Appl Environ Microbiol* **59**: 2657–2665.
20. Pillai SD, Josephson KL, Bailey RL, Gerba CP, Pepper IL (1991) Rapid method for processing soil samples for polymerase chain reaction amplification of specific gene sequences. *Appl Environ Microbiol* **57**: 2283–2286.
21. Tsai Y-L, Olson BH (1991) Rapid method for direct extraction of DNA from soil and sediments. *Appl Environ Microbiol* **57**: 1070–1074.
22. Shizuya H, Birren B, Kim UJ, Mancino V, Slepak T, Tachiiri Y, Simon M (1992) Cloning and stable maintenance of 300-kilobase-pair fragments of human DNA in *Escherichia coli* using an F-factor-based vector. *Proc Natl Acad Sci USA* **89**: 8794–8797.
23. Zimmer R, Verrinder-Gibbins AM (1997) Construction and characterization of a large-fragment chicken bacterial artificial chromosome library. *Genomics* **42**: 217–226.
24. Sternberg N, Tiemeier D, Enquist L (1977) In vitro packaging of a λ *Dam* vector containing *Eco*RI DNA fragments of *Escherichia coli* and phage P1. *Gene* **1**: 255–280.
25. Sternberg NL (1992) Cloning high molecular weight DNA fragments by the bacteriophage P1 system. *Trends Genet* **8**: 11–16.

26. Schleper C, DeLong EF, Preston CM, Feldman RA, Wu KY, Swanson RV (1998) Genomic analysis reveals chromosomal variation in natural populations of the uncultured psychrophilic archaeon *Cenarchaeum symbiosum*. *J Bacteriol* **180**: 5003–5009.
27. Rondon MR, Raffel SJ, Goodman RM, Handelsman J (1999) Toward functional genomics in bacteria: Analysis of gene expression in *Escherichia coli* from a bacterial artificial chromosome library of *Bacillus cereus*. *Proc Natl Acad Sci USA* **96**: 6451–6455.
28. Giovannoni SJ, DeLong EF, Schmidt TM, Pace NR (1990) Tangential flow filtration and preliminary phylogenetic analysis of marine picoplankton. *Appl Environ Microbiol* **56**: 2572–2575.
29. Krsek M, Wellington EMH (1999) Comparison of different methods for the isolation and purification of total community DNA from soil. *J Microbiol Methods* **39**: 1–16.
30. Kozdrój J, van Elsas JD (2000) Application of polymerase chain reaction-denaturing gradient gel electrophoresis for comparison of direct and indirect extraction methods of soil DNA used for microbial community fingerprinting *Biol Fertil Soils* **31**: 372–378
31. Miller DN, Bryant JE, Madsen EL, Ghiorse WC (1999) Evaluation and optimization of DNA extraction and purification procedures for soil and sediment samples. *Appl Environ Microbiol* **65**: 4715–4724.
32. Roose-Amsaleg CL, Garnier-Sillam E, Harry M (2001) Extraction and purification of microbial DNA from soil and sediment samples. *Appl Soil Ecol* **18**: 47–60.
33. Tien CC, Chao CC, Chao WL (1999) Methods for DNA extraction from various soils: a comparison. *J Appl Microbiol* **86**: 937–943.
34. Birren B, Lai E (1993) *Pulsed Field Gel Electrophoresis: A Practical Guide*. Academic Press, San Diego.
35. Zhou J, Brun MA, Tiedje JM (1996) DNA recovery from soils of diverse composition. *Appl Environ Microbiol* **62**: 316–322.
36. Zhu H, Dean RA (1999) A novel method for increasing the transformation efficiency of *Escherichia coli*-application for bacterial artificial chromosome library construction. *Nucleic Acids Res* **27**: 910–911.
37. Silhavy RJ, Berman ML, Enquist LW (1984) *Experiments with Gene Fusion*. Cold Spring Harbor Press, Cold Spring Harbor, NY.
38. Atrazhev AM, Elliott JF (1996) Simplified desalting of ligation reactions immediately prior to electroporation into *E coli*. *BioTechniques* **21**: 1024.
39. Beja O, Aravind L, Koonin *et al.* (2000) Bacterial rhodopsin: evidence for a new type of phototrophy in the sea. *Science* **289**: 1902–1906.
40. Beja O, Koonin EV, Aravind L, Taylor LT, Seitz H, Stein JL, Bensen DC, Feldman RA, Swanson RV, DeLong EF (2002) Comparative genomic analysis of archaeal genotypic variants in a single population and in two different oceanic provinces. *Appl Environ Microbiol* **68**: 335–345.
41. Brady SF, Chao CJ, Handelsman J, Clardy J (2001) Cloning and heterologous expresson of a natural product biosynthetic gene cluster from eDNA. *Org Lett* **3**: 1981–1984.
42. Brady SF, Clardy J (2000) Long-chain *N*-acyl amino acid antibiotics isolated from heterologously expressed environmental DNA. *J Am Chem Soc* **122**: 12903–12904.
43. Cottrell MT, Moore JA, Kirchman DL (1999) Chitinases from uncultured marine microorganisms. *Appl Environ Microbiol* **65**: 2553–2557.
44. Entcheva P, Liebl W, Johann A, Hartsch T, Streit WR (2001) Direct cloning from enrichment cultures, a reliable strategy for isolation of complete operons and genes from microbial consortia. *Appl Environ Microbiol* **67**: 89–99.
45. Henne A, Daniel R, Schmitz RA, Gottschalk G (1999) Construction of

environmental DNA libraries in *Escherichia coli* and screening for the presence of genes conferring utilization of 4-hydroxybutyrate. *Appl Environ Microbiol* **65**: 3901–3907.
46. Majernik A, Gottschalk G, Daniel R (2001) Screening of environmental DNA libraries for the presence of genes conferring Na+(Li+)/H+ antiporter activity on *Escherichia coli*: characterization of the recovered genes and the corresponding gene products. *J Bacteriol* **183**: 6645–6653.
47. MacNeil IA, Minor TC, August PR *et al.* (2001) Expression and isolation of antimicrobial small molecules from soil DNA libraries. *J Mol Microbiol Biotechnol* **3**: 301–308.
48. Preston CM, Wu KY, Molinski TF, DeLong EF (1996) A psychrophilic crenarchaeon inhabits a marine sponge: *Cenarchaeum symbiosum* gen nov, sp. nov. *Proc Natl Acad Sci USA* **93**: 6241–6246.
49. Schleper C, Swanson RV, Mathur EJ, DeLong EF (1997) Characterization of a DNA polymerase from the uncultivated psychrophilic archaeon *Cenarchaeum symbiosum*. *J Bacteriol* **179**: 7803–7811.
50. Gillespie DE, Brady SF, Bettermann AD, Cianciotto NP, Liles MR, Rondon MR, Clardy J, Goodman RM, Handelsman J (2000) Isolation of antibiotics turbomycin A and B from a metagenomic library of soil microbial DNA. *Appl Environ Microbiol* **68**: 4301–4306.
51. Wang G-Y-S, Graziani E, Waters B, Pan W, Li X, McDermott J, Meurer G, Saxena G, Andersen J, Davies J (2000) Novel natural products from soil DNA libraries in a Streptomycete host. *Org Lett* **2**: 2401–2404.
52. Piel J (2002) A polyketide synthase-peptide synthetase gene cluster from an uncultured bacterial symbiont of *Paederus* beetles. *Proc Natl Acad Sci USA* **99**: 14002–14007.
53. Diaz-Torres ML, McNab R, Spratt DA, Villedieu A, Hunt N, Wilson M, Mullany P (2003) Novel tetracycline resistance determinant from the oral metagenome. *Antimicrob Agents Chemother* **47**: 1430–1432.

Protocol 11.1

Construction of a large insert library in a BAC vector is described.

MATERIALS

Reagents (duplicated reagents are listed only once)

Reagents for plugs:	Low melting temperature agarose, preparative quality
	Plug molds or 1 ml syringes
	Buffer A (50 mM Tris-HCl, pH 8.0, 1 M NaCl)
	Buffer B (50 mM Tris-HCl, pH 8.0, 100 mM EDTA, pH 8.0, 100 mM NaCl, 0.2% sodium deoxycholate, 0.5% 20 cetyl ether (Brij-58), 0.5% N-lauryl sarcosine, 5 mg/ml lysozyme)
	Buffer C (50 mM Tris-HCl, pH 8.0, 500 mM EDTA, pH 8.0, 100 mM NaCl , 0.5% N-lauryl sarcosine, 0.2 mg/ml proteinase K)
	TE (10 mM Tris-HCl, pH 8.0, 1 mM EDTA, pH 8.0)
	100 mM PMSF (phenylmethylsulfonylfluoride) stock in 100% isopropanol (caution: PMSF is toxic)
	Wide-bore pipette tips
Reagents for digestion:	Selected restriction enzyme
	Restriction enzyme buffer, lacking Mg^{2+}
	100 mM $MgCl_2$ solution
	0.5 M EDTA
Reagents for electrophoresis:	TAE (50 × stock solution 2 M Tris, 25 mM EDTA, adjusted to pH 7.5–7.8 with acetic acid)
	UV-detectable stain of choice [for example, EtBr at 0.5 mg/ml or SYBR® Green I (Molecular Probes, Eugene, OR, USA) at 1:10 000 dilution of provided stock]
Reagents for direct isolation:	Extraction buffer (100 mM Tris-HCl, pH 8.0, 100 mM EDTA, 100 mM sodium phosphate pH 8.5, 1.5 M NaCl, 1% CTAB (hexadecyltrimethylammonium bromide), 100 µg/ml proteinase K)

20% SDS (w/v in H_2O)

Chloroform:iso-amyl alcohol (24:1, v:v)

Isopropanol

70% ethanol (v/v in H_2O)

Reagents for vector preparation:

Purified undigested BAC [for example, pCC1BAC™ (Epicenter, Madison, WI, USA)]

Selected restriction enzyme

Alkaline phosphatase, preferably a phosphatase that can be heat-inactivated

Phenol:chloroform:isoamyl alcohol (25:24:1)

10 M ammonium acetate (pH does not need to be adjusted)

T4 DNA ligase (3 IU/ml); 10 × buffer

EtBr staining solution

Commercial kit for isolating DNA from agarose

Reagents for ligation:

T4 DNA Ligase (3 IU/μl)

Ligation buffer (10 × stock 300 mM Tris-HCl (pH 7.8), 100 mM $MgCl_2$; 100 mM DTT; 10 mM ATP)

Reagents for transformation:

tRNA at 3 mg/ml

Ice-cold EtOH

10% sterile glycerol (v/v in H_2O)

Electroporation-competent cells [for example, DH10B (F^- *mcr*A Δ (*mrr-hsd*RMS-*mcr*BC) Φ80*lac*ZΔM15 Δ*lac*X74 *rec*A1 *end*A1 *ara*Δ139 Δ(*ara, leu*)7697 *gal*U *gal*K λ^- *rps*L (Str^R) *nup*G]

SOC (2% bacto-tryptone, 0.5% bacto-yeast extract, 0.05% NaCl, 2.5 mM KCl, 10 mM $MgCl_2$, 10 mM $MgSO_4$, 20 mM glucose) $MgSO_4$ and glucose should be added separately, after autoclaving

LB plates (1% bacto-tryptone, 0.5% bacto-yeast extract, 1% NaCl, 1.5% bacto-agar)

Chloramphenicol (stock concentration 12.5 mg/ml in 95% EtOH)

Isopropyl-beta-D-thiogalactopyranoside (IPTG) (Stock concentration 100 mM in H_2O)

5-Bromo-4-chloro-3- indolyl-beta-D-galactopyranoside (X-gal) (Stock concentration 200 mg/ml in dimethyl formamide)

Equipment

Pulsed field gel electrophoresis (PFGE) unit, highly recommended

Standard agarose electrophoresis unit

UV wavelength box for viewing agarose gels

Spectrophotometer (optional)

Centrifuge, capable of 12 000 *g* for tubes containing >10 ml

Microcentrifuge

Water bath/heating block for restriction enzyme digestion

14°C incubation chamber for ligations

–20°C freezer

Electroporation cuvettes

Electroporation apparatus

37°C shaker

37°C incubator for plates

Adjustable temperature water baths

Tangential flow filtration apparatus and intake filter (for sea water only)

PREPARATION OF HIGH MOLECULAR WEIGHT INSERT DNA

1. Preparation of insert DNA from agarose plugs

Preparation of very high molecular weight DNA is easiest if cells are first immobilized in agarose plugs. This may be difficult due to contaminants in some environmental samples, such as particulates, eukaryotic cells, and humic acids. For the purposes of this protocol it is assumed that reasonably purified cell preparations can be obtained. Details of buffer composition are described in the Reagents section. This protocol is adapted from Rondon *et al.* (27) for lysis of *Bacillus cereus* cells, and is expected to be suitable for lysis of most prokaryotic cells.

For all steps in which isolated environmental DNA is pipetted, use wide-bore pipette tips to minimize shearing.

1A. Extraction of cells from an environmental sample

Extraction methods differ depending on the environmental source. This protocol for extraction of cells from seawater is from Giovannoni *et al.* (28) and Schmidt *et al.* (1). Comparisons of methods for extraction of cells from soils are described elsewhere (29–33).

1. Collect seawater through a 10 μm pore-size intake filter.
2. Concentrate cells with a tangential flow filtration apparatus (28) equipped with a 10 ft^2 0.1 μm pore fluorocarbon membrane. The sizes of the intake filter and tangential flow filter bracket the sizes of the targeted microbes, while allowing a high flow rate through the system.
3. Pellet the concentrated cells by centrifugation at (16 300 *g*).

1B. Preparation of metagenomic DNA in agarose plugs

Day 1

1. Resuspend cell preparation to a concentration of 2×10^9 to 1×10^{10} cells per ml in buffer A at room temperature. No cell clumps should be visible.
2. Mix cells (cells should be at room temperature) with an equal volume of molten (65°C) 1.6% preparative quality low melting temperature agarose dissolved in water. (Note: 2% agarose may be preferred if using 1 ml syringes.)
3. Pipette the mixture into agarose plug molds (these molds are generally provided with pulsed field gel electrophoresis (PFGE) equipment but may be purchased separately). Alternatively, draw the mixture into a 1 ml syringe as described in Stein *et al.* (7).
4. Solidify plugs on ice.
5. Remove plugs from molds or from syringe. When removing agarose from syringe, cut into 100 μl plugs. Transfer plugs to 2 × plug volume (200 μl) of buffer B for lysozyme and detergent disruption of membranes.
6. Incubate for 24 h at 37°C.

Day 2

1. Remove buffer B.
2. Add 200 ml of buffer C , which contains proteinase K.

3. Incubate for 24 h at 50°C.

Day 3

1. Remove buffer C and repeat day 2, steps 2 and 3.

Day 4

1. Wash plugs three times for 10 min each time with 1 ml TE.
2. Inactivate residual proteinase K with PMSF by washing individual plugs two times for 1 h each time in 1 ml TE with 1 mM PMSF.
3. Plugs are ready to use. Store at 4°C in TE for short-term use, or in 10 mM Tris-HCl (pH 8.0), 50 mM EDTA, (pH 8.0) for long-term storage.

1C. Digestion of metagenomic DNA in agarose plugs

Conditions for digestion will vary depending on the sample, restriction enzyme, and desired size of digested DNA. For construction of metagenomic libraries, partial digestion of the sample DNA is preferred to minimize cloning bias. Varying the restriction enzyme concentration, varying the time of digestion, or using methylation competition may be used to limit the rate of digestion (34). However, the most common method, described here, is to limit the amount of Mg^{2+} in the reaction (34). The amount of enzyme and Mg^{2+} used must be determined empirically for each batch of DNA in agarose plugs, since DNA concentration and purity will affect the digestion conditions. It is usually preferable to perform a series of test digestions with small plug fragments (i.e. one-quarter of a plug per test condition). For these tests, it is important that the plug volume be kept constant. A negative control, containing all components of the digestion except the restriction enzyme, is also recommended. This control is necessary to detect any degradation of DNA caused by non-specific nucleases.

The following protocol is for partial digestion of the DNA sample while embedded in agarose. Complete digestion can be achieved by using 2 × plug volume of restriction enzyme buffer containing Mg^{2+} and incubating the reaction for 16 h or overnight.

1. Remove TE or storage buffer and incubate seven plugs in 2 × plug volume of restriction

enzyme buffer lacking Mg^{2+} for 5 h at room temperature or overnight at 4°C with one change of buffer to remove excess EDTA.

2. Place each test plug in a separate microcentrifuge tube and add 2 × plug volume of restriction enzyme buffer without Mg^{2+}. For each different test condition, the plugs and starting buffers should be identical.
3. To six test plugs, add the amount of restriction enzyme needed for a complete digestion of the DNA in 16 h. Generally, more restriction enzyme is needed to completely digest DNA that is embedded in agarose than for digestion of DNA in solution. Follow the manufacturer's guidelines for a specific enzyme of choice; in many cases, information is available regarding the amount of enzyme needed for digestion of agarose-embedded DNA.
4. Place all tubes at 4°C overnight to allow the restriction enzyme to diffuse evenly into the plugs.
5. Add 1, 3, 10, 30 or 100 µl of 100 mM $MgCl_2$ solution to five of the tubes containing agarose plugs and enzyme. The remaining tube will be incubated without $MgCl_2$.
6. Incubate the tubes at 37°C for 1 h.
7. Stop the digestion reaction by adding 10 µl of 0.5 M EDTA. Keep tubes on ice or at 4°C until gel is run.
8. Remove 1/4 plug from each test plug and run on a pulsed field gel to determine the extent of digestion in each plug.

As an alternative to the digestion method presented here, the agarose plug can be melted at 65°C, equilibrated to 37°C, and then digested in the molten state. The digested DNA can then be pipetted directly into the wells of a gel and allowed to solidify.

1D. Size separation and isolation of digested DNA

Conditions for size separation of digested DNA will vary according to the overall size distribution of the DNA, the size fraction desired, and the equipment used. PFGE provides the most consistent size separation, but rough separation is also possible by standard agarose gel electrophoresis. When using PFGE, it is important to experiment with several test runs, using commercially available size markers, to determine optimal electrophoresis conditions. Compression gel conditions, in which DNA at, and above, the targeted size is highly compressed into a small region of the lane (1–3 mm), are preferred, in order to obtain concentrated DNA for the ligation. The gel illustrated in *Figure 11.1* is a compression gel, run for 10 h in 1× TAE running buffer at 6 V/cm with a 120° switch angle. The switch times are ramped from 0.1 to 3 s, with a ramping factor of –1.5. Samples are commonly run in 1% preparative quality low melting temperature agarose, to facilitate subsequent agarose digestion during DNA recovery. Selection of the electrophoresis buffer (TBE or TAE) may depend on compatibility with the agarose-digesting enzyme used in Step 6.

1. Pour 1% gel with comb to generate wells that will accommodate agarose plugs.
2. Load plugs containing partially digested DNA into wells. Seal plug into well with a few drops of molten agarose. Load size markers in wells flanking samples to facilitate identification of the correct region to excise.
3. Size-fractionate DNA, using PFGE or standard agarose gel electrophoresis.
4. Cut off the marker lanes and stain in water or electrophoresis buffer with a UV-detectable stain of choice. Photograph the marker lanes, being sure to line up the gel slices with a ruler for the photograph.
5. Using the distance on the ruler as a guide, excise the desired region(s) from the unstained portion of the gel. It is advisable to excise additional slices surrounding the primary region of interest for back-up material.
6. Digest the agarose with a commercially available agarose-digesting enzyme, following the manufacturer's guidelines.
7. Run a small portion of the eluted agarose-digested samples on a gel to visualize the size and quantity of DNA recovered. This gel enables selection of the optimal samples to use for ligations.

2. Direct isolation of metagenomic DNA from environmental samples

If the environmental samples do not allow for cell separation and preparation of agarose plugs, the DNA may be extracted directly from the sample. This will generally yield DNA with a smaller average size than from the plug method but may avoid some of the bias inherent in separating cells from particulates in a sample such as soil. Numerous methods are available for direct extraction of DNA from environmental samples for purposes of PCR amplification of specific genes. We do not recommend following those protocols directly, since the size of the DNA that is isolated by those protocols is generally much too small to be useful for construction of large insert libraries. The following protocol is from Rondon *et al.* (2000) (11) and is adapted from Zhou *et al.* (35).

2A. Extraction of DNA.

1. Suspend sample (5 g) in 10 ml CTAB/proteinase K extraction buffer.
2. Incubate at 37°C for 30 min. One or more freeze–thaw cycles may be introduced at this step, if desired.
3. Add SDS to 2% final concentration; mix gently.
4. Incubate for 2 h at 60°C, occasionally inverting gently.
5. Centrifuge sample at low speed (10 min at 6000 ***g***). Released DNA will be in supernatant.
6. Remove supernatant to a new tube. Extract sample again with 5 ml extraction buffer plus SDS, if desired; pool supernatants.
7. Carefully add an equal volume of chloroform:isoamyl alcohol (24:1) to the pooled supernatants. Place centrifuge bottle on its side and rotate it gently, ~15 rpm, on shaker for 10 min.
8. Centrifuge sample at low speed (e.g. 10 min at 6000 ***g***).
9. Transfer the aqueous (upper) layer to a new tube.
10. Carefully add 0.6 vol. isopropanol to aqueous layer. Mix gently for 5 min as in step 7; let stand at room temperature for 20 min.

11. Centrifuge 10 min at 12 000 *g* to pellet DNA.
12. Wash pellet with 70% ethanol; air dry.
13. Resuspend DNA in 500 µl of TE. Pipette gently with wide-bore pipette to resuspend. Overnight incubation at 4°C may be necessary. (Note: if extracting DNA from soil, additional extractions with chloroform:isoamyl alcohol or phenol:chloroform:isoamyl alcohol may be helpful to clean the DNA preparation. However, Step 14 should separate the DNA from most contaminants.)
14. Separate the DNA by electrophoresis on a 1% low melting temperature gel with size markers in adjacent lanes. This can be done either using a conventional horizontal electrophoresis unit or a PFGE unit. The primary reason for this electrophoresis is to separate the DNA from humic acids and other impurities, so a short standard electrophoresis run at 60 V for 30 min is generally sufficient. As in the protocol above, use the markers as a guide for excision of the region of interest.
15. Excise the appropriate gel region and store slices in TE or in storage buffer as described in IC, above.

2B. Digestion of metagenomic DNA in gel slices

This can be done according to the protocol for agarose plugs, as described above.

2C. Size separation and isolation of digested DNA

This can be done according to the protocol for agarose plugs, as described above.

3. Preparation of linearized BAC vector

Commercial kits for BAC/plasmid isolation are available and are recommended for purifying BAC DNA prior to digestion. It is recommended that a specific protocol for isolation of BACs/cosmids/very low copy plasmids be used to maximize recovery and minimize chromosomal DNA contamination. Vector preparation is critical to the success of library construction, so special care should be taken to avoid introducing into the final library ligation any linearized vector that will self-ligate or any non-linearized vector.

1. Linearize 2 μg purified BAC with selected restriction enzyme.
2. Heat-inactivate restriction enzyme, according to manufacturer's guidelines.
3. Add alkaline phosphatase directly to inactivated reaction. Use manufacturer's guidelines to determine the appropriate units of phosphatase to be added.
4. Heat-inactivate the phosphatase, according to manufacturer's guidelines.
5. Add an equal volume of phenol:chloroform:isoamyl alcohol (25:24:1), vortex, and centrifuge at 14 000 rpm for 5 min. Transfer aqueous layer to new tube.
6. Add to aqueous layer 0.33 vol. 10 M ammonium acetate and 1.5 volume isopropanol to precipitate the DNA; centrifuge at 14 000 rpm for 5 min; rinse pellet with 70% EtOH and air dry.
7. Resuspend pellet in 200 μl of 1× ligation buffer. Add 3 IU T4 DNA ligase; incubate overnight at 14°C. During this step, any vector that has not been dephosphorylated will self-ligate.
8. Separate self-ligated from linearized vector by electrophoresis of the ligation reaction in preparative quality agarose. In adjacent lanes, load markers, undigested vector, and digested but not self-ligated vector as controls. Identify and excise the band containing linearized vector.
9. Use a commercially available gel isolation kit to isolate linearized vector DNA from the agarose.

4. Ligation of insert and vector DNA

These ligation volumes may be increased, if the insert DNA is too dilute for a small ligation volume.

1. Quantify vector and insert DNA by optical density or by electrophoresis and EtBr staining.
2. Mix vector (50 ng) and insert in a 10:1 molar ratio (vector: insert), basing molar calculation

of insert DNA on the average insert expected from the size selection.

3. Add sterile H_2O to give a total volume of 17 µl.
4. Heat to 65°C for 10 min; cool on ice.
5. Add 2 µl ligation buffer and 3 IU T4 DNA ligase.
6. Incubate at 14°C overnight.

5. Transformation

Steps 1–5 describe one protocol adapted from Zhu and Dean (36) for concentrating and desalting the ligation prior to transformation. Alternative protocols are described in Silhavy *et al.* (37) and Atrazhev and Elliott (38).

1. Add 3 µg tRNA to ligation.
2. Add 5 × volume ice-cold EtOH.
3. Place at –20°C for 30 min.
4. Centrifuge for 10 min at 14 000 rpm; rinse pellet with 70% EtOH; air dry.
5. Resuspend ligation in 10 µl 10% glycerol.
6. Add 2 µl of ligation to freshly thawed *E. coli* competent cells in chilled electroporation cuvette.
7. Incubate on ice for 10 min.
8. Electroporate at 100 Ω, 2 V, 25 µFD; immediately dilute in 1 ml SOC medium.
9. Incubate with shaking (200 rpm) at 37°C for 45 min.
10. Spread transformation mixture on LB plates containing chloramphenicol (or the appropriate antibiotic for the vector of choice), IPTG and X-gal (12.5, 32 and 32 µg ml^{-1}, respectively) at a range of dilutions. IPTG and X-gal are necessary only if the BAC cloning site lies within a *lacZ* expression cassette for blue/white selection of cloned inserts.
11. Incubate plates at 37°C overnight. A 2 day incubation may be necessary to visualize color development.

A molecular toolbox for bacterial ecologists: PCR primers for functional gene analysis

12

Michael J. Larkin, A. Mark Osborn and Derek Fairley

12.1 Introduction

Microbial ecology encompasses the study of the functional role and the diversity of microorganisms in their environment. This can be expressed in terms of the diversity of microbial species, the genetic potential encoded by the microbially encoded functional genes, and their expressed phenotype as defined by gene expression, protein production and cellular activity (1). Microbial communities are typically extremely complex and exhibit high diversity consisting of thousands of different bacterial species within small environmental samples (2). In addition to the high levels of species diversity, the bacteria that comprise environmental microbial communities may display a panoply of different functional roles. For many years our understanding of the diversity of microorganisms was limited to those organisms that we could physically culture. However, microscopic techniques revealed that these cultured organisms represented only ≤10% of the microorganisms within any given environment (3). The last two decades have seen a revolution in microbial ecology with the application of culture-independent methodologies typically involving analysis of nucleic acids extracted directly from an environmental sample that comprises the genomes of all microorganisms present within that sample. Approaches for isolation of nucleic acids are described in Chapter 1.

In order to progress towards an understanding of the structure and function of complex microbial systems, it is necessary to determine which nucleotide sequences are present, and also to investigate the relative abundance, distribution, and diversity of specific sequences in the system under study. The sequences of interest may be valuable primarily as phylogenetic markers for assessing microbial diversity (Chapter 2), or the so-called 'functional genes' which encode metabolic or catabolic functions, and define the genetic potential of these microorganisms (this chapter). Alternatively, they may in fact be metabolic gene transcripts, which effectively describe expression levels, metabolic activity, and viability. In practice, there is considerable overlap between these categories, and this is arguably where the answers to some of the most interesting questions in modern biology lie. It should also be noted that due to the considerable extent to which genes can be transferred between bacteria (lateral or horizontal gene transfer), often mediated by an array of mobile genetic elements, many protein-coding genes are found to have their own phylogenies, which are profoundly different from currently accepted 'organismal' phylogenies (or rRNA sequence-based phylogenies). Similarly, there is growing evidence that many microbial genomes are chimeric in nature (4). These important features of the microbial world can only be fully understood by adopting the 'reductionist' approach of sequence detection.

A variety of molecular techniques can be applied to address the central problems which surround sequence detection, with the most appropriate technique depending both on the system under study, and also on the nature of the question being asked (*Table 12.1*). Regardless of the method adopted, most of these techniques depend heavily on the use of short synthetic oligonucleotide 'primers', which allow selective amplification of the target sequence(s) (typically using PCR, or other enzymatic amplification methods). These primers can therefore be applied to detect sequences

Table 12.1 Detecting specific nucleotide sequences of interest

Objective	Technique	References
Detecting a sequence of interest in colonies or axenic culture	Colony probe hybridization	(17,18)
	Polymerase chain reaction (PCR) → probe hybridization or DNA sequencing	(19,20)
Detecting sequences of interest in a mixed culture/environmental sample	PCR → cloning → probe hybridization or DNA sequencing	(21,22)
	Nucleic acid sequence-based amplification (NASBA) → probe hybridization or PCR/real-time PCR (rt)PCR	(23,24)
	Ligase chain reaction (LCR)	(25)
Detecting a limited number of sequences, with high sensitivity and selectivity	rtPCR	(26,27)
Simultaneous detection of many related, but polymorphic, sequences in mixed culture/environmental DNA (e.g. 16S rDNA-based population profiling)	PCR → TGGE/DGGE	(7,8)
	PCR → reverse hybridization/DNA macroarray/microarray analysis	(28–30)
	Reverse hybridization without PCR	(31)
Detecting a sequence of interest in individual microbial cells	Oligonucleotide-primed *in situ* DNA synthesis (PRINS)	(32,33)
	In situ PCR	(34,35)
	Laser flow cytometry/*in situ* PCR	(36)
	RING-FISH: recognition of individual genes – fluorescence *in situ* hybridization	(37)
Detecting expression of sequence of interest in individual microbial cells	*In situ* RT-PCR	(38)

DGGE: denaturing gradient gel electrophoresis; TGGE: thermal gradient gel electrophoresis.

of interest, and are often used in combination with further DNA 'probe' molecules, which may be labeled with fluorescent, radioactive or antigenic tags, or can be immobilized for certain applications.

In many respects, PCR primers and DNA probes are the basic 'tools of the trade' for molecular microbial ecologists. To date, however, there are few centralized resources or online databases (containing primer and probe sequences) that are dedicated to supporting microbial molecular ecology research. In the sections that follow we have assembled information on resources that comprise a 'molecular toolbox', with an emphasis on providing access both to pertinent research papers and to relevant online databases. This toolbox focuses on the vast array of PCR primers that are available primarily for the study of functional genes, but also provides brief coverage of some of the more important primers used for analysis of bacterial 16S rRNA genes. Whilst this chapter does not include specific data on DNA probes that have been applied in microbial ecology, the PCR primer sets referred to in the accompanying tables can themselves be used to generate DNA probes for hybridization-based analysis of nucleic acids isolated from the environment. This chapter does, however, include discussion of some of the factors that are important to probe design, and the reader is also referred to Chapter 8 in which the application of DNA probes in microbial ecology is discussed at length.

Molecular microbial ecology is also reliant on the sequence databases and bioinformatics tools that enable specific genes to be targeted by experimental analysis in the laboratory. Chapter 15 provides a comprehensive review of the various bioinformatics tools available to the molecular microbial ecologist, and the reader is encouraged to visit regularly the links therein, and to become familiar with the numerous software tools which are available for manipulating and viewing biological sequence data (often at no cost).

This chapter therefore focuses primarily on the many PCR primer sets that have been developed to investigate functional genes in the environment. PCR primer sets to detect individual bacterial species and/or genes of particular interest to clinical microbiologists are beyond the scope of this current review; instead, we highlight PCR primers that can be used to study functional genes of interest to the microbial ecologist, especially to study genes that encode biodegradation functions, biogeochemical cycling functions, and those encoding important resistance phenotypes including resistances to heavy metals and antibiotics. It should also be noted that many of the PCR primers considered in this chapter can in principle be applied both in assessing genetic potential (i.e. detecting genes which encode particular metabolic or catabolic functions), and to investigate expression of these genes via RT-PCR (using complementary primer sequences, where appropriate; see Chapter 5).

12.2 Adopting a strategy to detect sequences from the environment

If appropriate primer or probe sequences are available from the literature (or from a database) then these can generally be applied to the detection of sequences of interest. Primers or probes of known specificity can be selected, for which published protocols and validation data are generally available. This facilitates rational experimental design, and also enables suitable controls to be chosen. In addition, comparative studies, or studies which extend previously published work, can be undertaken more easily.

However, if no suitable primer or probe sequences have been previously identified, then it may be necessary to design them *de novo*, based either on sequence data obtained from one of the major sequence databases (*Table 12.2* and Chapter 15) or on 'in house' sequence data. This approach enables organisms or genes for which no published primer or probe sequences are

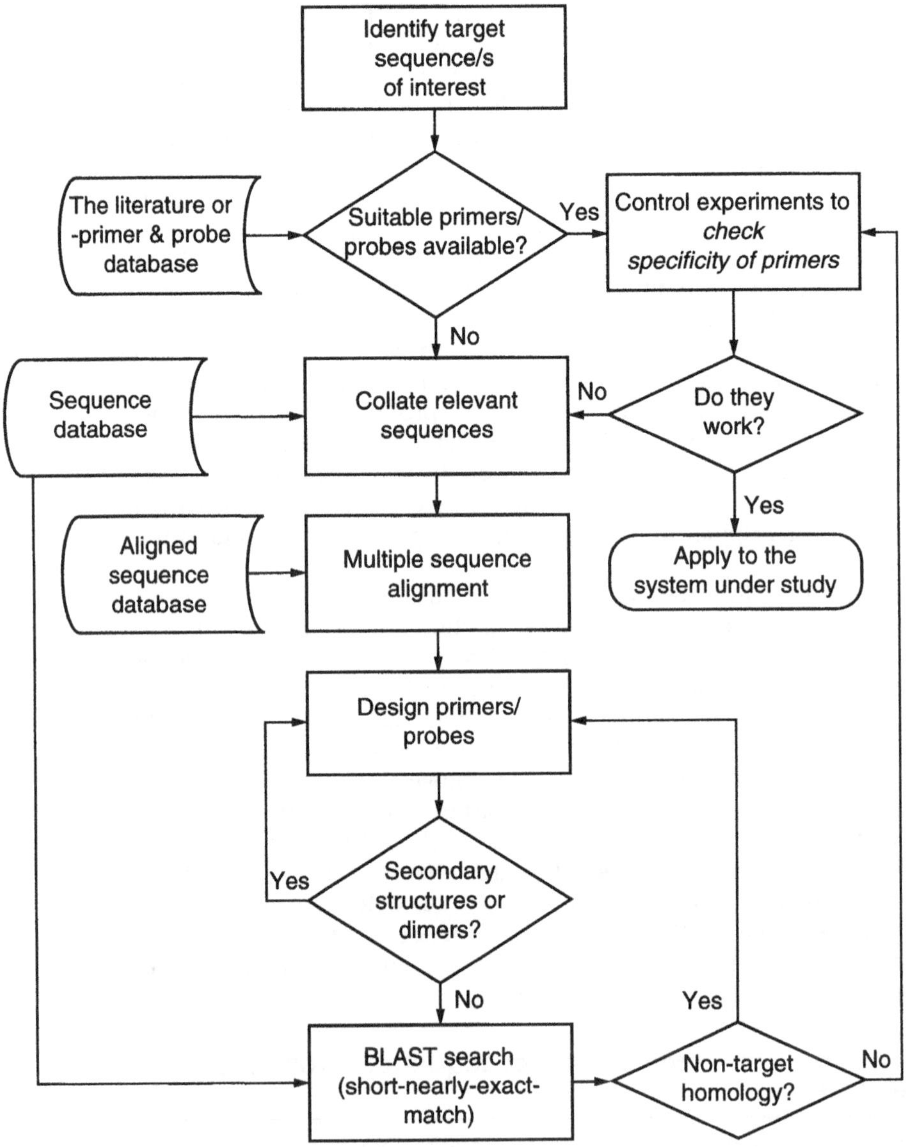

Figure 12.1

A primer and oligonucleotide probe design strategy.

Table 12.2 Selected online resources for molecular microbial ecology research (see also Chapter 15)

Resource type	Name	URL
Sequence databases	GenBank EMBL	http://www.ncbi.nlm.nih.gov/genbank/index.html http://www.ebi.ac.uk/embl.html
BLAST[a] searching	blastn and 'short nearly exact match' blast	http://www.ncbi.nlm.nih.gov/BLAST/
Web-based sequence alignment tools	Clustal TCoffee[b]	http://www.ebi.ac.uk/clustalw/ http://www.ch.embnet.org/software/TCoffee.html
Sequence alignment and phylogeny construction tools for download	Clustal PHYLIP	ftp://ftp.ebi.ac.uk/pub/software/dos/clustalw/ http://evolution.genetics.washington.edu/phylip.html

[a] 'Basic Local Alignment Search Tool'.
[b] The sequence alignment algorithm used by TCoffee may be more effective than the Clustal algorithm for aligning multiple sequences which are not very closely related (http://acer.gen.tcd.ie/mmm/tco.html).

available to be targeted, although various factors must be carefully considered (Section 12.3), and new primers and probes must be thoroughly validated with respect to their specificity. A strategy for selecting oligonucleotide primers and probes is presented in *Figure 12.1*.

Frequently, a combination of approaches is adopted. For example, well-established PCR primers may be used in combination with newly designed DNA probes to detect target sequences amongst PCR products. DNA probes for detecting metabolic genes, or their transcripts, may also be constructed from either cloned gene fragments or denatured PCR products (5).

12.3 Primer and probe design

A detailed consideration of oligonucleotide primer and probe design is beyond the scope of this chapter [see (6) for an overview]. However, a number of general principles apply, and are worth repeating. Oligonucleotide probes must not self-anneal, or form secondary structures (i.e. hairpins). In addition, paired oligonucleotides for use as PCR primers must not anneal to either themselves or each other (dimerise), and they must generally exhibit closely matched G + C contents and T_m values. Software tools for checking oligonucleotides are provided with most commercial molecular biology packages. Various online resources are also available (see: http://molbiol-tools.ca/PCR.htm).

Most PCR primers and DNA probes are designed to anneal to a specific target sequence with high specificity, although they may contain a limited number of ambiguous bases, to accommodate sequence variation in target sequences. However, to avoid unexpected cross-hybridization, their sequences must be meticulously checked for homology to non-target sequences (via a BLAST search, using the 'short nearly exact match' facility; *Table 12.2*).

For certain applications, longer DNA probes can be used to detect specific sequences of interest. These are typically either denatured PCR products which have been amplified from reference strains or restriction fragments isolated from plasmids carrying cloned genes or gene fragments (see Chapter 8).

Discriminating between sequences in a mixed community that are almost identical (i.e. target sequences which exhibit single base-pair mismatches) is often possible using techniques such as DGGE/TGGE (7,8; and see Chapter 3), or by using DNA probes which hybridize with very high specificity (for example using Applied BioSystems 'minor groove binding' nucleotides (9) or *Taq*man chemistry (see Chapter 6).

12.4 PCR primers to amplify 16S rRNA gene sequences

The main nucleotide sequence databases, e.g. GenBank and EMBL, contain an enormous and ever-expanding number of rRNA sequences, and are clearly of central importance as resources for designing PCR primers and DNA probes. Moreover, many laboratories which are involved in microbial ecology studies also have extensive in-house databases of rRNA sequence data, which can be used for the design of rRNA primers and probes that are directly relevant to the researcher's environment of interest. However, most researchers will utilize the three key resources for rRNA sequences where large datasets of pre-aligned rRNA sequences are available (*Table 12.3*). These resources enable the rapid identification of PCR primers or oligonucleotide probes that exhibit varying degrees of group specificity and are discussed in more detail in Chapter 15. PCR analysis of 16S rRNA sequences is discussed at length in Chapter 2. For such analysis, many different PCR primer sets have been designed both to amplify rRNA sequences at the domain level, and to target individual species. A suite of so-called 'universal' primers were described by Lane (10) (Table 12.4), with these primers finding widespread adoption in microbial ecology. More recently Marchesi *et al.* (11) developed an improved primer pair that permits amplification of

Table 12.3 rRNA sequence resources

Database	URL	Comments
Ribosomal Database Project (RDP)	http://rdp.cme.msu.edu	Large repository of aligned rRNA sequences with online analysis tools
WWW rRNA Server	http://www.psb.ugent.be/rRNA/index.html	Aligned rRNA sequence database and sequence visualization software
ARB Project	http://www.arb-home.de	Large database of aligned sequences, and a comprehensive software environment for managing aligned sequences, and for probe design and evaluation. Only available for UNIX/Linux platforms

Table 12.4 Key PCR primers for 16S rRNA sequence detection and analysis

Primer[a]	Sequence (5′ to 3′)	Reference
27F[b]	AGAGTTTGATCMTGGCTCAG	(10)
63F	CAGGCCTAACACATGCAAGTC	(11)
341F[c]	CCTACGGGAGGCAGCAG	(7)
357R	CTCCTACGGGAGGCAGCAG	(10)
519R	GWATTACCGCGGCKGCTG	(10)
534R	ATTACCGCGGCTGCTGG	(7)
926R	AAACTYAAAKGAATTGACGG	(10)
1387R	GGGCGGWGTGTACAAGGC	(11)
1392R	ACGGGCGGTGTGTRC	(10)
1492R	TACGGYTACCTTGTTACGACTT	(10)
1525R[b]	AAGGAGGTGWTCCARCC	(10)

[a]F and R designations indicate forward and reverse primers, respectively.
[b]27F and 1525R are sometimes referred to a pA and pH (39).
[c]Typically used with a GC clamp for DGGE analysis in combination with 534R (also referred to as primers 1 and 2 (7).

16S rRNA sequences from a wider range of bacterial taxa. Two recent reviews have assessed the applicability of domain-specific 16S rRNA gene primers (12), and also their application in fingerprinting methodologies (13). The application of universal and specific rRNA probes for studying bacterial communities and/or specific species via fluorescent *in situ* hybridization (FISH) is discussed in Chapter 9.

12.5 PCR primers for detecting functional genes

The core of this chapter is the toolbox of PCR primers for functional genes. This toolbox is presented as a series of tables (*Tables 12.5–12.11*) that have been divided on the basis of function. In particular we focus on functional genes involved in biodegradation (*Table 12.5*), bacterial resistance (*Tables 12.6 and 12.7*) and on biogeochemical cycling (*Tables 12.9* and *12.10*). The toolbox includes primers that are applicable to amplification of diverse genes, notably in the case of degenerate primers for amplification of catechol-2,3-dioxygenases (see *Table 12.5*), but also for specific detection of genes from individual species. In many instances these primer sets can also be used to derive DNA probes for hybridization experiments (Chapter 8), and indeed in some cases to investigate gene expression via RT-PCR (Chapter 5), or to quantify gene abundances (Chapter 6). In addition to the tables below we would also refer the reader to recent reviews covering the application and evaluation of molecular tools to three areas of environmental microbiology. Widada *et al.* (14) present an appraisal of recent developments in molecular techniques for studying biodegradation systems. In addition to highlighting PCR primer systems for biodegradation gene detection, they review other methodologies that can be used for studying biodegradation. In a second review, Throback *et al.* (15) evaluate the applicability of multiple primer sets for studying denitrification with an emphasis on nitrite and nitrous oxide reductases. This excellent review,

Table 12.5 PCR primers for biodegradation genes

Enzyme type and/or substrate	Target genes	Notes	Reference
Dioxygenases			
Catechol-2,3-dioxygenases	Multiple	Consensus primers designed from C-2,3-O genes	(40)
	Multiple	Consensus primers designed from C-2,3-O genes	(41)
	Multiple	Consensus primers designed from C-2,3-O genes	(42)
	edo genes	Designed from *Rhodococcus* *edo* genes encoding C-1,2-O genes and C-2,3-O genes	(43)
	Multiple	Degenerate primers for chlorocatechol 2,3-dioxygenase	(44)
	Multiple	Degenerate primers designed from multiple dioxygenase genes	(45)
	Multiple	Consensus primers designed from C-2,3-O genes from *Pseudomonas*	(46)
	Multiple	Consensus primers designed from C-2,3-O genes	(47)
Catechol-1,2-dioxygenases	Multiple	Degenerate primers for chlorocatechol 1,2-dioxygenase	(48)
Biphenyl	*bphA*	Specific primers for *Pseudomonas pseudoalcaligenes bphA*	(49)
	bphC	Consensus primers designed from *Pseudomonas* and *Burkholderia bphC* genes	(5)
4-Chlorobiphenyl	*pcbC*	Specific primers for *Pseudomonas* sp. *pcbC* genes	(50)
Chlorobenzoate	*cba*	*cba* gene from Tn*5271* from *Alcaligenes*	(51)
Naphthalene	*nagAc*	Detects *nagAc* from *Ralstonia.* Used for real-time PCR	(52)
	nahAc	Conserved primers designed from *Pseudomonas nahAc* genes	(21)
	ndoB	Specific primers for *P. putida* PaW 736 and degenerate primers for multiple *ndo* genes	(53)
	ndoB	Specific primers for *Pseudomonas putida ndoB*	(49)
Toluene	*todC*	Specific primers for *Pseudomonas putida todC1*	(49)
	xylE	Specific primers for *Pseudomonas putida xylE* gene	(54)
	xylE	Specific primers for *Pseudomonas putida xylE*	(49)
2,4-Dichlorophenoxyacetic acetic acid (2,4-D)	*tfdA*	Primers designed to conserved regions of *Alcaligenes* and *Burkholderia tfdA* genes	(55)
	tfdA, tfdB	Degenerate primers designed to *Alcaligenes* and *Burkholderia tfdA* genes Note: primers also designed to *tfdB* (2,4-dichlorophenol hydroxylase)	(56)

Table 12.5 *continued*

Monoxygenases			
Alkanes	*alkB*	Degenerate primers for *alkB*	(57)
	alkB, alkM	Multiple primer sets for diverse *alkB* and *alkM* sequences	(58)
	alkB	Degenerate primers for *alkB*	(59)
	alkB, alkM	Mutliple primers for *alkB* and *alkM* genes	(60)
Dehalogenases			
	deh	Degenerate primers designed to multiple *deh* genes	(61)
Chloroethene		Degenerate and specific PCR primers for chloroethene reductive dehalogenase genes	(62)
cis-1,3-dichloropropene	*dhlA*	Specific *dhlA* primers	(63)
Regulatory genes			
Naphthalene	*nahR, nahG*	Multiple sets of degenerate primers to amplify *nahR* and *nahG* transcriptional regulator genes	(64)
Miscellaneous degradation			
Carbofuran	*mcd*	Specific primers for *mcd* gene	(65)
Hydrocarbon desulfurization	*dszA, dszB* and *dszC*	Specific primers for *dsz* genes from *Rhodococcus erythropolis* IGTS8	(66)
Poly(3-hydroxybutyrate) (PHB)		Consensus primers designed from six fibronectin type III linker domain-encoding sequences of PHB depolymerases	(67)

Table 12.6 PCR primers for heavy metal resistance genes

Heavy Metal	Target genes	Notes	Reference
Arsenic	*arsA, arsB, arsC*	Target *E. coli* R773 *ars* genes	(68)
	arsC	Target *arsC* from Enterobacteriaceae. Primers used in Q-PCR assay	(69)
Cadmium	*cadA*	Target Gram-positive *cadA* genes	(70)
Cobalt, zinc, cadmium	*czc*	The *czc* gene encodes resistance to all three metals	(71)
Copper	*pco*	Multiple primer sets	(72)
Manganese	*mofA*	Target *Leptothrix* manganese oxidation genes	(73)
Mercury	*merRTP*	Gram-negative mercury resistance determinants	(74)
	mer (various)	Gram-negative mercury resistance determinants	(75)
	merRTP, merR	Primers for T-RFLP analysis of *mer* genes	(76)
	merA	Specific primers for pMERPH/R391 *merA* genes	(77)
	merA, merB	Specific primers for *Bacillus* sp. RC607 *mer* genes	(78)
	merA	Target Gram-positive mercury resistance determinants	(79)
	mer (various)	Target Gram-positive mercury resistance determinants	(80)

Table 12.7 PCR primers for antibiotic resistance genes

Antibiotic	Target genes	Notes	Reference
Ampicillin	*amoC*	Targets *ampC* genes from Proteobacteria	(81)
Erythromycin	*erm* genes (various)	Targets *erm* genes from Gram positive bacteria	(82)
Gentamicin	*aac* genes (various)	Multiple primers targeting various *aac* genes	(83)
Kanamycin	*nptII*		(84)
Methicillin	*mecA*	Targets *mecA* genes from *Staphylococcus* sp.	(85)
Streptomycin	*strA, strB1*	Target streptomycin resistance genes common to many bacteria	(86)
	ant, aph	Suite of primers to target streptomycin-modifying genes	(87)
Tetracycline	Various *tet* genes, *otr*	Comprehensive primer suite for amplification of tetracycline resistance genes	(88)
Vancomycin	*vanA, vanB*	Multiple vancomycin resistance genes from *Enterococcus* sp.	(89)

Table 12.8 PCR primers for detection of mobile genetic elements (MGE)

MGE (function)	Target genes	Notes	Reference
Gene cassettes	unknown	Permits amplification of (unknown) gene cassettes that are carried on class I integrons	(90)
Integrons	*intI1, intI2*	Two primer sets to amplify class I and class II integrase genes	(91)
Plasmids			
Bacillus RCR plasmids	*rep, mob*	Targets replication and mobilization genes in *Bacillus* rolling circle replicating plasmids	(92)
IncN-plasmids	*rep*	Target replication genes in IncN-related plasmids	(93)
IncP-1 plasmids	*trfA, korA*	Target replication and maintenance genes	(94)
IncP-9 plasmid	Various genes	Target replication, maintenance and transfer genes	(95)
Plasmid (maintenance)	*kikA, korA*	Target maintenance genes IncN, and P broad host range plasmids	(96)
Plasmid (replication)	*rep* genes, *ori* sequences	Target replicon sequences from IncN, P, Q and W broad host range plasmids	(96)
Plasmid (replication)	Various *rep* sequences	Target plasmid replication sequences from marine bacteria	(97,98)
Plasmid (transfer)	*oriT, traG, trwAB*	Target transfer genes from IncN, P, Q and W broad host range plasmids	(96)
Plasmid (transfer)	*trbB, traG*	Degenerate primers	(99)
IS/Transposons			
IS*1*			(100)
Class II transposons	*tnpA*	Tn*21*-, Tn*501*- and Tn*3*-related *tnpA* sequences	(101)
	tnpA, tnpR	Class II-related transposon sequences	(102)

Table 12.9 PCR primers for genes encoding nitrogen cycle functions

Functional group	Target genes	Notes	Reference
Ammonia oxidation	*amoA*	Consensus primers designed from *Nitrosomonas europaea amoA* genes	(103)
	amoA	Degenerate primers designed from alignment of multiple *amoA* sequences	(104)
	amoA	Modification of primers from	(104,105)
	amoA	Degenerate primers	(106)
Nitrate reduction	*narG*	Degenerate primers designed from *Bacillus subtilis* and *Escherichia coli narG* for nested PCR protocol	(107)
		Degenerate primers target *narG* from Bacteria and Archaea	(108)
	narH	Degenerate primers	(109)
	napA	Degenerate primers designed from *Paracoccus pantotrophus, Ralstonia eutropha* and *Escherichia coli napA* for nested PCR protocol	(110)
Nitrite reduction	*nirS*	Suites of primers designed from Proteobacteria *nirS* sequences	(111,112)
	nirS	Degenerate primers for nested PCR protocol	(113)
	nirS	Degenerate primers designed for DGGE application	(15)
	nirS	Degenerate primers	(114)
	nirK	Suite of primers designed from Proteobacteria *nirK* sequences	(111,112)
	nirK	Degenerate primers used for Q-PCR	(115)
	nirK	Degenerate primers designed for DGGE application	(15)
Nitric oxide reduction	*norB*	Suite of degenerate primers	(116)
Nitrous oxide reduction	*nosZ*	Suite of degenerate primers	(117)
	nosZ	Degenerate primers target *nosZ* from α-, β-, γ-Proteobacteria	(118)
Nitrate ammonification	*nrfA*	Degenerate primers target *nrfA* genes from Proteobacteriaceae	(119)
Nitrogen assimilation	*nasA, narB*	Target diverse nitrogen assimilation genes	(120)
Nitrogen fixation	*nifH*	Degenerate primers widely used since original application	(121)
	nifH	Used for RT-PCR amplification	(122)
	nifH	Used for RT-PCR amplification of *Azotobacter nifH* mRNA	(123)
	nifH	Suite of *nifH* primer sets for free-living nitrogen fixing soil bacteria	(124)

Table 12.10 PCR primers for genes encoding biogeochemical cycling functions

Functional group	Target genes	Notes	Reference
CO oxidation	*coxL*	Degenerate primers	(125,126)
H_2 oxidation	(NiFe) hydrogenase genes	Degenerate primers for Proteobacterial (NiFe) hydrogenase genes	(127)
Methanogenesis	*mcrA*	Targets α-subunit of the methyl coenzyme M reductase (MCR)	(22)
Methanotrophs	*mmo* genes	Note: these primers amplify both methane and ammonium monooxygenase genes	(128)
	mmo genes (various)	Designed from *mmo* genes of *Methylosinus trichosporium*	(129,130)
	mmoX, *mmoY*	Designed from *mmo* genes of *Methylococcus capsulatus*	(131)
	moxF	Designed from *moxF of Paracoccus denitrificans*	(130)
	pmoA	Both studies describe new reverse primers specific to *mmoA* genes that do not amplify *amoA* sequences	(132,133)
	pmoA	New primer set specific to *mmoA* genes	(134)
	pmoA	Primer set for Q-PCR analysis of *pmoA* genes	(135)
Sulfate reduction	*dsrAB*	Designed from *dsrA* sequences of *Archaeoglobus fulgidus* and *Desulfovibrio vulgaris*	(136)

Table 12.11 PCR primers for miscellaneous genes

Function	Target genes	Notes	Reference
Chitanase	*chiA*	Consensus primers designed to amplify *chiA* from diverse genera	(137)
	Group I chitinase genes	Degenerate primer set	(138)
Peptidase	*apr, npr, sub*	Suite of primers to amplify multiple extracellular peptidases	(139)
Type III secretion systems	*hrc*	Consensus primers designed to *hrc* genes from fluorescent Pseudomonads	(140)

in addition to being extremely valuable to researchers in this particular area, also highlights the importance of choosing primer systems that are appropriate to the study of the particular environmental system of interest, and underlines that empirical testing of primers should be considered essential. The final review that we highlight by Smalla and Sobecky (16) provides a discussion of the molecular approaches available to investigate the distribution and diversity of mobile genetic elements (MGE). MGE are central to adaptation in bacteria and provide the mechanism for dissemination of many of the key functional genes in bacteria.

In conclusion this chapter will, we hope, provide a useful starting point for the analysis of functional genes in environmental systems by PCR. Although there is now increasing emphasis on microarray approaches (see Chapters 8 and 10), the ability to target specific functional genes that are present at low abundances will continue to mean that PCR-based studies will remain an important methodology for the study of function in environmental systems.

References

1. Keller M, Zengler K (2004) Tapping into microbial diversity. *Nature Rev Microbiol* **2**: 141–150.
2. Torsvik V, Goksoyr J, Daae FL (1990) High diversity in DNA of soil bacteria. *Appl Environ Microbiol* **56**: 782–787.
3. Amann RI, Ludwig W, Schleifer KH (1995) Phylogenetic identification and *in situ* detection of individual microbial cells without cultivation. *Microbiol Rev* **59**: 143–169.
4. Ochman H, Lawrence JG, Groisman EA (2000) Lateral gene transfer and the nature of bacterial innovation. *Nature* **405**: 299–304.
5. Erb RW, Wagner-Döbler I (1993) Detection of polychlorinated biphenyl degradation genes in polluted sediments by direct DNA extraction and polymerase chain reaction. *Appl Environ Microbiol* **59**: 4065–4073.
6. Stahl DA, Amann R (1991) Development and application of nucleic acid probes. In: Stackebrandt E, Goodfellow M (eds), *Nucleic Acid Techniques in Bacterial Systematics*, pp. 205–248. John Wiley & Sons, Chichester.
7. Muyzer G, de Waal EC, Uitterlinden AG (1993) Profiling of complex microbial populations by denaturing gradient gel electrophoresis analysis of polymerase chain reaction-amplified genes coding for 16S rRNA. *Appl Environ Microbiol* **59**: 695–700.
8. Heuer HM, Krsek P, Baker K, Smalla EMH, Wellington EMH (1997) Analysis of actinomycete communities by specific amplification of genes encoding 16S rRNA and gel-electrophoretic separation in denaturing gradients. *Appl Environ Microbiol* **63**: 3233–3241.
9. Kutyavin IV, Afonina IA, Mills A *et al.* (2000) 3'-Minor groove binder-DNA probes increase sequence specificity at PCR extension temperatures. *Nucleic Acids Res* **28**: 655–661.
10. Lane DJ (1991) 16S/23S rRNA sequencing. In: Stackebrandt E, Goodfellow M (eds), *Nucleic Acid Techniques in Bacterial Systematics*, pp. 115–175. John Wiley & Sons, Chichester.
11. Marchesi JR, Sato T, Weightman AJ, Martin TA, Fry JC, Hiom SJ, Wade WG (1998) Design and evaluation of useful bacterium-specific PCR primers that amplify genes coding for bacterial 16S rRNA. *Appl Environ Microbiol* **64**: 795–799.
12. Baker GC, Smith JJ, Cowan DA (2003) Review and re-analysis of domain-specific 16S primers. *J Microbiol Meth* **55**: 541–555.
13. Watanabe K, Kodama Y, Harayama S (2001) Design and evaluation of PCR primers to amplify bacterial 16S ribosomal DNA fragments used for community fingerprinting. *J Microbiol Meth* **44**: 253–262.
14. Widada J, Nojiri H, Omori T (2002) Recent developments in molecular techniques for identification and monitoring of xenobiotic-degrading bacteria and their catabolic genes in bioremediation. *Appl Microbiol Biotechnol* **60**: 45–59.
15. Throback IN, Enwall K, Jarvis A, Hallin S (2004) Reassessing PCR primers targeting *nirS*, *nirK* and *nosZ* genes for community surveys of denitrifying bacteria with DGGE. *FEMS Microbiol Ecol* **49**: 401–417.

16. Smalla K, Sobecky PA (2002) The prevalence and diversity of mobile genetic elements in bacterial communities of different environmental habitats: insights gained from different methodological approaches. *FEMS Microbiol Ecol* **42**: 165–175.
17. Layton AC, Lajoie CA, Easter JP, Jernigan, R, Beck MJ, Sayler GS (1994) Molecular diagnostics for polychlorinated biphenyl degradation in contaminated soils. *Ann NY Acad Sci* **721**: 407–422.
18. Hosein SG, Millette D, Butler BJ, Greer CW (1997) Catabolic gene probe analysis of an aquifer microbial community degrading creosote-related polycyclic aromatic and heterocyclic compounds. *Microb Ecol* **34**: 81–89.
19. Osborn AM, Bruce KD, Ritchie DA, Strike P (1993) PCR-RFLP analysis shows divergence among *mer* determinants from Gram-negative bacteria indistinguishable by DNA-DNA hybridisation. *Appl Environ Microbiol* **59**: 4024–4030.
20. Osborn AM, Bruce KD, Ritchie DA, Strike P (1995) Sequence conservation between regulatory mercury resistance genes in bacteria from mercury polluted and pristine environments. *Syst Appl Microbiol* **18**: 1–6.
21. Herrick JB, Madsen EL, Batt CA, Ghiorse WC (1993) Polymerase chain-reaction amplification of naphthalene-catabolic and 16S ribosomal-RNA gene-sequences from indigenous sediment bacteria. *Appl Environ Microbiol* **59**: 687–694.
22. Hales BA, Edwards C, Ritchie DA, Hall G, Pickup RW, Saunders JR (1996) Isolation and identification of methanogen-specific DNA from blanket bog peat by PCR amplification and sequence analysis. *Appl Environ Microbiol* **62**: 668–675.
23. Compton J (1991) Nucleic-acid sequence-based amplification. Nature **350**: 91–92.
24. Gionata L, Schijndel H, Gemen B, Kramer FR, Schoen CD (1998) Molecular beacon probes combined with amplification by NASBA enable homogeneous, real-time detection of RNA. *Nucleic Acids Res* **26**: 2150–2155.
25. Wilson VL, Tatford BC, Yin XQ, Rajki SC, Walsh MM, LaRock P (1999) Species-specific detection of hydrocarbon-utilizing bacteria. *J Microbiol Methods* **39**: 59–78.
26. Suzuki, MT, Taylor LT, DeLong EF (2000) Quantitative analysis of small-subunit rRNA genes in mixed microbial populations via 5'-nuclease assays. *Appl Environ Microbiol* **66**: 4605–4614.
27. Gruntzig V, Nold SC, Zhou J, Tiedje JM (2001) *Pseudomonas stutzeri* nitrite reductase gene abundance in environmental samples measured by real-time PCR. *Appl Environ Microbiol* **67**: 760–768.
28. Kawasaki E, Saiki R, Erlich H (1993) Genetic-analysis using polymerase chain reaction-amplified DNA and immobilized oligonucleotide probes—reverse dot-blot typing method. *Enzymology* **218**: 369–381.
29. Levesque CA, Harlton CE, de Cock AWAM (1998) Identification of some oomycetes by reverse dot blot hybridization. *Phytopathology* **88**: 213–222.
30. Rhee S-K, Liu X, Wu L, Chong SC, Wan X, Zhou J (2004) Detection of genes involved in biodegradation and biotransformation in microbial communities by using 50-mer oligonucleotide microarrays. *Appl Environ Microbiol* **70**: 4303–4317.
31. Small J, Call DR, Brockman FJ, Straub TM, Chandler DP (2001) Direct detection of 16S rRNA in soil extracts by using oligonucleotide microarrays. *Appl Environ Microbiol* **67**: 4708–4716.
32. Komminoth P, Long AA (1993) *In-situ* polymerase chain reaction. *Virchows Arch B Cell Pathol* **64**: 67–73.
33. Jacobs D, Angles ML, Goodman AE, Neilan BA (1997) Improved methods for *in situ* enzymatic amplification and detection of low copy number genes in bacteria. *FEMS Microbiol Lett* **152**: 65–73.
34. Haase AT, Retzel EF, Staskus KA (1990) Amplification and detection of lentiviral DNA inside cells. *Proc Natl Acad Sci USA* **87**: 4971–4975.

35. Hodson RE, Dustman WA, Garg RP, Moran MA (1995). *In situ* PCR for visualization of microscale distribution of specific genes and gene products in prokaryotic communities. *Appl Environ Microbiol* **61**: 4074–4082.
36. Porter J, Pickup R, Edwards C (1995) Flow cytometric detection of specific genes in genetically modified bacteria using *in situ* polymerase chain reaction. *FEMS Microbiol Lett* **134**: 51–56.
37. Zwirglmaier K, Ludwig W, Schleifer K-H (2004) Recognition of individual genes in a single bacterial cell by fluorescence *in situ* hybridization–RING-FISH. Mol Microbiology **51**: 89.
38. Holmstrøm K, Tolker-Nielsen T, Molin S (1999) Physiological states of individual *Salmonella typhimurium* cells monitored by *in situ* reverse transcription-PCR. *J Bacteriol* **181**: 1733–1738
39. Edwards U, Rogall T, Blöcker H, Emde M, Böttger EC (1989) Isolation and direct complete nucleotide determination of entire genes. Characterization of a gene coding for 16S ribosomal RNA. *Nucleic Acids Res* **17**: 7843–7853.
40. Joshi B, Walia S (1996) PCR amplification of catechol 2,3-dioxygenase gene sequences from naturally occurring hydrocarbon degrading bacteria isolated from petroleum hydrocarbon contaminated groundwater. *FEMS Microbiol Ecol* **19**: 5–15.
41. Wikstrom P, Wiklund A, Andersson AC, Forsman M (1996) DNA recovery and PCR quantification of catechol 2,3-dioxygnase genes from different soil types. *J Biotechnol* **52**: 107–120.
42. Meyer S, Moser R, Neef A, Stahl U, Kampfer P (1999) Differential detection of key enzymes of polyaromatic-hydrocarbon-degrading bacteria using PCR and gene probes. *Microbiology* **145**: 1731–1741.
43. Kulakov LA, Delcroix VA, Larkin MJ, Ksenzenko VN, Kulakova AN (1998) Cloning of new *Rhodococcus* extradiol dioxygenase genes and study of their distribution in different *Rhodococcus* strains. *Microbiology* **144**: 955–963.
44. Sei K, Asano K, Tateishi N, Mori K, Ike M, Fujita M (1999) Design of PCR primers and gene probes for the general detection of bacterial populations capable of degrading aromatic compounds via catechol cleavage pathways. *Biosci Bioengng* **88**: 542–550.
45. Wilson MS, Bakermans C, Madsen EL (1999) *In situ*, real-time catabolic gene expression: Extraction and characterization of naphthalene dioxygenase mRNA transcripts from groundwater. *Appl Environ Microbiol* **65**: 80–87.
46. Mesarch MB, Nakatsu CH, Nies L (2000) Development of catechol 2,3-dioxygenase-specific primers for monitoring bioremediation by competitive quantitative PCR. *Appl Environ Microbiol* **66**: 678–683.
47. Yeates C, Holmes AJ, Gillings MR (2000) Novel forms of ring-hydroxylating dioxygenases are widespread in pristine and contaminated soils. *Environ Microbiol* **2**: 644–653.
48. Leander M, Vallaeys T, Fulthorpe RR (1998) Amplification of putative chlorocatechol dioxygenase gene fragments from alpha- and beta-proteobacteria. *Can J Microbiol* **44**: 482–486.
49. Luz AP, Pellizari VH, Whyte LG, Greer CW (2004) Survey of indigenous microbial hydrocarbon degradation genes in soils from Antarctica and Brazil. *Can J Microbiol* **50**: 323–333.
50. Lee SY, Song MS, You KM, Kim BH, Bang SH, Lee IS, Kim CK, Park YK (2002) Monitoring 4-chlorobiphenyl-degrading bacteria in soil microcosms by competitive quantitative PCR. *J Microbiol* **40**: 274–281.
51. Wyndham RC, Nakatsu C, Peel M, Cashore A, Ng J, Szilagyi F (1994) Distribution of the catabolic transposon Tn*5271* in a groundwater bioremediation system. *Appl Environ Microbiol* **60**: 86–93.
52. Dionisi HM, Chewning CS, Morgan KH, Menn FM, Easter JP, Sayler GS (2004)

Abundance of dioxygenase genes similar to *Ralstonia* sp strain U2 *nagAc* is correlated with naphthalene concentrations in coal tar-contaminated freshwater sediments. *Appl Environ Microbiol* **70**: 3988–3995.

53. Hamann C, Hegemann J, Hildebrandt A (1999) Detection of polycyclic aromatic hydrocarbon degradation genes in different soil bacteria by polymerase chain reaction and DNA hybridization. *FEMS Microbiol Lett* **173**: 255–263.
54. Hallier-Soulier S, Ducrocq V, Mazure N, Truffaut N (1996) Detection and quantification of degradative genes in soils contaminated by toluene. *FEMS Microbiol Ecol* **20**: 121–133.
55. Hogan DA, Buckley DH, Nakatsu CH, Schmidt TM, Hausinger RP (1997) Distribution of the *tfdA* gene in soil bacteria that do not degrade 2,4-dichlorophenoxyacetic acid (2,4-D). *Microbial Ecol* **34**: 90–96.
56. Vallaeys T, Fulthorpe RR, Wright AM, Soulas G (1996) The metabolic pathway of 2,4-dichlorophenoxyacetic acid degradation involves different families of *tfdA* and *tfdB* genes according to PCR-RFLP analysis. *FEMS Microbiol Ecol* **20**: 163–172.
57. Whyte LG, Greer CW, Inniss WE (1996) Assessment of the biodegradation potential of psychrotrophic microorganisms. *Can J Microbiol* **42**: 99–106.
58. Whyte LG, Schultz A, van Beilen JB, Luz AP, Pellizari V, Labbe D, Greer CW (2002) Prevalence of alkane monooxygenase genes in Arctic and Antarctic hydrocarbon-contaminated and pristine soils. *FEMS Microbiol Ecol* **41**: 141–150.
59. Kohno T, Sugimoto Y, Sei K, Mori K (2002) Design of PCR primers and gene probes for extensive detection of alkane-degrading bacteria. *Microb Environ* **17**: 114–121.
60. Smits TH, Balada SB, Witholt B, van Beilen JB (2002) Functional analysis of alkane hydroxylases from Gram-negative and Gram-positive bacteria. *J Bacteriol* **184**: 1733–1742.
61. Hill KE, Marchesi JR, Weightman AJ (1999) Investigation of two evolutionarily unrelated halocarboxylic acid dehalogenase gene families. *J Bacteriol* **181**: 2535–2547.
62. Regeard C, Maillard J, Holliger C (2004) Development of degenerate and specific PCR primers for the detection and isolation of known and putative chloroethene reductive dehalogenase genes. *J Microbiol Methods* **56**: 107–118.
63. Verhagen C, Smit E, Janssen DB, Van Elsas JD (1995) Bacterial dichloropropene degradation in soil; screening of soils and involvement of plasmids carrying the dhlA gene. *Soil Biol Biochem* **27**: 1547–1557.
64. Park W, Padmanabhan P, Padmanabhan S, Zylstra GJ, Madsen EL (2002) *nahR*, encoding a LysR-type transcriptional regulator, is highly conserved among naphthalene-degrading bacteria isolated from a coal tar waste-contaminated site and in extracted community DNA. *Microbiology* **148**: 2319–2329.
65. Parekh NR, Hartmann A, Fournier JC (1996) PCR detection of the *mcd* gene and evidence of sequence homology between the degradative genes and plasmids from diverse carbofuran-degrading bacteria. *Soil Biol Biochem* **28**: 1797–1804.
66. Duarte GF, Rosado AS, Seldin L, de Araujo W, van Elsas JD (2001) Analysis of bacterial community structure in sulfurous-oil-containing soils and detection of species carrying dibenzothiophene desulfurization (*dsz*) genes. *Appl Environ Microbiol* **67**: 1052–1062.
67. Sei K, Nakao M, Mori K, Ike M, Kohno T, Fujita M (2001) Design of PCR primers and a gene probe for extensive detection of poly(3-hydroxybutyrate) (PHB)-degrading bacteria possessing fibronectin type III linker type-PHB depolymerases. *Appl Microbiol Biotechnol* **55**: 801–806.
68. Saltikov CW, Olson BH (2002) Homology of *Escherichia coli* R773 *ars*A, *ars*B, and *ars*C genes in arsenic-resistant bacteria isolated from raw sewage and arsenic-enriched creek waters. *Appl Environ Microbiol* **68**: 280–288.

69. Sun YM, Polishchuk EA, Radoja U, Cullen WR (2004) Identification and quantification of *ars*C genes in environmental samples by using real-time PCR. *J Microbiol Methods* **58**: 335–349
70. Oger C, Berthe T, Quillet L, Barray S, Chiffoleau JF, Petit F (2001) Estimation of the abundance of the cadmium resistance gene *cad*A in microbial communities in polluted estuary water. *Res Microbiol* **152**: 671–678.
71. Trajanovska S, Britz ML, Bhave M (1997) Detection of heavy metal ion resistance genes in Gram-positive and Gram-negative bacteria isolated from a lead-contaminated site. *Biodegradation* **8**: 113–124.
72. Williams JR, Morgan AG, Rouch DA, Brown NL, Lee BTO (1993) Copper-resistant enteric bacteria from United Kingdom and Australian piggeries. *Appl Environ Microbiol* **59**: 2531-2537.
73. Siering PL, Ghiorse WC (1997) PCR detection of a putative manganese oxidation gene (*mof*A) in environmental samples and assessment of *mof*A gene homology among diverse manganese-oxidizing bacteria. *GeoMicrobiol J* **14**: 109–125.
74. Osborn AM, Bruce KD, Strike P, Ritchie DA (1993) Polymerase chain reaction-restriction fragment length polymorphism analysis shows divergence among mer determinants from Gram-negative soil bacteria indistinguishable by DNA-DNA hybridization. *Appl Environ Microbiol* **59**: 4024–4030.
75. Liebert CA, Wireman J, Smith T, Summers AO (1997) Phylogeny of mercury resistance (*mer*) operons of Gram-negative bacteria isolated from the fecal flora of primates. *Appl Environ Microbiol* **63**: 1066–1076.
76. Bruce KD (1997) Analysis of *mer* gene subclasses within bacterial communities in soils and sediments resolved by fluorescent-PCR-restriction fragment length polymorphism profiling. *Appl Environ Microbiol* **63**: 4914–4919.
77. Osborn AM, Bruce KD, Ritchie DA, Strike P (1996) The mercury resistance operon of the IncJ plasmid pMERPH exhibits structural and regulatory divergence from other Gram-negative *mer* operons. *Microbiology* **142**: 337–345.
78. Nakamura K, Silver S (1994) Molecular analysis of mercury-resistant *Bacillus* isolates from sediment of Minamata Bay, Japan. *Appl Environ Microbiol* **60**: 4596–4599.
79. Hart MC, Elliott GN, Osborn AM, Ritchie DA, Strike P (1998) Diversity amongst *Bacillus mer* A genes amplified from mercury resistant isolates and directly from mercury polluted soil. *FEMS Microbiol Ecol* **27**: 73–84.
80. Narita M, Chiba K, Nishizawa H, Ishii H, Huang CC, Kawabata Z, Silver S, Endo G (2003) Diversity of mercury resistance determinants among *Bacillus* strains isolated from sediment of Minamata Bay. *FEMS Microbiol Lett* **S223**: 73–82.
81. Schwartz T, Kohnen W, Jansen B, Obst U (2003) Detection of antibiotic-resistant bacteria and their resistance genes in wastewater, surface water, and drinking water biofilms. *FEMS Microbiol Ecol* **43**: 325–335.
82. Jensen LB, Agerso Y, Sengelov G (2002) Presence of *erm* genes among macrolide-resistant Gram-positive bacteria isolated from Danish farm soil. *Environment Int* **28**: 487–491.
83. Heuer H, Krogerrecklenfort E, Wellington EMH *et al.* (2002) Gentamicin resistance genes in environmental bacteria: prevalence and transfer. *FEMS Microbiol Ecol* **42**: 289–302.
84. Smalla K, van Overbeek LS, Pukall R, van Elsas JD (1993) Prevalence of *npt*II and Tn5 in kanamycin-resistant bacteria from different environments. *FEMS Microbiol Ecol* **13**: 47–58.
85. Murakami K, Minamide W, Wada K, Nakamura E, Teraoka H, Watanabe S (1991) Identification of methicillin-resistant strains of staphylococci by polymerase chain reaction. *J Clin Microbiol* **29**: 2240–2244.

86. Tolba S, Egan S, Kallifidas D, Wellington EMH (2002) Distribution of streptomycin resistance and biosynthesis genes in streptomycetes recovered from different soil sites. *FEMS Microbiol Ecol* **42**: 269–276.
87. van Overbeek LS, Wellington EMH, Egan S, Smalla K, Heuer H, CollardJ-M, Guillaume G, Karagouni AD, Nikolakopoulou TL, Dirk van Elsas JD (2002) Prevalence of streptomycin-resistance genes in bacterial populations in European habitats. *FEMS Microbiol Ecol* **42**: 277–288.
88. Aminov RIN, Garrigues J, Mackie RI (2001) Molecular ecology of tetracycline resistance: development and validation of primers for detection of tetracycline resistance genes encoding ribosomal protection proteins. *Appl Environ Microbiol* **67**: 22–32.
89. Uhl J, Kohner P, Hopkins M, Cockerill FR (1997) Multiplex PCR detection of *vanA, vanB, vanC-1,* and *vanC-2/3* genes in enterococci. *J Clin Microbiol* **35**: 703–707.
90. Stokes HW, Holmes AJ, Nield BS, Holley MP, Nevalainen KMH, Mabbutt BC, Gillings MR (2001) Gene cassette PCR: sequence-independent recovery of entire genes from environmental DNA. *Appl Environ Microbiol* **67**: 5240–5246.
91. Barlow RS, Pemberton JM, Desmarchelier PM, Gobius KS (2004) Isolation and characterization of integron-containing bacteria without antibiotic selection. *Antimicrob Agents Chemother* **48**: 838–842.
92. Mason VP, Syrett N, Hassanali T, Osborn AM (2002) Diversity and linkage of replication and mobilisation genes in *Bacillus* rolling circle replicating plasmids from diverse geographical origins. *FEMS Microb Ecol* **42**, 235–241.
93. Osborn AM, Pickup RW, Saunders JR (2000) Development and application of molecular tools in the study of IncN-related plasmids from lakewater sediments. *FEMS Microbiol Lett* **186**: 203–208.
94. Thomas CM, Thorsted P (1994) PCR probes for promiscuous plasmids. *Microbiology* **140**: 2–3.
95. Krasowiak R, Smalla K, Sokolov S, Kosheleva I, Sevastyanovich Y, Titok M, Thomas CM (2002) PCR primers for detection and characterisation of IncP-9 plasmids. *FEMS Microbiol Ecol* **42**: 217–225.
96. Gotz A, Pukall R, Smit E, Tietze E, Prager R, Tschape H, van Elsas JD, Smalla K (1996) Detection and characterization of broad-host-range plasmids in environmental bacteria by PCR. *Appl Environ Microbiol* **62**: 2621–2628.
97. Sobecky PA, Mincer TJ, Chang MC, Toukdarian A, Helinski DR (1998) Isolation of broad-host-range replicons from marine sediment bacteria. *Appl Environ Microbiol* **64**: 2822–2830.
98. Cook MA, Osborn AM, Bettandorff J, Sobecky PA (2001) Endogenous isolation of replicon probes for assessing plasmid ecology of marine sediment microbial communities. *Microbiology* **147**: 2089–2101.
99. Disque-Kochem C, Battermann A, Stratz M, Dreiseikelmann B (2001) Screening for *trbB*- and *traG*-like sequences by PCR for the detection of conjugative plasmids in bacterial soil isolates. *Microbiological Res* **156**: 159–168.
100. Lawrence JG, Ochman H, Hartl DL (1992) The evolution of insertion sequences within enteric bacteria. *Genetics* **131**: 9–20.
101. Dahlberg C, Hermansson M (1995) Abundance of Tn*3*, Tn*21*, and Tn*501* transposase (*tnpA*) sequences in bacterial community DNA from marine environments. *Appl Environ Microbiol* **61**: 3051–3056.
102. Pearson AJ, Bruce KD, Osborn AM, Ritchie DA, Strike P (1996) Distribution of class II transposase and resolvase genes in soil bacteria and their association with mer genes. *Appl Environ Microbiol* **62**: 2961–2965.
103. Sinigalliano CD, Kuhn DN, Jones RD (1995) Amplification of the *amoA* gene from diverse species of ammonium-oxidizing bacteria and from an indigenous bacterial population from seawater. *Appl Environ Microbiol* **61**: 2702–2706.

104. Rotthauwe JH, Witzel KP, Liesack W (1997) The ammonia monooxygenase structural gene *amoA* as a functional marker: molecular fine-scale analysis of natural ammonia-oxidizing populations. *Appl Environ Microbiol* **63**: 4704–4712.
105. Stephen JR, Chang Y-J, Macnaughton SJ, Kowalchuk GA, Leung KT, Flemming CA, White DC (1999) Effect of toxic metals on the indigenous β-subgroup ammonia oxidizer community structure and protection by inoculated metal-resistant bacteria. *Appl Environ Microbiol* **65**: 95–101.
106. Mendum TA, Sockett RE, Hirsch PR (1999) Use of molecular and isotopic techniques to monitor the response of autotrophic ammonia-oxidizing populations of the beta subdivision of the class proteobacteria in arable soils to nitrogen fertilizer. *Appl Environ Microbiol* **65**: 4155–4162.
107. Gregory LG, Karakas-Sen A, Richardson DJ, Spiro S (2000) Detection of genes for membrane-bound nitrate reductase in nitrate-respiring bacteria and in community DNA. *FEMS Microbiol Lett* **183**: 275–279.
108. Philippot L, Piutti S, Martin-Laurent F, Hallet S, Germon JC (2002) Molecular analysis of the nitrate-reducing community from unplanted and maize-planted soils. *Appl Environ Microbiol* **68**: 6121–6128.
109. Petri R, Imhoff JF (2000) The relationship of nitrate reducing bacteria on the basis of *narH* gene sequences and comparison of *narH* and 16S rDNA based phylogeny. *Syst Appl Microbiol* **23**: 47–57.
110. Flanagan DA, Gregory LG, Carter JP, Karakas-Sen A, Richardson DJ, Spiro S (1999) Detection of genes for periplasmic nitrate reductase in nitrate respiring bacteria and in community DNA. *FEMS Microbiol Lett* **177**: 263–270.
111. Braker GA, Fesefeldt A, Witzel KP (1998) Development of PCR primer systems for amplification of nitrite reductase genes (*nirK* and *nirS*) to detect denitrifying bacteria in environmental samples. *Appl Environ Microbiol* **64**: 3769–3775.
112. Hallin S, Lindgren PE (1999) PCR detection of genes encoding nitrite reductase in denitrifying bacteria. *Appl Environ Microbiol* **65**: 1652–1657.
113. Michotey V, Mejean V, Bonin P (2000) Comparison of methods for quantification of cytochrome cd_1-denitrifying bacteria in environmental marine samples. *Appl Environ Microbiol* **66**: 1564–1571.
114. Jayakumar DA, Francis CA, Naqvi SWA, Ward BB (2004) Diversity of nitrite reductase genes (*nirS*) in the denitrifying water column of the coastal Arabian Sea. *Aquatic Microbial Ecol* **34**: 69–78.
115. Henry S, Baudoin E, Lopez-Gutierrez JC, Martin-Laurent F, Brauman A, Philippot L (2004) Quantification of denitrifying bacteria in soils by *nirK* gene targeted real-time PCR. *J Microbiol Methods* **59**: 327–335.
116. Braker G, Tiedje JM (2003) Nitric oxide reductase (*norB*) genes from pure cultures and environmental samples. *Appl Environ Microbiol* **69**: 3476–3483.
117. Scala DJ, Kerkhof LJ (1998) Nitrous oxide reductase (*nosZ*) gene-specific PCR primers for detection of denitrifiers and three *nosZ* genes from marine sediments. *FEMS Microbiol Lett* **162**: 61–68.
118. Delorme S, Philippot L, Edel-Hermann V, Deulvot C, Mougel C, Lemanceau P (2003) Comparative genetic diversity of the *narG*, *nosZ*, and 16S rRNA genes in fluorescent pseudomonads. *Appl Environ Microbiol* **69**: 1004–1012.
119. Mohan SB, Schmid M, Jetten M, Cole J (2004) Detection and widespread distribution of the *nrfA* gene encoding nitrite reduction to ammonia, a short circuit in the biological nitrogen cycle that competes with denitrification. *FEMS Microbiol Ecol* **49**: 433–443.
120. Allen AE, Booth MG, Frischer ME, Verity PG, Zehr JP, Zani S (2001) Diversity and detection of nitrate assimilation genes in marine bacteria. *Appl Environ Microbiol* **67**: 5343–5348.
121. Zehr JP, McReynolds LA (1989) Use of degenerate oligonucleotides for

amplification of the *nifH* gene from the marine cyanobacterium *Trichodesmium* spp. *Appl Environ Microbiol* **55**: 2522–2526
122. Zani S, Mellon MT, Collier JL, Zehr JP (2000) Expression of *nifH* genes in natural microbial assemblages in Lake George, New York, detected by reverse transcriptase PCR. *Appl Environ Microbiol* **66**: 3119–3124.
123. Burgmann H, Widmer F, Sigler WV, Zeyer J (2003) mRNA extraction and reverse transcription-PCR protocol for detection of *nifH* gene expression by Azotobacter vinelandii in soil. *Appl Environ Microbiol* **69**: 1928–1935.
124. Burgmann H, Widmer F, Von Sigler W, Zeyer J (2004) New molecular screening tools for analysis of free-living diazotrophs in soil. *Appl Environ Microbiol* **70**: 240–247.
125. King GM (2003) Uptake of carbon monoxide and hydrogen at environmentally relevant concentrations by mycobacteria. *Appl Environ Microbiol* **69**: 7266–7272.
126. Dunfield KE, King GM (2004) Molecular analysis of carbon monoxide-oxidizing bacteria associated with recent Hawaiian volcanic deposits. *Appl Environ Microbiol* **70**: 4242–4248.
127. Lechner S, Conrad R (1997) Detection in soil of aerobic hydrogen-oxidizing bacteria related to *Alcaligenes eutrophus* by PCR and hybridization assays targeting the gene of the membrane-bound (NiFe) hydrogenase FEMS. *Microbiol Ecol* **22**: 193–206.
128. Holmes AJ, Costello AM, Lidstrom ME, Murrell JC (1995) Evidence that particulate methane monooxygenase and ammonia monooxygenase may be evolutionarily related. *FEMS Microbiol Lett* **132**: 203–208.
129. Mohan KS, Walia SK (1994) Detection of soluble methane monooxygenase producing methylosinus-trichosporium OB3B by polymerase chain-reaction. *Can J Microbiol* **40**: 969–973.
130. McDonald IR, Kenna EM, Murrell JC (1995) Detection of methanotrophic bacteria in environmental samples with the PCR. *Appl Environ Microbiol* **61**: 116–121.
131. Miguez CB, Bourque D, Sealy JA, Greer CW, Groleau D (1997) Detection and isolation of methanotrophic bacteria possessing soluble methane monooxygenase (sMMO) genes using the polymerase chain reaction (PCR). *Microbiol Ecol* **33**: 21–31.
132. Costello AM, Lidstrom ME (1999) Molecular characterization of functional and phylogenetic genes from natural populations of methanotrophs in lake sediments. *Appl Environ Microbiol* **65**: 5066–5074.
133. Bourne DG, McDonald IR, Murrell JC (2001) Comparison of *pmoA* PCR primer sets as tools for investigating methanotroph diversity in three Danish soils. *Appl Environ Microbiol* **67**: 3802–3809.
134. Steinkamp R, Zimmer W, Papen H (2001) Improved method for detection of methanotrophic bacteria in forest soils by PCR. *Curr Microbiol* **42**: 316–322.
135. Kolb S, Knief C, Stubner S, Conrad R (2003) Quantitative detection of methanotrophs in soil by novel *pmoA*-targeted real-time PCR assays. *Appl Environ Microbiol* **69**: 2423–2429.
136. Wagner M, Roger AJ, Flax JL, Brusseau GA, Stahl DA (1998) Phylogeny of dissimilatory sulfite reductases supports an early origin of sulfate respiration. *J Bacteriol* **180**: 2975–2982.
137. Ramaiah N, Hill RT, Chun J, Ravel J, Matte MH, Straube WL, Colwell RR (2000) Use of a *chiA* probe for detection of chitinase genes in bacteria from the Chesapeake Bay. *FEMS Microbiol Ecol* **34**: 63–71.
138. LeCleir GR, Buchan A, Hollibaugh JT (2004) Chitinase gene sequences retrieved from diverse aquatic habitats reveal environment-specific distributions. *Appl Environ Microbiol* **70**: 6977–6983.

139. Bach HJ, Hartmann A, Schloter M, Munch JC (2001) Primers and functional probes for amplification and detection of bacterial genes for extracellular peptidases in single strains and in soil. *J Microbiol Methods* **44**: 173–182.
140. Mazurier S, Lemunier M, Siblot S, Mougel C, Lemanceau P (2004) Distribution and diversity of type III secretion system-like genes in saprophytic and phytopathogenic fluorescent pseudomonads. *FEMS Microbiol Ecol* **49**: 455–467.

Molecular detection of fungal communities in soil

13

Eric Smit, Francisco de Souza and Renske Landeweert

13.1 Introduction

Fungi are an important group of microorganisms inhabiting a diverse range of ecosystems (1). The majority of fungi are multicellular organisms. They can have a sexual and/or asexual lifecycle with numerous species forming fruiting bodies and producing large numbers of spores. Due to their filamentous growth, fungi are highly efficient in mineralizing virtually any organic substrate. Saprotrophic fungi colonize dead substrates, while necrotrophic and mutualistic fungal species colonize living tissues by means of highly specialized infection structures. Several fungal species have antagonistic properties that prevent the development of other potentially pathogenic fungi and bacteria.

In the past, detection and identification of fungi relied on classification of the fruiting bodies or on cultivation-based methods to distinguish hyphal growth, infection structures and resting spores. Since many fungi do not produce fruit bodies, are refractory to cultivation or do not form observable sexual structures, these identification methods have limited the number of species currently recognized (2). The development of molecular techniques provides new possibilities for fungal species identification. Sequencing or molecular typing of fungal genes provides researchers with new tools to identify fungal isolates. The small and large subunits of ribosomal RNA (rDNA) and the internal transcribed spacer (ITS) regions of fungi can be targeted for sequencing using a suite of general eukaryotic primers (3,4). However, in order to obtain sequences of sufficient quality the isolates have to be cultured first, which limited the application of general primer sets to culturable fungi only.

Alternatively, the use of species-specific primers allows DNA to be extracted and identified from less pure sources and even from environmental sources directly. This approach detects fungi with sequences complementary to the primers. However, the disadvantage of this approach is that divergent (unknown) sequences will not be amplified, potentially biasing fungal community and diversity analyses.

To fully explore the natural fungal diversity within the environment, a better approach might be to use general fungal primers covering all known fungal taxonomic diversity. By analogy to developments in molecular ecology of bacterial communities (5), researchers started to analyze fungal communities in the environment using molecular methods. In this chapter we give an overview of the various molecular strategies that are currently being used to analyze fungal diversity.

13.2 Molecular strategies for fungal community analyses

Molecular methods for analyzing fungal diversity in environmental samples are based on extraction of total DNA from the sample. Total DNA extracts contain DNA from all lysed cells and hence reflects the composition of all organisms that were present in the substrate sampled. After purification of the DNA extracts, specific seqeunces are PCR-amplified using either group-specific or universal fungal community primers. The various primer sets that have been used in studies on fungal communities will be evaluated in this chapter. Each primer set targets a particular gene or sequence and the choice of these target gene(s) and regions will be determined by the specific aim of the study. The ribosomal genes are the most suitable target genes since they contain both conserved and variable regions. The conserved regions (5.8S, 18S and 28S gene) can be used to construct primers distinguishing a wide range of fungal taxa while the variable regions that amplify the internal transcribed spacer regions (ITS1 and ITS2) can be used to specifically distinguish certain fungal species. The detection sensitivity or specificity can also be increased by applying a nested PCR approach, i.e. subjecting the amplicons of the first reaction to a second amplification cycle using primers that anneal within the first fragment.

Generally amplification of community DNA targeting ribosomal genes will yield a collection of different DNA amplicons of similar sizes. Only the size of the various ITS regions can differ enough to allow size analysis for these amplicons (6). Since size variation of the 18S rDNA is very limited, size is not the appropriate characteristic to distinguish different 18S rDNA fragments. However, fragments that have the same size can be distinguished on the basis of their differing base pair compositions using denaturing gradient gel electrophoresis (DGGE) or temperature gradient gel electrophoresis (TGGE) (see Chapter 3). To date, a number of studies have analyzed fungal community diversity using DGGE or TGGE (7–14).

Another technique to separate equally sized DNA amplicons is to digest the PCR product with various restriction enzymes. This technique is known as amplified ribosomal DNA restriction analysis (ARDRA) (15). When visualized on an agarose gel, the fragments will form a species (or community)-specific banding pattern that can in principle be compared to a database consisting of banding patterns from known microorganisms.

Three other molecular techniques currently used to characterize fungal populations are single strand conformation polymorphism (SSCP), automated ribosomal intergenic spacer analysis (ARISA) and terminal restriction fragment length polymorphism (T-RFLP). For SSCP analysis (16,17) single DNA strands are created by enzymatic digestion and then, following denaturing, are electrophoresed on native polyacrylamide gels (see Chapter 3). ARISA and T-RFLP are related techniques based on the terminal labeling of the amplified DNA using a fluorescently labeled primer. In ARISA the amplified and labeled ITS fragments are analyzed directly on a sequencer to determine their size in base pairs (18). For T-RFLP, the amplified fragments are first subjected to restriction enzyme digestion and then their size is determined on a sequencer (19). These methods will in principle yield one fluorescently labeled fragment per organism, while ARDRA might yield more bands per organism.

Both ARISA and T-RFLP will yield a fingerprint of the fungal community but these methods are not particularly suited to identify fungi to species level. To accurately identify fungi from a community to species/genus level, the DNA amplicons need to be sequenced. To this end the generated amplicons from a mixed community can be cloned (9,20–22) or alternatively bands can be cut out of a DGGE/TGGE or an SSCP gel for sequencing (see Chapter 3), and they can be identified by comparing them to sequence databases. At present, only a limited number of molecular studies investigating fungal diversity in the environment have been performed.

13.3 DNA extraction

To obtain microbial DNA from environmental samples, several different extraction protocols have been developed for soil, sediment, plant tissue, wood and even art objects. For an accurate assessment of the total fungal community, DNA from all fungal groups present in the substrate needs to be extracted in detectable and representative amounts. Successful extraction therefore implies that cells from all fungi are lysed with equal efficiency. Most DNA extraction procedures involve three steps:

(i) cells are disrupted in a lysis step;
(ii) cell fragments and debris are removed;
(iii) nucleic acids are precipitated and purified.

For optimal DNA extraction and purification, each step in the extraction protocol needs to be adjusted to the particular properties of each sample and this has led to the development of many different protocols. Many authors mention the successful use of commercially available DNA extraction kits (19,23) to extract DNA from soil as well as from other environmental substrates (11,24). In general, small sample amounts (10–500 mg) are used for soil DNA extraction when using these commercially available extraction kits (17,23,25–27). However, other extraction procedures may require larger sample sizes (28).

In the lysis step, cells are disrupted by chemical, mechanical or enzymatic methods, and their contents, including the DNA, are released. This step is critical since it will determine what percentage of the cells actually present in the sample can be detected using molecular methods (29). Direct cell lysis methods, whereby cells are lysed directly within the environmental matrix, in general give good yields of DNA and the extracted DNA seems to represent the microbial as well as nonmicrobial community of the sample (30). Most DNA extraction protocols involve suspension of the sample in an extraction buffer followed by the addition of lysozyme and a further cell disruption step. Other protocols for DNA extraction from soil are based on mechanical lysis of cells using a bead-beater (14). In general, bead-beating gives a good DNA yield, but the extraction method typically results in smaller sheared DNA fragments than are generated by extraction methods using freeze–thawing or chemical lysis (31). Potential problems associated with sheared DNA are discussed in Chapters 1 and 2.

After cell lysis the sample is repeatedly extracted with chloroform-isoamyl alcohol and finally the DNA can be precipitated. The efficiency of DNA extractions depends on efficient cell lysis as well as on the chemical properties of the sample. Often DNA extracted from soil needs to be purified to remove co-extracted contaminants. Besides phenolic compounds, humic acids can inhibit PCR reactions by chelating Mg^{2+} ions, therefore purification of DNA extracted from soil is critical (30). Commercially available DNA purification kits are often used for this purpose. Purification using agarose gel electrophoresis in combination with a gel extraction kit has also been reported in the literature (23). Dilution of the DNA or the addition of sequestrating agents to the PCR mixture may also help to overcome PCR inhibition by these co-extracted substances. Yet dilution of the template DNA should be done with caution as very low DNA concentrations may also influence PCR efficiency (32).

Although some extraction protocols have been shown to reproducibly generate identical results when used on replicate soil samples (26), no study has yet extensively compared the efficiency of different extraction protocols when assessing fungal communities. Work done on soil bacterial communities suggests that the extraction method largely determines the amount, quality and source of the DNA recovered (29,30,32–34). Recovery rates however, for fungal as well as bacterial DNA, remain difficult to determine (32). Obviously, this important point requires future investigation.

13.4 Target genes

There are a large number of nonspecific primers available that target the 18S rRNA, 28S rRNA, 5.8S rRNA or the ITS1 and ITS2 regions. Although these primers can be used for amplification of DNA from environmental samples, they are not necessarily fungal specific. The use of these nonspecific primers might also lead to the amplification of plant DNA (7,11,16) and animal DNA (13,35). Therefore it is recommended to use such primers only in combination with a fungal species or group-specific primer.

Choosing the most appropriate target genes for fungal primer development is of great importance. In a number of ecological studies, primers for the 18S rRNA gene were used (7–11,15,16,21,23). Since this gene has a limited sequence divergence among fungi, the resolution to distinguish sequences of different species is also limited. For instance, Leeflang *et al.* (21) could not distinguish between a number of different *Fusarium* species based on a part of their 18S rRNA sequence. However, the advantage of using the 18S rRNA gene is that there are a relatively large number of sequences present in the database enabling researchers to develop primers and to identify sequences originating from the environment (9).

The ITS1 and ITS2 region exhibits much greater sequence divergence than the 18S rRNA gene and will therefore provide a higher resolution to discriminate between relatively closely related fungi. On the other hand it has been reported that there is substantial within-species sequence diversity of the ITS which can lead to overestimation of fungal diversity (10,36).

The ITS-based primers ITS1F–ITS4B, originally developed for the characterization of ecto-mycorrhizal root tips (4), were recently applied to study Basidiomycete diversity in forest soil (22). These primers were found to be phylum- specific since all obtained clone sequences belonged to the Basidiomycetes. Jasalavich *et al.* (37) compared the primer pair ITS1F–ITS4 and ITS1F–ITS4B on a collection of fungal isolates. The first primer pair, which is fungal specific, amplified all brown and white rot Basidiomycetes and the wood-inhabiting Ascomycetes, while the Basidiomycete-specific primer pair ITS1F–ITS4B amplified only the Basidiomycete species.

Until now, only ribosomal genes or intergenic spacer regions have been used for the characterization of fungal communities. The advantage of using these sequences is the presence of conserved and variable regions that allow development of primers for broad taxonomic groups. In bacterial ecology, researchers have employed a range of primers for functional genes (see Chapter 12). The application of primer sets for fungal functional genes will no doubt also increase, the recent PCR detection of fungal ligninolytic peroxidase genes and mRNAs being a notable example (38). The future combination of ribosomal-based community analysis with analysis of functional genes and quantitative PCR should yield exciting data in the coming years.

13.5 Primers for fungal populations

The number of molecular studies on fungal diversity in soil is currently limited. However, a relatively large number of different primers for specific fungal communities have been developed. The development and analysis of primers can be greatly facilitated by using several publicly available sequence databases or programs. The ribosomal database located at: http://rdp.cme.msu.edu/html/ can be used to download software and alignments of ribosomal sequences and is extremely convenient for analyzing the taxonomic coverage of primers. The NCBI web site provides access to a wealth of sequence information. This site is located at: http://www.ncbi.nlm.nih.gov/. The Entrez program at the NCBI site can be used to retrieve sequences and the BLAST program is very useful for testing *in silico* primer specificity. Information on ribosomal sequences can be found at http://www.psb.ugent.be/rRNA/index.html and http://rdp.cme.msu.edu/html/ (see also Chapter 15). In the past, the design of group primers was done manually and required considerable time and patience. There are, however, a number of software programs which can aid researchers in this task. Most notable is the ARB software package containing the PROBE DESIGN program (http://arb-home.de) although the software can only be installed on Unix- or Linux-based computers. More recently, Ashelford *et al.* (39) developed the program PRIMROSE for designing group-specific primers which can be run on Microsoft Windows 95/NT/2000.

Most primers used the target SSU (18S rDNA), primarily because more SSU sequences than LSU sequences are available. The primer pairs that have been applied in community studies are listed in *Table 13.1*. The

Table 13.1 Primers used for the analysis of fungal communities in environmental samples

Code		Position[a]	Sequence (5′–3′)	Target	Taxon[b]	Reference
NS1	f	19–38	GTAGTCATATGCTTGTCTC	SSU	Euk	(3)
NS8	r	1789–1769	TCCGCAGGTTCACCTACGGA	SSU	Euk	(3)
NS3	f	553–573	GCAAGTCTGGTGCCAGCAGCC	SSU	Euk	(3)
NS31		not S.c.	TTGGAGGGCAAGTCTGGTGCC	SSU	Glom	(56)
2234C	f	1766–1785	GTTTCCGTAGGTGAACCTGC	SSU	Euk	(6)
3126T	r	not 18S	ATATGCTTAAGTTCAGCGGGT	LSU	Euk	(6)
NS1GC	f	17–38	CCAGTAGTCATATGCTTGTC	SSU	Euk	(7)
NS2+10	r	583–564	GAATTACCGCGGCTGCTGGC	SSU	Euk	(7)
EF4	f	195–214	GGAAGGG[G/A]TGTATTTATTAG	SSU	Fung	(9)
EF3	r	1747–1727	TCCTCTAAATGACCAAGTTTG	SSU	Fung	(9)
Fung5	r	747–729	GTAAAAGTCCTGGTTCCCC	SSU	Fung	(9)
NS26	f	305–323	CTGCCCTATCAACTTTCGA	SSU	Euk	(11)
518r	r	581–565	ATTACCGCGGCTGCTGG	SSU	Euk	(11)
ITS1	f	1769–1787	TCCGTAGGTGAACCTGCGG	SSU–ITS	Euk	(4)
ITS1F	f	1731–1752	CTTGGTCATTTAGAGGAAGTAA	SSU–ITS	Fung	(4)
ITS4A	r	not 18S	CGCCGTTACTGGGGCAATCCCTG	LSU–ITS	Asc	(57)
ITS4B	r	not 18S	CAGGAGACTTGTACACGGTCCAG	LSU–ITS	Bas	(4)
SSU0817	f	794–817	TTAGCATGGAATAATRRAATAGGA	SSU	Fung	(23)
SSU1196	r	1215–1196	TCTGGACCTGGTGAGTTTCC	SSU	Fung	(23)
SSU1536	r	1555–1336	ATTGCAATGCYCTATCCCCA	SSU	Fung	(23)
FR1	r	1664–1949	AICCATTCAATCGGTAIT	SSU	Fung	(10)
FF390	f	1317–1334	CGATAACGAACGAGACCT	SSU	Fung	(10)
PN3		not 18S	CCGTTGGTGAACCAGCGGAGGGATC	LSU		(58)
PN34		not 18S	TTGCCGCTTCACTCGCCGTT	LSU		(58)
U1		not 18S	GTGAAATTGTTGAAAGGGAA	LSU	Fung	(58)
U2		not 18S	GACTCCTTGGTCCGTGTT	LSU	Fung	(58)
VANS1		not S.c.	GTCTAGTATAATCGTTATACAGG	SSU	Glom	(56)
AM1		not S.c.	GTTTCCCGTAAGGCGCCGAA	SSU	Glom	(20)

[a]Primer positions are given on the SSU rRNA for *Saccharomyces cerevisiae*, except where stated.
[b]**Euk**: Eukarya; **Fung**: Fungi; **Asc**: Ascomycetes; **Bas**: Basidiomycetes; **Glom**: Glomales.

performance of some of these primers was evaluated in a number of studies (11,13,19,23,35). Primers can be tested in various ways:

(i) their taxonomic coverage can be analyzed by computer, e.g. by performing a BLAST search on the NCBI database;
(ii) they can be tested on a collection of known microorganisms;
(iii) they can be analyzed by sequencing a clone library made from amplified DNA from environmental samples;
(iv) their bias should be checked by amplifying a known mix of microorganisms.

Each method can yield slightly different results, and the ultimate test is the analysis of a clone library. There are several important primer parameters that should be checked, namely that:

(i) the taxonomic coverage that will determine which groups will be amplified;
(ii) the resolution of the amplicons to determine which particular taxonomic level can be distinguished, e.g. species level, genus level;

(iii) the primers are sufficiently specific, to insure that a strong signal is produced from DNA from the environment.

The latter is obviously also dependent upon the concentration (abundance) of the target template DNA in the environmental sample. Additionally, one should consider if the amplified DNA is of suitable length and quality for the method of subsequent analysis, e.g. fingerprinting or sequencing.

Schabereiter-Gurtner *et al.* (11) constructed a fungal 18S rDNA clone library (60 clones) from samples collected from glass from a historic church window and tested the performance of several primer sets by DGGE analysis. The use of nonspecific primers to amplify DNA from the glass yielded clones of nonfungal origin. The primer set NS1GC/NS2+10 or NS1/NS2+10CG did not produce sharp DGGE bands. Primer set EF4/518rGC failed to amplify one clone and DGGE analysis followed by sequencing of these amplicons revealed nine different clusters. The primer pair NS26/518rGC similarly failed to amplify one phylotype with DGGE analysis followed by sequencing of these amplicons revealing eight different clusters. However, the clusters identified using the different primer sets consisted of different clones. On this basis the authors recommended the use of two different primer sets to obtain maximum resolution and coverage of the fungal community (11). Another primer comparison study showed that the primer sets nu-SSU0817/nu-SSU1196, nu-SSU0817/nu-SSU1536 with EF4/EF3 and EF4/fung5, when tested on a collection of fungal isolates, generated amplicons from species from all fungal divisions and that only the EF-fung primer pairs failed to amplify a few species (23). Vainio and Hantula (10) developed several primers for investigation of wood-inhabiting fungi by DGGE analysis. They compared several combinations of the primers FR1, FF390, FF700 and FF1100 on a collection of fungal species and found that a 1650 bp fragment amplified by FR1/NS1 had the highest resolution for distinguishing the different species using DGGE analysis. This is remarkable since it is generally accepted to use short (<500 bp) DNA fragments for DGGE (5). Klamer *et al.* (19) tested the amplification range of ITS1–ITS4, ITS1F–ITS4, ITS1F–ITS4A, ITS1F–ITS4B and ITS1–ITS4B on a collection of DNA extracts from plants and Ascomycetes and Basidiomycetes. These primer pairs target the end of the 18S rRNA and the beginning of the 28S rRNA and will amplify the ITS1, the 5.8S rDNA and ITS2 regions. While the ITS1–ITS4 pair amplified plant and fungal DNA, the ITS1F–ITS4 primers were specific to fungi only. However, the ITS1F–ITS4A primer pair, which is supposed to be specific for Ascomycetes, did allow amplification of some Basidiomycete sequences and failed to cover all of the Ascomycetes. The ITS1–ITS4B combination that was designed to be specific for Basidiomycetes appeared to be highly specific for this group and worked better than the combination ITS1F–ITS4B. Sequence variation between ITS fragments of different fungal species are greater then differences between their 18S rRNA. Thus amplified ITS fragments give a higher resolution than for 18S rRNA gene sequences, enabling closely related species to be distinguished. Current knowledge of the performance of the various primer combinations does not allow the selection of one method or primer pair best suited to analyze the fungal community is soil. Clearly, no single primer system will satisfy all identification purposes and a regular iterative evaluation of each system is necessary as more reference sequence information and a better understanding of the genetic diversity of the target group becomes available (40).

13.6 *In silico* analysis of 18S rRNA primers

The large number of fungal primers available at present makes it virtually impossible to test and compare them all in one laboratory study. It is therefore difficult to determine which primer set is best suited to amplify all or specific fungal species.

Yet, an *in silico* analysis of the specificity of fungal primers would shed light on their taxonomic coverage. For this purpose a Probe Match analysis was performed on the Ribosomal Database Project (RDP) database (41) with most of the presently available primers for the 18S rDNA. Primers targeting the 28S rDNA were not analyzed since the number of 28S rDNA entries in the LSU database of the RDP is relatively small in comparison. The results of the 18S rDNA primer search are given in *Table 13.2*. For each of the primers the number of entries matching the primer sequence is given. The primers were analyzed on a set of aligned sequences of the small subunit ribosomal sequences only. Nonaligned sequences were omitted since they contain many partial sequences. However, interpretation of these results should be treated with caution since the coverage remains biased by the inevitable presence of partial sequences in the aligned database. Additionally, the ends of the 18S rDNA are relatively under-represented in the database and primers that anneal close to the ends of the 18S rDNA may appear to have a poor taxonomic coverage compared to primers that anneal towards the centre of the 18S rRNA for which more sequence information is available. To estimate the relative specificity of the primers we calculated the ratio between the number of matches to fungal sequences and the number of matches to sequences of other Eukaryotes. In *Table 13.2* the ratio between the number of hits for the Fungi and the Eukaryotes (F/E) and the ratio between the Ascomycota and Basidiomycota was calculated. The ratios demonstrate that the general Eukaryotic primers such as NS1, NS2+10, NS3 and NS26 potentially amplify a number of nonfungal organisms (*Table 13.2*). For the nonspecific primers the ratios between the Fungi and Eukaryotes range between 0.22 and 0.42, while for the fungal-specific primers the ratios range between 0.80 and 0.99. This clearly illustrates the necessity to use fungal-specific primers when assessing fungal diversity. Yet, for fungal-specific primers, the Basidiomycota to Ascomycota ratios reveal that the various primers have different biases for these taxa, since the ratios range from 0.5 to 1.38. The ITS1 and ITS1F primers preferably amplify the Basidiomycota, whereas the general primers NS1 and NS3 preferentially amplify the Ascomycota. This is definitely a point to consider when choosing primers. The primers that have been developed more recently, such as the SSU primers, seem to amplify a substantially higher number of fungal species than those developed in the past such as ITS1 and EF3 and EF4. This can be explained by the fact that the number of 18S rDNA sequences in database increases rapidly, allowing researchers to design primers more accurately.

Table 13.2 *In silico* determination of fungal primer specificity

Taxon	NS1	NS3	ITS1	ITS1F	FR1	EF3	EF4	NS26	SSU0817	SSU1196	SSU1536	NS2+10
Ascomycota	186	689	134	196	559	319	390	614	691	562	443	674
Basidiomycota	93	382	147	271	366	241	245	349	355	394	318	567
Zygomycota	1	73	6	52	55	53	19	33	66	37	23	73
Chitridiomycota	1	57	3	6	27	10	40	50	61	51	6	53
Other Eukaryotes	979	2950	430	9	255	7	133	1463	243	145	3	2788
Ratio F/E[a]	0.22	0.29	0.39	0.98	0.80	0.99	0.84	0.42	0.83	0.88	0.99	0.30
Ratio A/B[b]	0.50	0.55	1.09	1.38	0.65	0.76	0.63	0.57	0.51	0.70	0.72	0.80

[a]Ratio of the number of hits generated for the Fungi, to number of hits generated for the Eukaryotes.
[b]Ratio of the number of hits generated for the Ascomycetes, to number of hits generated for the Basidiomycetes.
The number of hits generated with each primer using a probe match analysis of the RDP database of aligned 18S rDNA sequences was used to estimate the taxonomic coverage of a number of fungal-specific primers targeting the 18S rDNA.

13.7 Specific application of molecular identification techniques for studying mycorrhizal fungi

13.7.1 Analyzing arbuscular mycorrhizal fungal diversity

An ecologically important group of symbiotic soil fungi are the arbuscular mycorrhizal fungi (AMF) (42). Unfortunately, studies of AMF ecology, genetics and evolution have been hampered by the inability to culture AMF and by the difficulty of identifying them (40). However, the recent application of molecular biological techniques for characterization of AMF has led to important advances in AMF ecology (12,20,40,43–45).

The AMF form one monophyletic group, making them a relatively suitable group for molecular phylogenetic studies. The SSU (18S) rRNA gene has become the most widely used gene for detection and identification of AMF (40). Several PCR-based strategies targeting rRNA genes have been developed to detect AMF in DNA extracted from roots, soil or spores (12,20,40,43,46). Two PCR-based strategies are currently in use to assess full AMF diversity and to monitor changes in AMF field populations. The first strategy is based on the use of group-specific primers to amplify AMF sequences. Amplicon diversity can then be assessed by analyzing clone libraries. Clones in such a library can be selected by characterizing them using RFLP analysis or DGGE analysis. Representatives of the different types or groups can then be sequenced. This strategy of cloning and RFLP analysis was successfully used by Helgason *et al.* (20) and Daniell *et al.* (47) using the Eukaryotic primer NS31 and the fungal-specific primer AM1. Chelius and Triplett (25) used the VANS1 primer as the AMF-specific primer. However, the VANS1 primer seems to amplify only a small number of AMF species (40,48). A DGGE approach to study AMF diversity in soil was used by Kowalchuk *et al.* (12), using the primers of Helgason *et al.* (20) with a GC clamp attached to NS31. Several family- and/or genus-specific AMF primers have been developed to cover all known AMF diversity (49). The use of several primer sets should be more accurate for investigating AMF diversity, but this of course increases both the number and cost of analyses. The application of molecular-based detection and characterization methods offers interesting possibilities to study AMF ecology. However, to date there is insufficient knowledge about the sequence variability and genetic organization of the ribosomal genes and ITS regions within species or individual organisms to interpret the variability of ecological data (40).

13.7.2 Analyzing ectomycorrhizal fungal diversity

Another group of symbiotic fungi are the ectomycorrhizal (EcM) fungi. EcM fungi constitute a significant part of the soil fungal community, colonizing the roots of many trees in most boreal and temperate forests and several tropical forests. In contrast to AMF, the EcM fungi are not a monophyletic group. Instead, EcM fungi are found across all phyla of true fungi (50), making molecular detection using a general EcM primer impossible. Diversity analyses of EcM fungal communities are currently based on identification of fruiting bodies or EcM root tips.

Common methods to identify EcM fungi on root tips are based on morphological root tip grouping and identification, often complemented with molecular identification of single root tips. DNA is extracted from a single root tip by manual grinding of the sample followed by purification and subsequent amplification by PCR. In general, universal primer pairs (ITS1 and ITS4) (3,51) as well as fungal or Basidiomycete-specific primer pairs (ITS1 or ITS1F and ITS4B) (4) for amplification of the ITS regions of the rDNA are used (50). However, the ITS region in some EcM fungal species may show intra-specific and even intra-individual variation. Although many EcM fungal species have not yet been identified on the basis of their root tips or fruiting bodies, the rapid accumulation of ITS sequence data offers opportunities for future identification and classification (52).

Whilst the EcM root tips function as exchange sites for nutrients and carbohydrates, the EcM mycelium presents another relevant feature of the symbiosis. The extraradical mycelium actively takes up water and nutrients and is involved in the mobilization of nutrients from various substrates. To truly understand the function of EcM in ecosystems, the mycelial distribution of individual fungal species in soil needs to be revealed. At this level, molecular techniques should play an important role in improving our understanding of these complex systems. A few recent studies have focused on the analysis of extracts of total EcM fungal DNA from soil samples (22,53,54). Using Basidiomycete-specific primers, EcM hyphae were detected and identified at several distances from tree roots in a wet alder forest (53) and at several depths in Swedish forest soil (22). The recent development of real-time PCR assays for quantification of EcM fungi (55) now offers the opportunity to describe ECM communities both compositionally and quantitatively. Hence whilst providing new insights into the taxonomic classification of EcM fungi, the potential of molecular identification and quantification techniques in ecological studies of EcM fungi is vast and will greatly enhance our understanding of EcM functioning and community dynamics.

13.8 Conclusions

To date, molecular methods have been applied in a number of studies to describe fungal communities. Whilst these methods seem promising, there remains, as with all molecular approaches, several aspects requiring evaluation. Lysis efficiency and amplification biases of the current methods used for fungi are still largely unknown. Moreover, we do not have information concerning the proportion of dead fungal cellular material that is present in the environment and how this affects DNA-extraction-based methods. For some fungi the sequence heterogeneity of the rRNA or ITS copies may also pose problems. As discussed above, evaluation of primers and PCR amplification bias is essential in order for ecologically meaningful data to be generated. Nevertheless, despite these caveats molecular approaches afford significant advantages for the future study of fungi in complex environments.

References

1. Domsch KH, Gams W, Anderson T-H (1980) *Compendium of Soil Fungi*, Vol. 1. Academic Press, London.
2. Bridge P, Spooner B (2001) Soil fungi: diversity and detection. *Plant and Soil* **232**: 147–154.
3. White TJ, Burns T, Lee S, Taylor J (1990) Amplification and direct sequencing of fungal ribosomal RNA genes for phyologenetics. In: Innes MA, Gelfand DH, Sninsky JJ, White TJ (eds), *PCR Protocols*, pp. 315–322. Academic Press, San Diego.
4. Gardes M, Bruns TD (1993) ITS primers with enhanced specificity for Basidiomycetes—application to the identification of mycorrhizae and rusts. *Mol Ecol* **2**: 113–118.
5. Muyzer G, De Waal EC, Uitterlinden AG (1993) Profiling of complex microbial populations by denaturing gradient gel electrophoresis analysis of polymerase chain reaction–amplified gene coding for 16S rRNA. *Appl Environ Microbiol* **59**: 695–700.
6. Ranjard L, Poly F, Lata JC, Mougel C, Thioulouse J, Nazaret S (2001) Characterization of bacterial and fungal soil communities by automated ribosomal intergenic spacer analysis fingerprints: biological and methodological variability. *Appl Environ Microbiol* **67**: 4479–4487.
7. Kowalchuk GA, Gerards S, Woldendorp JW (1997) Detection and characterization of fungal infections of *Ammophila arenaria* (Marram grass) roots by denaturing gradient gel electrophoresis of specifically amplified 18S-rDNA. *Appl Environ Microbiol* **63**: 3858–3865.
8. Kowalchuk G (1999) Fungal community analysis using denaturing gradient gel electrophoresis (DGGE). In: Akkermans ADL, Van Elsas JD, De Bruijn FJ (eds), *Molecular Microbial Ecology Manual*. Kluwer Academic Publishers, The Hague.
9. Smit E, Leeflang P, Glandorf B, Van Elsas JD, Wernars K (1999) Analysis of fungal diversity in the wheat rhizosphere by sequencing of cloned PCR-amplified genes encoding 18S rRNA and temperature gradient gel electrophoresis. *Appl Environ Microbiol* **65**: 2614–2621.
10. Vainio EJ, Hantula J (2000) Direct detection of wood inhabiting fungi using deaturing gradient gel electrophoresis of amplified ribosomal DNA. *Mycol Res* **104**: 927–936.
11. Schabereiter-Gurtner C, Pinar G, Lubitz W, Rolleke S (2001) Analysis of fungal communities on historical church window glass by denaturing gradient gel

electrophoresis and phylogenetic 18S rDNA sequence analysis. *J Microbiol Method* **47**: 345–354.
12. Kowalchuk GA, de Souza FA, van Veen JA (2002) Community analysis of arbuscular mycorrhizal fungi associated with *Ammophila arenaria* in Dutch coastal sand dunes. *Mol Ecol* **11**: 571–581.
13. Zuccaro A, Schulz B, Mitchell JI (2003) Molecular detection of ascomycetes associated with *Fucus serratus*. *Mycol Res* **107**: 1451–1466.
14. Girvan MS, Bullimore J, Ball AS, Pretty J, Osborn AM (2004) Response of active bacterial and fungal communities in soils under winter wheat to differing fertiliser and pesticide regimes. *Appl Environ Microbiol* **70**: 2692–2701.
15. Glandorf DCM, Verheggen P, Jansen T *et al.* (2001) Effect of genetically modified *Pseudomonas putida* WCS358r on the fungal rhizosphere microflora of field grown wheat. *Appl Environ Microbiol* **67**: 3371–3378.
16. Peters S, Koschinsky S, Schwieger F, Tebbe CC (2000) Succession of microbial communities during hot composting as detected by PCR-single strand conformation polymorphism-based genetic profiles of small-subunit rRNA genes. *Appl Environ Microbiol* **66**: 930–936.
17. Lowell JL, Klein DA (2001) Comparative single-strand conformation polymorphism (SSCP) and microscopy-based analysis of nitrogen cultivation interactive effects on the fungal community of a semiarid steppe soil. *FEMS Microb Ecol* **36**: 85–92.
18. Ranjard L, Poly F, Lata JC, Mougel C, Thiolouse J, Nazaret S (2001) Characterization of bacterial and fungal soil communities by automated ribosomal intergenic spacer analysis fingerprints: biological and methodological variability. *Appl Environ Microbiol* **67**: 4479–4487.
19. Klamer M, Roberts MS, Levine LH, Drake BG, Garland JL (2002) Influence of elevated CO_2 on the fungal community in a coastal scrub oak forest soil investigated with terminal restriction fragment polymorphism analysis. *Appl Environ Microbiol* **68**: 4370–4376.
20. Helgason T, Daniell TJ, Husband R, Fitter AH, Young JPW (1998) Ploughing up the wood-wide web? *Nature* **384**: 431.
21. Leeflang P, Smit E, Glandorf DCM, Van Hannen EJ, Wernars K (2002) Effects of *Pseudomonas putida* WCS358r and its genetically modified phenazine producing derivative on the *Fusarium* population in a field experiment, as determined by 18S rDNA analysis. *Soil Biol Biochem* **34**: 1021–1025.
22. Landeweert R, Leeflang P, Kuyper TW, Hoffland E, Rosland A, Wernars K, Smit E (2003) Beyond the roots: Molecular identification of ectomycorrhizal mycelium in soil. *Appl Environ Microbiol* **69**: 327–333.
23. Bornemann J, Hartin RJ (2000) Primers that amplify fungal rRNA genes from environmental samples. *Appl Environ Microbiol* **66**: 4356–4360.
24. Buchan A, Newell SY, Moreta JIL, Moran MA (2002) Analysis of Internal Transcribed Spacer (ITS) regions of rRNA genes in fungal communities in a Southeastern US salt marsh. *Microb Ecol* **43**: 329–340.
25. Chelius MK, Triplett, EW (1999) Rapid detection of arbuscular mycorrhizae in roots and soil of an intensively managed turfgrass system by PCR amplification of small subunit rDNA. *Mycorrhiza* **9**: 61–64.
26. Pennanen T, Paavolainen L, Hantula J (2001) Rapid PCR-based method for the direct analysis of fungal communities in complex environmental samples. *Soil Biol Biochem* **33**: 697–699.
27. Adair S, Kim SH, Breuil C (2002) A molecular approach for monitoring of decay basidiomycetes in wood chips. *FEMS Microbiol Lett* **211**: 117–122.
28. Viaud M, Pasquier A, Brygoo Y (2000) Diversity of soil fungi studied by PCR–RFLP of ITS. *Mycol Res* **104**: 1027–1032.
29. Frostegard A, Courtois S, Ramisse V, Clerc S, Bernillon D, Le Gall F, Jeannin P,

Nesme X, Simonet P (1999) Quantification of bias related to the extraction of DNA directly from soils. *Appl Environ Microbiol* **65**: 5409–5420.
30. Roose-Amsaleg CL, Garnier Sillam E, Harry M (2001) Extraction and purification of microbial DNA from soil and sediment samples. *Appl Soil Ecol* **18**: 47–60.
31. Van Elsas JD, Mäntynen V, Wolters AC (1997) Soil DNA extraction and assessment of the fate of *Mycobacterium chlorophenolicum* strain PCP-1 in different soils by 16S ribosomal gene sequence based most-probable-number PCR and immunofluorescence. *Biol Fertil Soils* **24**: 188–195.
32. Von Wintzingerode F, Gobel UB, Stackebrandt E (1997) Determination of microbial diversity in environmental samples: pitfalls of PCR-based rRNA analysis. *FEMS Microbiol Rev* **21**: 213–229.
33. Porteous LA, Seidler RJ, Watrud LS (1997) An improved method for purifying DNA from soil for polymerase chain reaction amplification and molecular ecology applications. *Mol Ecol* **6**: 787–791.
34. Martin-Laurent F, Philippot L, Hallet S, Chaussod R, Germon JC, Soulas G, Catroux G (2001) DNA extraction from soils: old bias for new microbial diversity analysis methods. *Appl Environ Microbiol* **67**: 2354–2359.
35. Anderson IC, Campbell CD, Prosser JI (2003) Specificity of fungal 18S rDNA and ITS PCR primers for estimating fungal biodiversity in the soil. *Environ Microbiol* **5**: 36–48.
36. Fatehi J, Bridge P (1998) Detection of multiple rRNA–ITS regions in isolates of Ascomychyta. *Mycol Res* **98**: 614–618.
37. Jasalavich CA, Ostrofsky A, Jellison J (2000) Detection and identification of decay fungi in spruce wood by restriction fragment length polymorphism analysis of amplified genes encoding rRNA. *Appl Environ Microbiol* **66**: 4725–4734.
38. Stuardo M, Vasquez M, Vicuna R, Gonzalez B (2004) Molecular approach for analysis of model fungal genes encoding lignolytic peroxidases in wood-decaying soil systems. *Lett Appl Microbiol* **38**: 43–49.
39. Ashelford KE, Weightman AJ, Fry JC (2002) PRIMROSE: a computer program for generating and estimating the phylogenetic range of 16SrRNA oligonucleotide probes in conjunction with the RDP-II database. *Nucleic Acids Res* **30**: 3481–3489.
40. Clapp JP, Helgason T, Daniell TJ, Young JPW (2002) Genetic studies of the structure and diversity of arbuscular mycorhizal fungal communities. In: van der Heijden MGA and Sanders IR (eds), *Mycorrhizal Ecology. Ecological Studies Analysis and Synthesis*, Vol. 157, pp. 201–224. Springer Verlag, Berlin.
41. Maidak BL, Olsen GJ, Larson N, Overbeek R, McCaughey MJ, Woese CR (1997) The RDP (Ribosomal Database Project). *Nucleic Acids Res* **25**: 106–111.
42. Van der Heijden MGA, Sanders IR (eds), *Mycorrhizal Ecology. Ecological Studies Analysis and Synthesis*, Vol. 157. Springer Verlag, Berlin.
43. Van Tuinen D, Jacquot E, Zhao B, Gollotte A, Gianinazzi-Pearson V (1998) Characterization of root colonization profiles by a microcosm community of arbuscular mycorrhizal fungi using 25S rDNA-targeted nested PCR. *Mol Ecol* **7**: 879–887.
44. Helgason T, Fitter AH, Young JPW (1999) Molecular diversity of arbuscular mycorrhizal fungi colonising *Hyacinthoides non-scripta* (Bluebell) in a semi-natural woodland. *Mol Ecol* **8**: 659–666.
45. Turnau K, Ryszka P, Gianinazzi-Pearson V, Van Tuinen D (2001) Identification of arbuscular mycorrhizal fungi in soils and roots of plants colonizing zinc wastes in southern Poland. *Mycorrhiza* **10**: 169–174.
46. Jaquot E, Van Tuinen D, Gianinazzi S, Gianinazzi-Pearson V (2000) Monitoring species of arbuscular mycorrhizal fungi in plants and in soil by nested PCR: application to the study of the impact of sewage sludge. *Plant and Soil* **226**: 179–188.

47. Daniell TJ, Husband R, Fitter AH, Young JPW (2001) Molecular diversity of arbuscular mycorrhizal fungi colonizing arable crops. *FEMS Microb Ecol* **36**: 203–209.
48. Clapp JP, Fitter AH, Young JPW (1999) Ribosomal small subunit sequence variation within spores of an arbuscular mycorrhizal fungus, *Scutellospora* sp. *Mol Ecol* **8**: 915–921.
49. Redecker D (2000) Specific PCR primers to identify arbuscular mycorrhizal fungi within colonized roots. *Mycorrhiza* **10**: 73–80.
50. Horton TR, Bruns TD (2001) The molecular revolution in ectomycorrhizal ecology: peeking into the black-box. *Mol Ecol* **10**: 1855–1871.
51. Gardes M, White TJ, Fortin JA, Bruns TD, Taylor JW (1991) Identification of indigenous and introduced symbiotic fungi in ectomycorrhizae by amplification of nuclear and mitochondrial ribosomal DNA. *Can J Bot* **69**: 180–190.
52. Buscot F, Munch JC, Charcosset JY, Gardes M, Nehls U, Hampp R (2000) Recent advances in exploring physiology and biodiversity of ectomycorrhizas highlight the functioning of these symbioses in ecosystems. *FEMS Microbiol Rev* **24**: 601–614.
53. Baar J, Bastiaans T, van de Coevering MA, Roelofs JGM (2002) Ectomycorrhizal root development in wet Alder carr forests in response to desiccation and eutrophication. *Mycorrhiza* **12**: 147–151.
54. Smit E, Veenman C, Baar J (2003) Molecular analysis of ectomycorrhizal basidiomycete communities in a *Pinus sylvestris* L stand reveals long-term increased diversity after removal of litter and humus layers. *FEMS Microb Ecol* **45**: 49–57.
55. Landeweert R, Veenman C, Kuyper TW, Fritze H, Wernars K, Smit E (2003) Quantification of ectomycorrhizal mycelium in soil by real-time PCR compared to conventional quantification techniques. *FEMS Microb Ecol* **45**: 283–292.
56. Simon L, Lalonde M, Bruns TD (1992) Specific amplification of 18S fungal ribosomal genes from vesicular-arbuscular endomycorrhizal fungi colonising roots. *Appl Environ Microbiol* **58**: 291–295.
57. Larana I, Elwood HJ, Gonzales V, Julian MC, Rubio V (1999) Design of a primer for ribosomal DNA internal transcribed spacer with enhanced specificity for Ascomycetes. *J Bio Technology* **75**: 187–194.
58. Möhlenhoff P, Muller L, Gorbushina AA, Petersen K (2001) Molecular approach to the characterisation of fungal communities: methods for DNA extraction, PCR amplification and DGGE analysis of painted art objects. *FEMS Microbiol Lett* **195**: 169–173.

Protocol 13.1: DGGE analysis of ectomycorrhizal communities in soil

This procedure has been successfully applied to study ectomycorrhizal community structure in forest soil rich in organic matter (22,53,55). A sufficient number of samples should be taken to cover the diversity and heterogeneity of ectomycorrhiza in forest soil. Root tips should be separated from the soil containing the mycorrhizal hyphae.

MATERIALS

Reagents

Fast DNA Spin kit for Soil (Bio101)

Wizard DNA Clean up System (Promega)

Primers: ITS1F (5′-CTTGGTCATTTAGAGGAAGTAA-3′), and GC-clamped ITS4B-GC (5′-CGCCCGCCGCG CCCCGCGCCCGGCCCGCCGCCCCCGCCCCAGG AGACTTGTACACGGTCCAG-3′)

Expand Long template polymerase kit containing enzyme and buffers (Roche, Manheim, Germany)

dNTP stock solution (2.5 mM each)

Sterile ultra-pure water

PCR tubes

Reagents for running DGGE gels (see Chapter 3)

Sybr Gold stain (Molecular Probes) or EtBr

Equipment

Ribolyzer (Hybaid)

PCR machine

Agarose gel electrophoresis tank and power pack

DGGE electrophoresis equipment and power pack

UV transilluminator (for EtBr staining)

Blue-light illuminator (Clare Chemical) in combination with a red filter (for Sybr Gold staining)

Gel documentation system

METHODS

DNA extraction from soil

DNA can be extracted and purified from soil using the DNA Spin kit for soil (Bio 101, Vista, CA, USA). Typically 0.2–0.5 g of freshly collected soil is used, and the protocol provided by the manufacturer is followed. To facilitate cell lysis, tubes are shaken in a ribolyzer (Hybaid or comparable) for 5–30 s. After the purification procedure, the extract is further purified using the Wizard DNA Clean-up system (Promega Corporation, Madison, WI, USA). The DNA is eluted from the column using 50 μl of ultra-pure water (if necessary this purification step can be repeated).

PCR amplification of ITS regions

To detect most ectomycorrhizal species in forest soil, primers have been used that specifically amplify Basidiomycetes (4). Dilution of the DNA extract by 10–50-fold prior to amplification may be necessary to alleviate polymerase inhibition caused by co-extracted contaminants. To amplify Basidiomycetes for DGGE analysis, the following primers can be used: ITS1F and ITS4B-GC.

1. Prepare the PCR master mix (depending on the number of samples) to achieve a final reaction volume of 50 μl upon addition of DNA.

Expand long template PCR buffer 2 (10 ×)	5 μl
dNTP stock solution (2.5 mM)	4 μl
ITS1F primer (10 μM)	1 μl
ITS4B-GC primer (10 μM)	1 μl
PCR grade ultra-pure water	37.5 μl
Expand long template polymerase (2.5 IU)	0.5 μl

2. Add 49 μl of PCR mix in PCR tubes

3. Add 1 μl of DNA extract (diluted if necessary)
4. Run the PCR using a programme of: 94°C 3 min; 40 × (94°C for 1 min, 48°C for 1 min, 72°C for 1 min); 72°C for 10 min; 4°C.
5. Check for amplification of PCR products using agarose gel electrophoresis.

DGGE analysis

Prepare a DGGE gel with a gradient of 20–60% denaturant. Run the gel for 17 h at 80 V. Stain the gel for 30 min in SybrGold (diluted 1:10 000). Stained gels can be viewed on a blue-light illuminator (Clare Chemical) in combination with a red filter.

Environmental assessment: bioreporter systems

14

Steven A. Ripp and Gary S. Sayler

14.1 Introduction

Bioreporters refer to intact living cells that have been genetically engineered to produce a measureable signal transcriptionally induced in response to a specific chemical or physical agent in their environment. Bioreporters contain three essential genetic elements: a promoter sequence, a regulatory gene, and a reporter gene. In the wild-type cell, the promoter gene is transcribed upon exposure to an inducing agent, leading to subsequent transcription of downstream genes that encode for proteins that aid the cell in either adapting to or combating the agent to which it has been exposed. In the bioreporter, the downstream genes, or portions thereof, have been removed and replaced with a reporter gene. Consequently transcription of the promoter gene activates the reporter gene, reporter proteins are produced, and some type of measurable signal is generated. These signals can be categorised as either colorimetric, fluorescent, luminescent, chemiluminescent, electrochemical or amperometric. Although each functions differently, their end product always remains the same: a measurable signal that is, ideally, proportional to the concentration of the specific chemical or physical agent to which they have been exposed.

Bioreporters can also be constructed without such inherent specificity. These bioreporters rely on reporter genes that are induced by a group of substances rather than just one or a few. Their primary use is for the detection of toxic substances, which, upon exposure to the bioreporter, induce a stress response gene that is fused to a reporter gene. Thus, an increase in signal intensity indicates toxicity, but the substance that initiated the signal cannot be uniquely identified. Reporter systems can also be designed to operate in reverse, where a decrease in signal intensity indicates toxicity. These bioreporters contain a constitutively expressed reporter gene that always remains on. Upon toxin exposure, the bioreporters either die or their metabolic activities are severely reduced, thereby causing a reduction in signal strength.

Although all of the data generated by a bioreporter can be obtained much more accurately using conventional analytical techniques such as gas chromatography or mass spectrometry, bioreporters offer a distinct advantage in that they report not only on the presence of any given chemical but also on its bioavailability and the overall effect of the chemical on a living system. Bioreporters are also significantly cheaper, faster and easier to use than analytical methods. Additionally, for some select bioreporter systems, the bioassay can be performed continuously, on-line and in real-time. There are some excellent reviews on bioreporters (1–6).

14.2 Reporter systems

14.2.1 β-Galactosidase (*lacZ*)

The *lacZ* gene derived from *Escherichia coli* encodes a β-galactosidase (β-gal) that catalyzes the hydrolysis of β-galactosides. Traditional *lacZ* bioreporters are assayed colorimetrically using the method of Miller (7). The substrate of *o*-nitrophenyl-β-galactoside (ONPG) is added to permeabilized bioreporter cells after inducer exposure to generate a yellow by-product whose intensity correlates with β-gal activity to provide an estimate of target chemical concentration. The assay is simple and highly reliable, and has become integral to commercially available genotoxicity test kits such as the SOS Chromotest (8). Due to low sensitivities and narrow dynamic ranges, however, the colorimetric test is largely being replaced by other detection methods. By simply using different β-galactoside substrates, fluorescent (9), luminescent (10) or chemiluminescent (11) assays are possible. A major disadvantage remains, however, in that the reporter cells must be lysed or undergo a membrane disruption step in order to quantify β-gal activity. Thus, data are obtained only incrementally and results are delayed, sometimes by several hours, in relation to the time required to complete the assay. Newer electrochemical (12) and amperometric (13) assays are beginning to solve this problem by measuring β-gal activity either directly or indirectly in an online, near real-time format. However, the endogenous presence of β-gal in natural environments and its potentially high background activity must always be taken into account when performing any of these assays. *Table 14.1* provides a comprehensive list of available *lacZ* bioreporters for environmental monitoring.

14.2.2 Chloramphenicol acetyltransferase (CAT)

CAT from *E. coli* catalyzes the transfer of the acetyl group from acetyl-coenzyme A to the antibiotic chloramphenicol. Fluorescent substrates are available for measuring CAT activity (14). Most CAT bioreporter systems are mammalian-based since no endogenous CAT activity is present in eukaryotic cells. For example, sensor systems for the detection of toxicological stressors (15) and environmental chemicals with estrogenic activity (16) are available.

14.2.3 Catechol 2,3-dioxygenase (*xylE*)

The *xylE* gene that encodes a catechol 2,3-dioyxgenase is part of the pWWO plasmid of *Pseudomonas putida* and is involved in the degradation of aromatic compounds (17). Catechol 2,3-dioxygenase catalyzes the cleavage of colorless catechol to produce the yellow compound 2-hydroxymuconic

Table 14.1 *lacZ*-based bioreporters

Analyte	Reporter	Time for induction	Concentration	Reference
Acylhomoserine lactones (AHL)	*traG, luxI, lasB,*	24–48 h	Unknown	61
	traI	not specified	≥3 nM	62
Anaerobicity	ANR	<1 h	2×10^2 Pa	63
Antibacterial agents	*cspA, rpoH, ipb*	1–3 h	1 µg Polymyxin B	64
Antimonite, arsenic	*arsR*	30 min	10^{-15} M	65
Biphenyls	*bphA*	3 h	1 mM	66
Cadmium	*zntR*	<1 h	25 nM	67
Chlorocatechol	*clcR*	5 min	10^{-8} M	68
Chromate	*chr*	8 h	1 µM	69
Copper	*pcoE*	1 h	0.01 mM	70
	CUP1	25 min	0.5–2 mM	13
DNA damaging agents (genotoxicity tests)	*E.coli sulA*	1.5 h	Various	71
	Salmonella sulA	2 h	0.025 µg ml^{-1}	72
	dinA, B, D	<30 min	1 µg ml^{-1}	73
	umuC	3 h	0.05 µg ml^{-1}	74
	recA	<10 min	1 µg ml^{-1}	75
	sfiA	2 h	<1 ng ml^{-1}	8,76
Endocrine disruptors	RSP5, SPT3	4–24 h	10^{-12} to 10^{-4} M	77
Growth regulation	*osmY*	1 h	Not applicable	12
Mercury	*mer*	4 h	0.2 ng ml^{-1}	68,78
			<0.5 µM	79
Nickel	*cnr*	8 h	128 µM	69
Oxidative stress	*katB*	1 h	50 mM H_2O_2	80
Pesticide toxicity	*HSP104*	1.5 h	0.1 mg l^{-1}	81
Phenols	*dmpR*	4 h	0.5 mM	82
Sucrose	*scrY*	2.5 h	10^{-5} M	83
Tetracyclines	*tetR*	3 h	0.01 µg ml^{-1}	84
	tetA	1.5 h	0.05 ng ml^{-1}	85
Zinc	*smtA*	2 h	≤12 µM	86

semialdehyde, forming the basis of this reporter system (18). *xylE* reporter systems have been applied primarily for studying gene regulation, and their utility as environmental reporters is rather limited to the tagging of microorganisms destined for environmental release (19,20). *xylE* serves well in this regard since its endogenous activity in environmental systems is extremely low, as compared to *lacZ* (20).

14.2.4 β-Lactamase (*bla*)

β-Lactamase cleaves β-lactam rings in certain antibiotics. Synthetic substrates have been developed that can also be cleaved by β-lactamase to form colorimetric or fluorescent products (21,22). As with catechol 2,3-dioxygenase, β-lactamase is routinely used for gene regulation studies but rarely as a reporter for environmental assessment. β-Lactamase-derived reporters for mercury (23), arsenic (24) and cadmium (25) are available but have been tested only in a gene regulatory sense. Therefore their operational capacity under environmental conditions is unknown.

14.2.5 β-Glucorinidase (*gusA, gurA, uidA*)

β-Glucorinidase is routinely used in plant cell systems (26) but its application in whole cell biosensing is scarce, usually as a visual tag for tracking environmentally released microorganisms (27) or assessing bacterial/plant interactions (28).

14.2.6 Ice nucleation (*inaZ*)

The *inaZ* gene from *Pseudomonas syringae* encodes for ice nucleaction activity (29). When *inaZ* is expressed the InaZ protein is incorporated into the outer membrane and catalyzes ice formation at temperatures between –2 and –10°C. Ice nucleation activity can be quantitatively related to intracellular InaZ concentrations using a droplet-freezing assay performed within as little as 5 min (30). Miller *et al.* (31) developed an *inaZ*-based bioreporter for determining sucrose bioavailability in bean phyllospheres and demonstrated 35-fold greater sensitivity than *lacZ*- and *gfp*-based sucrose bioreporters at low (14.5 μM) concentrations. Although rapid and sensitive, few *inaZ*-based bioreporters have been constructed. One is available for the determination of iron bioavailability in various plant rhizospheres and phyllospheres (32–34) and another for tryptophan bioavailability in a grass root rhizosphere (35). *inaZ* has also been applied as a visual reporter for assessing bacterial growth dynamics on soybean leaves (36). Expression of the *inaZ* gene in anaerobic (37) and halophilic bacteria (38), however, indicates potential application of this reporter in a wide variety of environmental systems.

14.2.7 Green fluorescent protein (GFP)

Green fluorescent protein (GFP) is a photoprotein isolated and cloned from the jellyfish *Aequorea victoria* (39). Variants have also been isolated from the sea pansy *Renilla reniformis.* GFP converts blue light to produce a green fluorescent signal without the required addition of an exogenous substrate. All that is required is a blue or UV light source to activate the fluorescent properties of the photoprotein. This ability to autofluoresce makes GFP highly desirable in biosensor assays since it can be used online and in real-time to monitor intact living cells. Additionally the ability to alter GFP by amino acid substitutions to produce alternative light emissions (i.e. cyan, red and yellow) allows it to be used as a multianalyte detector. Consequently GFP has been incorporated into bioreporters for the detection of various analytes (*Table 14.2*). However, it has been used more often as a visual tag within bacterial, yeast, nematode, plant and mammalian hosts for monitoring purposes (*Table 14.3*). More detailed information on GFP can be found in the reviews by Errampalli *et al.* (40) and Jansson (6).

Table 14.2 GFP-based reporters

Analyte	Reporter	Time for induction	Concentration	Reference
Acylhomoserine lactones	*luxI*	2 h	1 nM	87
L-Arabinose	*araC*	3.5 h	5×10^{-7} M	88
Arsenic	*arsR*	6 h	1 ppb	89
Biocides	*TEF*	25 min	100 μg ml^{-1}	90
DNA damaging agents (genotoxicity tests)	*katG*	6 h	4–78 ppb	91
	lacZ	2 h	0.68 μM	92
	recA	5 h	12 nM	93
	rad54, rnr2	4 h	10^{-3}% methyl methane sulfonate	94 95
	umuC	12 h	1 μg ml^{-1}	96
	umuC	24 h	325 μg ml^{-1} methyl methane sulfonate	
Fructose, sucrose	*fruB*	1 h	0.15 pg	97
Iron	*pvd*	Unknown	10^{-4} M	98
Lactose	*bfp2*	3 h	10^{-6} M	99
Mercury	*Mer*	16 h	<50 ng ml^{-1}	72
Octane	*alkB*	1-2.5 h	0.01–0.1 μM	100
Tetracyclines	*tetR*	50 min	<10 ng ml^{-1}	84
UV radiation exposure	CMV	48 h	10 J m^{-2} UVC	101

Table 14.3 Representative examples of green fluorescent protein used as a visual tag

Application	Environmental matrix	Reference
Bacterial adhesion	Wastewater flocs	102
Fungal colonization	Leaf surfaces	103
Horizontal gene transfer (HGT)	Bush bean phyllosphere	104
HGT (TOL plasmid)	Agar surface	105
Monitoring *Alcaligenes faecalis*	Phenol-contaminated soil	106
Monitoring *Psuedomonas fluorescens*	Soil, plant roots	107
	Tomato seedling roots	108
Monitoring *Pseudomonas putida*	Activated sludge	109
Monitoring *Pseudomonas* sp.	PAH-contaminated soil[†]	110
Plasmid transfer	Alfalfa sprouts	111
Protozoan grazing of *Moraxella* sp.	Liquid culture	112
Screening metagenomic libraries	Groundwater metagenomes	113
Survival of *Moraxella* sp.	*p*-Nitrophenol-contaminated soil	114
TOL plasmid expression	Biofilm	115
Transport of *Pseudomonas putida*	Groundwater	116
Viable but non culturable *Salmonella typhi*	Groundwater	117
Visualizing *Rhizobium meliloti*	Plant-root rhizosphere	118
Visualizing *Xanthomonas campestris*	Cabbage plants	119

[†] PAH-polyaromatic hydrocarbon

14.2.8 Aequorin

Aequorin is a photoprotein also isolated from the bioluminescent jellyfish *Aequora victoria*. Upon addition of calcium ions and coelenterazine, a reaction occurs whose end product is the generation of blue light in the 460–470 nm range. In *A. victoria* this blue light is converted by GFP to give the characteristic green fluorescence. Aequorin-based whole-cell reporters have generally only been developed for the detection of intracellular calcium and calcium-related stressors (41) and for investigating calcium-mediated responses to alkaline stress in *Saccharomyces cerevisiae* (42). Aequorin has proven more useful when engineered into B-cell lines to produce detectors for pathogenic bacteria and viruses in what is referred to as the CANARY assay (cellular analysis and notification of antigen risks and yields) (43). The B-cells are genetically engineered to produce aequorin. Upon exposure to antigens of different pathogens, the recombinant B-cells emit light as a result of activation of an intracellular signaling cascade that releases calcium ions inside the cell. This technology has also been incorporated into reporters for biological warfare reagents (44).

14.2.9 Uroporphyrinogen (Urogen) III methyltransferase (UMT)

UMT catalyzes a reaction that yields two fluorescent products which produce a red-orange fluorescence in the 590–770 nm range when illuminated with UV light (45), and as with GFP no addition of exogenous substrate is required. UMT has been used for whole-cell sensing of antimonite, arsenite and arsenate (46), but is more often used as a marker for gene transcription in bacterial, yeast and mammalian cells (47).

14.2.10 Luciferases

Luciferase is a generic name for an enzyme that catalyzes a light-emitting reaction. Luciferases can be found in bacteria, algae, fungi, jellyfish, insects, shrimp and squid, and the resulting light that these organisms produce is termed bioluminescence. The use of luciferases as bioreporters has been reviewed by Greer and Szalay (48).

14.2.10.1 Insect luciferase (*luc*)

Firefly luciferase catalyzes a reaction that produces visible light in the 550–575 nm range. A click beetle luciferase is also available that produces light at a peak closer to 595 nm. Both luciferases require the addition of an exogenous substrate (luciferin) for the light reaction to occur. Several *luc*-based bioreporters have been constructed for the detection of inorganic and organic compounds of environmental concern (*Table 14.4*).

14.2.10.2 Bacterial luciferase (*lux*)

In bacteria the genes responsible for the light-emitting reaction (the *lux* genes) have been isolated and used extensively in the construction of bioreporters that emit a blue-green light with a maximum intensity at 490 nm

Table 14.4 *luc*-based biosensors

Analyte	Reporter	Time for induction	Concentration	Reference
Arsenite	*ars*	2 h	10 nM	47
Arsenite, antimonite, cadmium	*ars*	2 h	33 nM (antimonite)	120
Benzene, toluene, xylene	*xylR*	30 min	3 μM (xylene)	121
Cadmium, lead, antimony	*cadA*	2–3 h	1 nM (antimony)	122
Environmental estrogens	ERE	10–12 h	10^{-7} M (DDT)	123
Mercury	*mer*	2 h	100 nM	47
Organomercurials	*mer*	2 h	0.2 nM (methylmercury chloride)	124
Solvents	*sep*	Not stated	1 mM (various solvents)	125

(49). Three variants of *lux* are available, one that functions at <30°C, another at <37°C and a third at <45°C. The *lux* genetic system consists of five genes *luxA-E*, and depending on the combination of the genes used, several different types of bioluminescent reporters can be constructed. The application of *lux* bioreporters in the environment has been reviewed by Nunes-Halldorson and Duran (50).

14.2.10.3 *luxAB*

luxAB bioreporters contain only the *luxA* and *luxB* genes, which together are responsible for generating the light signal. However, to fully complete the light-emitting reaction, a substrate must be supplied to the cell. Typically this occurs through the addition of the chemical decanal at some point during the bioassay. Numerous *luxAB* bioreporters have been constructed within bacterial, yeast, insect, nematode, plant and mammalian cell systems. *Table 14.5* lists some of the various applications of *luxAB*-based bioreporters.

14.2.10.4 *luxCDABE*

Instead of containing only the *luxA* and *luxB* genes, bioreporters can contain all five genes of the *lux* cassette, thereby allowing for a completely independent light-generating system that requires no substrate addition, nor any excitation by an external light source. In these bioassays the bioreporter is simply exposed to a target analyte and a quantitative increase in bioluminescence results, often in <1 h. Due to the rapidity and ease of use, along with the ability to perform the bioassay repetitively in real-time and online, makes *luxCDABE* bioreporters extremely attractive for environmental monitoring. Additionally, the development of microluminometers for detecting the bioluminescent signal reduces this assay down

Table 14.5 Environmental applications of *luxAB*-based bioreporters

Application	Reference
Antibiotic effectiveness	126
Bacterial biofilms	127
Bacterial biomass	128
Bacterial motility	129
Bacterial stress response	130
Bacterial transport	131
Bioremediation process monitoring	132
Environmental contaminants	133–136
Gene transfer	137
Mercury	138
Mutagenicity tests	139
Naphthalene	140
Quorum sensing	141
Toxicity assays	142
Tributyltin	143,144
Visual tagging	117,145,146

to a miniaturized format (51). *Table 14.6* illustrates the widespread application of *luxCDABE*-based bioreporters.

14.2.10.5 *luxCDABE*-based non-specific reporters

Non-specific *lux* bioreporters are typically used for the detection of chemical toxins. They are usually designed to continuously bioluminesce. Upon exposure to a chemical toxin, either the cell dies or its metabolic activity is retarded, leading to a decrease in bioluminescence light levels. Their most familiar application is in the Microtox® assay where, following a short exposure to several concentrations of the query sample, decreased bioluminescence can be correlated to relative levels of toxicity (52). The Vitotox test operates similarly (53).

Table 14.6 *luxCDABE*-based bioreporters

Analyte	Reporter	Time for induction	Concentration	Reference
2,3-Dichlorophenol	*recA* (stress promoter)	2 h	50 mg l^{-1}	147
2,4,6-Trichlorophenol	*recA* (stress promoter)	2 h	10 mg l^{-1}	147
2,4-D	*tfdRP*	20–60 min	2 μM–5 mM	148
3-Xylene	*xyl*	Hours	3 μM	149
4-Chlorobenzoate	*fcbA*	1 h	380 μM–6.5 mM	150
4-Nitrophenol	*recA* (stress promoter)	2 h	0.25 mg l^{-1}	147
Aflatoxin B1	Various stress promoters	45 min	1.2 ppm	151
Alginate production	*algD*	1 h	50–150 mM NaCl	152
Ammonia	*hao*	30 min	20 μM	153

Table 14.6 *continued*

BTEX (benzene, toluene, ethylbenzene, xylene)	*tod*	1–4 h	0.03–50 mg l^{-1}	154
Cadmium	*cupS*	4 h	19 mg kg^{-1}	155
Chlorodibromomethane	*recA* (stress promoter)	2 h	20 mg l^{-1}	147
Chloroform	*recA* (stress promoter)	2 h	300 mg l^{-1}	147
Chromate	*chrA*	1 h	10 μM	69
Cobalt	*cnr*	4–6 h	9 μM	156
Copper	Not specified	1 h	1 μM–1 mM	157
DNA damage (cumene hydroperoxide)	Various stress promoters	50 min	6.25 mg l^{-1}	158
DNA damage (mitomycin)	*recA, uvrA, alkA*	1 h	0.032 mg l^{-1}	159
Gamma irradiation	*recA, grpE, katG*	1.5 h	1.5–200 Gy	160
Heat shock	*grpE*	20 min	Various depending on chemical inducer used	161,162
Hydrogen peroxide	*katG*	20 min	0.1 mg l^{-1}	163
Iron	*pupA*	Hours	10 nM–1 μM	164
	fepA		15 nM	165
Isopropyl benzene	*ipb*	1-4 h	1-100 μM	166
Lead	*pbr*	4 h	4.036 g kg^{-1}	155
Mercury	*mer*	70 min	0.025 nM	167
N-Acyl homoserine lactones	*luxI, lasI, rhiI*	4 h	Not specified	168
Naphthalene	*nahG*	8–24 min	12–120 μM	169
Nickel	*cnr*	4-6 h	0.1 μM	156
Nitrate	*narG*	4 h	0.05–50 μM	170
Organic pesticides	*katG*	20 min	Not specified	163
Oxidative stress	*pqi-5, soda, katG*	Not specified	Not specified	171
PCBs	*bph*	1–3 h	0.8 μM	172
p-Chlorobenzoic acid	*fcbA*	40 min	0.06 g l^{-1}	150
p-Cymene	*cym*	<30 min	60 ppb	173
Pentachlorophenol	*recA* (stress promoter)	2 h	0.008 mg l^{-1}	147
Phenol	*recA* (stress promoter)	2 h	16 mg l^{-1}	147
	Not specified			174
Salicylate	*nahG*	15 min	36 μM	169
Silver	*zntAP*	1 h	0.1 μM	175
Tetracycline	*tet*	50 min	<10 ng ml^{-1}	84
Trichloroethylene	*tod*	1–1.5 h	5–80 μM	176
Trinitrotoluene	Not specified	Not specified	Not specified	177
Ultrasound	Various stress promoters	1 h	500 W cm^{-2}	178
UV light	*recA*	1 h	2.5–20 J m^{-2}	179
Zinc	*smtA*	4 h	0.5–4 μM	180

14.3 Mini-transposons as genetic tools in biosensor construction

A transposon is a discrete mobile genetic element capable of translocating from a donor site within the DNA molecule into one of many nonhomologous target sites, mediated by the action of a transposase enzyme (54). The use of transposons as reporter elements was first applied in gene regulation studies using a phage Mu transposable element containing a promoterless *lac* gene (55). This was followed by similar constructs primarily using the Tn*5* (*Figure 14.1A*) and Tn*10* family of transposons as well as a variety of others such as Tn*3*/Tn*1*, γ^{δ} (Tn*10*) and the conjugative transposon Tn*916*/Tn*917*. Although powerful mutagenic tools, natural transposons had several disadvantages, especially in environmental applications; they required an antibiotic-resistant marker for selection, and the presence of inverted repeats at their termini promoted unwanted genetic rearrangements and inherent instability (secondary transposition) (56). Transposons were also in some cases large and difficult to work with genetically and were subject to transposition immunity which prevents multiple transposon insertion into the same strain, severely limiting their cloning value (57). The development of mini-transposons solved many of these problems (58). Mini-transposons are shortened hybrids of natural transposons, usually Tn*5* and Tn*10*, in which the transposase gene is placed outside the boundaries of the inverted repeats. In this formation, the mobile element undergoes insertion into the target site but the transposase gene does not, thus preventing any further rearrangements. Mini-transposons are also not

Table 14.7 Representative examples of mini-transposons/reporter gene constructs

Designation	Application	Reference
Mini-Tn*5 lacZ1*	*lacZl* fusions	181
Mini-Tn*5 lacZ2*	*lacZl* fusions	181
Mini-Tn*5 luxAB1*	*luxAB* fusions	181
Mini-Tn*5 luxCDABE*	*Photorhabdus luminescens lux*	182
Mini-Tn*5 phoA*	*phoA* fusions	181
Mini-Tn*5 xylE*	*xylE* fusions	181
Mini-Tn*5 gfp-km*	Promoterless fusion to *gfp* with antibiotic resistance	183
mTn*10phoA*	*phoA* fusions	184
mTn*5gfp-pgusA*	Bifunctional *gfp/gusA* reporter	185
mTn*5gusA21*	β-Glucorinidase reporter	186
mtn*Yl1*	GFP expression in yeast	187
pAG408	GFP-based reporter	188
pUT mini-Tn*5 luxCDABE*	Ecotoxicity reporter using *V. fischi lux*	189
pUTK214	*Tod-V. fischi lux* bioreporter for BTEX	154
Tn*10d-bla*	*bla fusions*	190
Tn*Max10*	*xylE* fusions	191
Tn*Max11*	*lacZ* fusions	191
Tn*Max6*	*phoA* fusions	191
Tn*Max7*	*blaM fusions*	191
Tn*MaxErCm*	*cat* transcriptional fusions	192

affected by transposition immunity, thereby allowing for multiple insertions of foreign inserts into the same strain, provided that each insert has its own selectable marker. Additionally, mini-transposons typically maintain an origin of replication that allows for delivery into a broad range of hosts, e.g. pUT (*Figure 14.1B*). Various mini-transposons customized with reporter genes have been developed for simplified construction of bioreporter organisms. Representative examples are listed in *Table 14.7*. By inserting a genetic promoter element into a unique cloning site within the

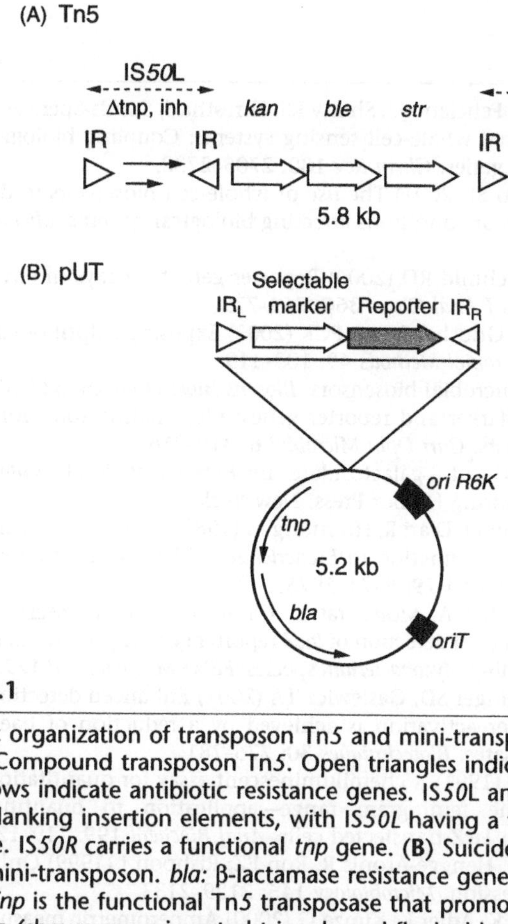

Figure 14.1

Schematic organization of transposon Tn*5* and mini-transposons-suicide vector pUT. **(A)** Compound transposon Tn*5*. Open triangles indicate inverted repeats. Open arrows indicate antibiotic resistance genes. IS*50*L and IS*50R* are non-identical flanking insertion elements, with IS*50L* having a truncated transposase (*tnp*) gene. IS*50R* carries a functional *tnp* gene. **(B)** Suicide vector pUT carrying the Tn*5* mini-transposon. *bla:* β-lactamase resistance gene for selection during cloning. *Tnp* is the functional Tn*5* transposase that promotes transposition of the transposon (shown as an inverted triangle and flanked by inverted repeats). The transposon also contains unique restriction sites into which selectable marker genes (open arrow) and the bioreporter genes (hatched arrow) can be inserted. pUT has the narrow host range origin of replication from R6K, and also an *oriT* to enable mobilization of the plasmid by broad host range plasmids, e.g. RK2. Following transformation/electroporation/conjugation of pUT constructs into the recipient cell, selection is made for the selectable marker within the transposon, but without selection for β-lactamase activity. Transposition of the transposon is mediated by the vector-borne *tnp* gene with the transposon carrying the bioreporter transposing into the chromosome. However, the pUT vector is unable to replicate in the recipient as it lacks host factors essential for replication from the *ori* R6K.

mini-transposon vector, one can theoretically engineer any of the bioreporter classes discussed above. Furthermore, the ability to stably insert the mini-transposon into the host chromosome makes these systems ideal for environmental applications, since the necessity for antibiotic selection, that would be necessary for a plasmid-borne reporter, can be reduced. Newer mini-transposons based on heavy metal resistance determinants make antibiotic selection obsolete (59). Methods for constructing and using mini-transposons are expertly described by de Lorenzo and Timmis (60) and Berg *et al.* (56).

References

1. Daunert S, Barrett G, Feliciano JS, Shetty RS, Shrestha S, Smith-Spencer W (2000) Genetically engineered whale-cell sensing systems: Coupling biological recognition with reporter genes. *Chem Rev* **100**: 2705–2738.
2. Hansen LH, Sorensen SJ (2001) The use of whole-cell biosensors to detect and quantify compounds or conditions affecting biological systems. *Microb Ecol* **42**: 483–494.
3. Kohler S, Belkin S, Schmid RD (2000) Reporter gene bioassays in environmental analysis. *Fresenius J Anal Chem* **366**: 769–779.
4. Keane A, Phoenix P, Ghoshal S, Lau PCK (2002) Exposing culprit organic pollutants: a review. *J Microbiol Methods* **49**: 103–119.
5. D'Souza SF (2001) Microbial biosensors. *Biosens Bioelectron* **16**: 337–353.
6. Jansson JK (2003) Marker and reporter genes: illuminating tools for environmental microbiologists. *Curr Opin Microbiol* **6**: 310–316.
7. Miller JH (1972) Assay of β-galactosidase. In: *Experiments in Molecular Genetics* pp. 352–355. Cold Spring Harbor Press, New York.
8. Quillardet P, Huisman O, D'ari R, Hoffnung M (1982) Chromotest, a direct assay of induction of an SOS function in *Escherichai coli* K12 to measure genotoxicity. *Proc Natl Acad Sci USA* **79**: 5971–5975.
9. Rowland B, Purkayastha A, Monserrat C, Casart Y, Takiff H, McDonough KA (1999) Fluorescence-based detection of *lacZ* reporter gene expression in intact and viable bacteria including *Mycobacterium* species. *FEMS Microbiol Lett* **179**: 317–325.
10. Nazarenko DA, Dertinger SD, Gasiewicz TA (2001) Enhanced detection of beta-galactosidase reporter activation is achieved by a reduction of haemoglobin content in tissue lysates. *Biotechniques* **30**: 776–781.
11. Jain VK, Magrath IT (1991) A chemiluminescent assay for quantitation of beta-galactosidase in the femtogram range—application to quantification of beta-galactosidase in *lacZ*-transfected cells. *Anal Biochem* **199**: 119–124.
12. Biran I, Klimentiy L, Hengge-Aronis R, Ron EZ, Rishpon J (1999) Online monitoring of gene expression. *Microbiology* **145**: 2129–2133.
13. Lehmann M, Riedel K, Adler K, Kunze G (2000) Amperometric measurement of copper ions with a deputy substrate using a novel *Saccharomyces cerevisiae* sensor. *Biosens Bioelectron* **15**: 211–219.
14. Sode K, Hatano N, Tatara M (1996) Cloning of a marine cyanobacterial promoter for foreign gene expression using a promoter probe vector. *Appl Biochem Biotechnol* **59**: 349–360.
15. Todd MD, Lee MJ, Williams JL, Nalezny JM, Gee P, Benjamin MB, Farr SB (1995) The Cat-Tox (L) Assay—a sensitive and specific measure of stress induced transcription in transformed human liver cells. *Fundam Appl Toxicol* **28**: 118–128.
16. Tully DB, Cox VT, Mumtaz MM, Davis VL, Chapin RE (2000) Six high priority organochlorine pesticides, either singly or in combination, are nonesterogenic in transfected HeLa cells. *Reprod Toxicol* **14**: 95–102.

17. Franklin FCH, Bagdasarian M, Bagdasarian MM, Timmis KN (1981) Molecular and functional analysis of the TOL plasmid pWWO from *Pseudomonas putida* and cloning of genes for the entire regulated aromatic ring meta-cleavage pathway. *Proc Natl Acad Sci USA* **78**: 7458–7462.
18. Zukowski MM, Gaffney DF, Speck D, Kauffmann M, Findeli A, Wisecup A, Lecocq JP (1983) Chromogenic identification of genetic regulatory signals in *Bacillus subtilis* based on expression of a cloned *Pseudomonas* gene. *Proc Natl Acad Sci USA* **80**: 1101–1105.
19. Han HY, Kim CK, Park YK, Ka JO, Lee BJ, Min KH (1996) Construction of genetically engineered microorganisms for overexpression of *xylE* gene encoding catechol 2,3-dioxygenase and the functional stability of the recombinant plasmid pSW3a containing *xylE* in aquatic environments. *J Microbiol* **34**: 341–348.
20. De Leij FAA, Sutton EJ, Whipps JM, Fenlon JS, Lynch JM (1995) Impact of field release of genetically modified *Pseudomonas fluorescens* on indigenous microbial populations of wheat. *Appl Environ Microbiol* **61**: 3443–3453.
21. Zlorarnik G, Negulescu PA, Knapp TE, Mere L, Burres N, Feng LX, Whitney M, Roemer K, Tsien RY (1998) Quantitation of transcription and clonal selection of single living cells with beta-lactamase as reporter. *Science* **279**: 84–88.
22. Cohenford MA, Abraham J, Medeiros AA (1988) A colorimetric procedure for measuring beta lactamase activity. *Anal Biochem* **168**: 252–258.
23. Chu L, Mukhopadhyay D, Yu H, Kim KS, Misra TK (1992) Regulation of the *Staphylococcus aureus* plasmid pI258 mercury resistance operon. *J Bacteriol* **174**: 7044–7047.
24. Ji G, Silver S (1992) Regulation and expression of the arsenic resistance operon from *Staphylococcus aureus* plasmid pI258. *J Bacteriol* **174**: 3684–3694.
25. Yoon KP, Misra TK, Silver S (1991) Regulation of the *cadA* cadmium resistance determinant of *Staphylococcus aureus* plasmid pI258. *J Bacteriol* **173**: 7643–7649.
26. Jefferson RA (1989) The GUS reporter system. *Nature* **342**: 837–838.
27. Hirsch PR, Mendum TA, Pühler A, Selbitschka W (2000) The field release and monitoring of GUS-marked rhizobial strain CT0370. In: Jansson JK, van Elsas JD, Bailey M (eds) *Tracking Genetically Engineered Micro-organisms: Method Development from Microcosms to the Field*, pp. 145–151. RG Landes Co., Austin, Texas.
28. Doohan FM, Smith P, Parry DW, Nicholson P (1998) Transformation of *Fusarium culmorum* with the B-D-glucuronidase (GUS) reporter gene: a system for studying host pathogen relationships and disease control. *Physiol Mol Plant Pathol* **53**: 17–37.
29. Margaritis A, Bassi AS (1991) Principles and biotechnological applications of bacterial ice nucleation. *Crit Rev Biotechnol* **11**: 277–295.
30. Lindow SE (1990) Bacterial ice nucleation measurements In: Sands D, Klement Z, Rudolf K (eds), *Methods in Phytobacteriology*, pp. 428–434. Akademia Kiado, Budapest.
31. Miller WG, Brandl MT, Quiñones B, Lindow SE (2001) Biological sensor for sucrose availability: relative sensitivities of various reporter genes. *Appl Environ Microbiol* **67**: 1308–1317.
32. Loper JE, Henkels MD (1997) Availability of iron to *Pseudomonas fluorescens* in rhizosphere and bulk soil evaluated with an ice nucleation reporter gene. *Appl Environ Microbiol* **63**: 99–105.
33. Loper JE, Lindow SE (1994) A biological sensor for iron available to bacteria in their habitats on plant surfaces. *Appl Environ Microbiol* **60**: 1934–1941.
34. Loper JE, Henkels MD (1999) Utilization of heterologous siderophores enhances levels of iron available to *Pseudomonas putida* in the rhizosphere. *Appl Environ Microbiol* **65**: 5357–5363.
35. Jaeger CH III, Lindow SE, Miller W, Clark W, Firestone MK (1999) Mapping of

sugar and amino acid availability in soil around roots with bacterial sensors of sucrose and tryptophan. *Appl Environ Microbiol* **65**: 2685–2690.
36. Rukayadi Y, Suwanto A, Tjahjono B, Harling R (2000) Survival and epiphytic fitness of a nonpathogenic mutant of *Xanthomonas campestris* pv Glycines. *Appl Environ Microbiol* **66**: 1183–1189.
37. Drainas C, Vartholomatos G, Panopoulos NJ (1995) The ice nucleation gene from *Pseudomonas syringae* as a sensitive gene reporter for promoter analysis in *Zymomonas mobilis. Appl Environ Microbiol* **61**: 3821–3825.
38. Arvanitis N, Vargas C, Tegos G, Perysinakis A, Nieto JJ, Ventosa A, Drainas C (1995) Development of a gene reporter system in moderately halophilic bacteria employing the ice nucleation gene of *Pseudomonas syringae. Appl Environ Microbiol* **61**: 3821–3825.
39. Misteli T, Spector DL (1997) Applications of the green fluorescent protein in cell biology and biotechnology. *Nature Biotechnol* **15**: 961–964.
40. Errampalli D, Leung K, Cassidy MB, Kostrzynska M, Blears M, Lee H, Trevors JT (1999) Applications of the green fluorescent protein as a molecular marker in environmental microorganisms—review. *J Microbiol Meth* **35**: 187–199.
41. Batiza AF, Schulz T, Masson PH (1996) Yeast respond to hypotonic shock with a calcium pulse. *J Biol Chem* **271**: 23357–23362.
42. Viladevall, L, Serrano R, Ruiz A, Domenech G, Giraldo J, Barcelo A, Arino J (2004) Characterization of the calcium-mediated response to alkaline stress in *Saccharomyces cerevisiae. J Biol Chem* **279**: 43614–43624.
43. Rider TH, Petrovick MS, Nargi FE, Harper JD, Schwoebel ED, Mathews RH, Blanchard DJ, Bortolin LT, Young AM, Chen JZ, Hollis MA (2003) A B cell-based sensor for rapid identification of pathogens, *Science* **301**: 213–215.
44. Pescovitz D (2000) Bioagent chip: a sensor to detect a biological warfare attack in seconds. *Sci Amer* **282**: 35.
45. Sattler I, Roessner CA, Stolowich NJ, Hardin SH, Harris-Haller LW, Yokubaitis NT, Murooka Y, Hashimoto Y, Scott AI (1995) Cloning, sequencing and expression of the uroporphyrinogen III methyltransferase *cobA* gene of *Propionibacterium freudenreichii* (*shermanii*). *J Bacteriol* **177**: 1564–1569.
46. Feliciano J, Liu Y, Ramanathan S, Daunert S (2000) Fluorescence-based sensing system for antimonite and arsenite using *cobA* as the reporter gene. 219th ACS national meeting San Francisco, CA, (Poster No. 58).
47. Wildt S, Deuschle U (1999) *cobA*, a red fluorescent transcriptional reporter for *Escherichia coli*, yeast, and mammalian cells. *Nat Biotech* **17**: 1175–1178.
48. Greer LF, Szalay AA (2002) Imaging of light emission from the expression of luciferases in living cells and organisms: a review. *Luminescence* **17**: 43–74.
49. Meighen E (1994) Genetics of bacterial luminescence. *Annu Rev Genet* **28**: 117–139.
50. Nunes-Halldorson VD, Duran NL (2003) Bioluminescent bacteria: *lux* genes as environmental biosensors. *Braz J Microbiol* **34**: 91–96.
51. Simpson ML, Sayler GS, Patterson G, Nivens DE, Bolton EK, Rochelle JM, Arnott JC, Applegate BM, Ripp S, Guillorn MA (2001) An integrated CMOS microluminometer for low-level luminescence sensing in the bioluminescent bioreporter integrated circuit. *Sens Actuators B Chem* **72**: 134–140.
52. Hermens J, Busser F, Leeuwangh P, Musch A (1985) Quantitative structure-activity relationships and mixture toxicity of organic chemicals in *Photobacterium phosphoreum*: the Microtox test. *Ecotoxicol Environ Saf* **9**: 17–25.
53. Verschaeve L, Van Gompel J, Thilemans L, Regniers L, Vanparys P, van der Lelie D (1999) VITOTOX bacterial genotoxicity and toxicity test for the rapid screening of chemicals. *Environ Mol Mutagen* **33**: 240–248.
54. Dyson PJ (1999) Isolation and development of transposons. In: Smith MCM, Sockett RE (eds) *Methods in Microbiology: Genetic Methods for Diverse Prokaryotes*, Vol. 20, pp. 133–167. Academic Press, San Diego, CA.

55. Casadaban M, Cohen S (1979) Lactose genes fused to exogenous promoters in one step using a Mu-*lac* bacteriophage: in vivo probe for transcriptional control sequences. *Proc Natl Acad Sci USA* **76**: 4530–4533.
56. Berg CM, Berg DE, Groisman EA (1989) Transposable elements and the genetic engineering of bacteria. In: Berg DE, Howe MM (eds) *Mobile DNA*, pp. 879–925. American Society for Microbiology, Washington, DC.
57. Wallace LJ, Ward JM, Bennett PM, Robinson MK, Richmond MH (1981) Transposition immunity. *Cold Spring Harb Symp Quant Biol* **45**: 183–188.
58. de Lorenzo V, Herrero M, Sanchez JM, Timmis KN (1998) Mini-transposons in microbial ecology and environmental biotechnology. *FEMS Microbiol Ecol* **27**: 211–224.
59. Taghavi S, Delanghe H, Lodewyckx C, Mergeay M, van der Lelie D (2001) Nickel-resistance-based minitransposons: new tools for genetic manipulation of environmental bacteria. *Appl Environ Microbiol* **67**: 1015–1019.
60. de Lorenzo V, Timmis KN (1994) Analysis and construction of stable phenotypes in Gram-negative bacteria with Tn*5* and Tn*10*-derived minitransposons. *Methods Enzymol* **235**: 386–405.
61. Cha C, Gao P, Chen YC, Shaw PD, Farrand SK (1998) Production of acyl-homoserine lactone quorum-sensing signals by gram-negative plant-associated bacteria. *Mol Plant Microbe Interact* **11**: 1119–1129.
62. McLean RJC, Whiteley M, Stickler DJ, Fuqua WC (1997) Evidence of autoinducer activity in naturally-occurring biofilms. *FEMS Microbiol Lett* **154**: 259–263.
63. Hojberg O, Schnider U, Winteler HV, Sorensen J, Haas D (1999) Oxygen-sensing reporter strain of *Pseudomonas fluorescens* for monitoring the distribution of low-oxygen habitats in soil. *Appl Environ Microbiol* **65**: 4085–4093.
64. Bianchi AD, Baneyx F (1999) Stress responses as a tool to detect and characterize the mechanism of action of antibacterial agents. *Appl Environ Microbiol* **65**: 5023–5027.
65. Ramanathan S, Shi W, Rosen BP, Daunert S (1998) Bacteria-based chemiluminescence sensing system using β-galactosidase under the control of the ArsR regulatory protein of the *ars* operon. *Anal Chim Acta* **369**: 189–195.
66. Master ER, Mohn WW (2001) Induction of *bphA*, encoding biphenyl dioxygenase, in two polychlorinated biphenyl-degrading bacteria, psychrotolerant *Pseudomonas* strain Cam-1 and mesophilic *Burkholderia* strain LB400. *Appl Environ Microbiol* **67**: 2669–2676.
67. Biran I, Babai R, Levcov K, Rishpon J, Ron EZ (2000) Online and in situ monitoring of environmental pollutants: electrochemical biosensing of cadmium. *Environ Microbiol* **2**: 285–290.
68. Guan X, Ramanathan S, Garris JP, Shetty RS, Ensor M, Bachas LG, Daunert S (2000) Chlorocatechol detection based on a *clc* operon/reporter gene system. *Anal Chem* **72**: 2423–2427.
69. Peitzsch N, Eberz G, Nies DH (1998) *Alcaligenes eutrophus* as a bacterial chromate sensor. *Appl Environ Microbiol* **64**: 453–458.
70. Rouch DA, Parkhill J, Brown NL (1995) Induction of bacterial mercury responsive and copper responsive promoters—functional differences between inducible systems and implications for their use in gene fusions for *in vivo* metal biosensors. *J Ind Microbiol* **14**: 349–353.
71. McDaniels AE, Reyes AL, Wymer LJ, Rankin CC, Stelma GN Jr (1993) Genotoxic activity detected in soils from a hazardous waste site by the Ames test and an SOS colorimetric test. *Environ Mol Mutagen* **22**: 115–122.
72. El Mzibri M, De Meo MP, Laget M, Guiraud H, Seree E, Barra Y, Dumenil G (1996) The *Salmonella* sulA-test: a new *in vitro* system to detect genotoxins. *Mutat Res* **369**: 195–208.
73. Oh TJ, Lee CW, Kim IG (1999) The damage-inducible (*din*) genes of *Escherichia*

coli are induced by various genotoxins in a different way. *Microbiol Res* **154**: 179–183.
74. Aoki T, Nakamura S, Oda Y (1989) Improvement of the SOS/*umu* test for preliminary screening of mutagens. *Anal Lett* **22**: 2463–2469.
75. Nunoshiba T, Nishioka H (1991) 'Rec-lac test' for detecting SOS-inducing activity of environmental genotoxic substances. *Mutat Res* **254**: 71–77.
76. Fish F, Lampert I, Halachmi A, Riesenfeld G, Herzberg M (1987) The SOS Chromotest kit: a rapid method for the detection of genotoxicity. *Toxicity Assess* **2**: 135–147.
77. Gaido KW, Leonard LS, Lovell S, Gould J, Babai D, Portier C, McDonnell DP (1997) Evaluation of chemicals with endocrine modulating activity in a yeast-based steroid hormone receptor gene transcription assay. *Toxicol Appl Pharmacol* **143**: 205–212.
78. Hansen LH, Sørensen SJ (2000) Versatile biosensor vectors for detection and quantification of mercury. *FEMS Microbiol Lett* **193**: 123–127.
79. Schaefer JK, Letowski J, Barkay T (2002) *mer*-mediated resistance and volatilization of Hg(II) under anaerobic conditions. *Geomicrobiol J* **19**: 87–102.
80. Elkins JG, Hassett DJ, Stewart PS, Schweizer HP, McDermott TR (1999) Protective role of catalase in *Pseudomonas aeruginosa* biofilm resistance to hydrogen peroxide. *Appl Environ Microbiol* **65**: 4594–4600.
81. Fujita K, Iwahashi H, Kawai R, Komatsu Y (1998) Hsp104 expression and morphological changes associated with disinfectants in *Saccharomyces cerevisiae*: environmental bioassay using stress response. *Wat Sci Technology* **38**: 237–243.
82. Wise AA, Kuske CR (2000) Generation of novel bacterial regulatory proteins that detect priority pollutant phenols. *Appl Environ Microbiol* **66**: 163–169.
83. Miller WG, Brandl MT, Quinones B, Lindow SE (2001) Biological sensor for sucrose availability: relative sensitivities of various reporter genes. *Appl Environ Microbiol* **67**: 1308–1317.
84. Hansen LH, Sørensen SJ (2000) Detection and quantification of tetracyclines by whole-cell biosensors. *FEMS Microbiol Lett* **190**: 272–278.
85. Chopra I, Hacker K, Misulovin Z, Rothstein DM (1990) Sensitive biological detection method for tetracyclines using a *tetA-lacZ* fusion system. *Antimicrob Agents Chemother* **34**: 111–116.
86. Huckle JW, Morby AP, Turner JS, Robinson NJ (1993) Isolation of a prokaryotic metallothionein locus and analysis of transcriptional control by trace metal ions. *Mol Microbiol* **7**: 177–187.
87. Andersen JB, Heydorn A, Hentzer M, Eberl L, Geisenberger O, Christensen BB, Molin S, Givskov M (2001) *gfp*-based N-acyl homoserine-lactone sensor systems for detection of bacterial communication. *Appl Environ Microbiol* **67**: 575–585.
88. Shetty S, Ramanathan IH, Badr A, Wolford J, Daunert S (1999) Green fluorescent protein in the design of a living biosensing system for L-Arabinose. *Anal Chem* **71**: 763–768.
89. Roberto FF, Barnes JM, Bruhn DF (2002) Evaluation of a GFP reporter gene construct for environmental arsenic detection. *Talanta* **58**: 181–188.
90. Webb JS, Barratt SR, Sabev H, Nixon M, Eastwood IM, Greenhalgh M, Handley PS, Robson GD (2001) Green fluorescent protein as a novel indicator of antimicrobial susceptibility in *Aureobasidium pullulans*. *Appl Environ Microbiol* **67**: 5614–5620.
91. Mitchell R, Gu MB (2004) An *Escherichia coli* biosensor capable of detecting both genotoxic and oxidative damage. *Appl Microbiol Biotechnol* **64**: 46–52.
92. Rabbow E, Rettberg P, Baumstark-Khan C, Horneck G (2002) SOS-LUX- and LAC-FLUORO-TEST for the quantification of genotoxic and/or cytotoxic effects of heavy metal salts. *Anal Chim Acta* **456**: 31–39.

93. Kostrzynska M, Leung KT, Lee H, Trevors JT (2002) Green fluorescent protein based biosensor for detecting SOS-inducing activity of genotoxic compounds. *J Microbiol Meth* **48**: 43–51.
94. Afanassiev V, Sefton M, Anantachaiyong T, Barker MG, Walmsley RM, Wölfl S (2000) Application of yeast cells transformed with GFP expression constructs containing the RAD54 or RNR2 promoter as a test for the genotoxic potential of chemical substances. *Mut Res: Genet Toxicol Environ Mutagen* **464**: 297–308.
95. Arai R, Makita Y, Oda Y, Nagamune T (2001) Construction of green fluorescent protein reporter genes for genotoxicity test (SOS/*umu*-test) and improvement of mutagen-sensitivity. *J Biosci Bioeng* **92**: 301–304.
96. Justus T, Thomas SM (1999) Evaluation of transcriptional fusions with green fluorescent protein versus luciferase as reporters in bacterial mutagenicity tests. *Mutagenesis* **14**: 351–356.
97. Leveau JH, Lindow SE (2001) Appetite of an epiphyte: quantitative monitoring of bacterial sugar consumption in the phyllosphere. *Proc Natl Acad Sci USA* **98**: 3446–3453.
98. Joyner DC, Lindow SE (2000) Heterogeneity of iron bioavailability on plants assessed with a whole-cell GFP-based bacterial biosensor. *Microbiology* **146**: 2435–2445.
99. Shrestha S, Shetty RS, Ramanathan S, Daunert S (2001) Simultaneous detection of analytes based on genetically engineered whole cell sensing systems. *Anal Chim Acta* **444**: 251–260.
100. Jaspers MCM, Meier C, Zehnder AJB, Harms H, van der Meer JR (2001) Measuring mass transfer processes of octane with the help of an *alkS-alkB::gfp*-tagged *Escherichia coli*. *Env Microbiol* **3**: 512–524.
101. Baumstark-Khan C, Hellweg CE, Palm M, Horneck G (2001) Enhanced green fluorescent protein (EGFP) for space radiation research using mammalian cells in the International Space Station. *Phys Med* **17** Suppl 1: 210–214.
102. Olofsson AC, Zita A, Hermansson M (1998) Floc stability and adhesion of green-fluorescent-protein-marked bacteria to flocs in activated sludge. *Microbiology* **144**: 519–528.
103. Vanden-Wymelenberg AJ, Cullen D, Spear RN, Schoenike B, Andrews JH (1997) Expression of green fluorescent protein in *Aureobasidium pullulans* and quantification of the fungus on leaf surfaces. *BioTechniques* **23**: 686–690.
104. Normander B, Christensen BB, Molin S, Kroer N (1998) Effect of bacterial distribution and activity on conjugal gene transfer on the phylloplane of the Bush Bean (*Phaseolus vulgaris*). *Appl Environ Microbiol* **64**: 1902–1909.
105. Christensen BB, Sternberg C, Molin S (1996) Bacterial plasmid conjugation on semi-solid surfaces monitored with the green fluorescent protein (Gfp) from *Aequorea victoria* as a marker. *Gene* **173**: 59–65.
106. Bastos AE, Cassidy MB, Trevors JT, Lee H, Rossi A (2001) Introduction of green fluorescent protein gene into phenol-degrading *Alcaligenes faecalis* cells and their monitoring in phenol-contaminated soil. *Appl Microbiol Biotechnol* **56**: 255–260.
107. Tombolini R, Unge A, Davey ME, de Bruijn FJ, Jansson JK (1997) Flow cytometric and microscopic analysis of GFP-tagged *Pseudomonas fluorescens* bacteria. *FEMS Microbiol Ecol* **22**: 17–28.
108. Bloemberg GV, O'Toole GA, Lugtenberg BJJ, Kolter R (1997) Green fluorescent protein as a marker for *Pseudomonas* spp. *Appl Environ Microbiol* **63**: 4543–4551.
109. Eberl L, Schulze R, Ammendola A, Geisenberger O, Erhart R, Sternberg C, Molin S, Amann R (1997) Use of green fluorescent protein as a marker for ecological studies of activated sludge communities. *FEMS Microbiol Lett* **149**: 77–83.
110. Errampalli D, Okamura H, Lee H, Trevors JT, van Elsas JD (1998) Use of green fluorescent protein (GFP) as a reporter system to detect phenanthrene

mineralizing *Pseudomonas sp* UG14r in the environment. *FEMS Microbiol Ecol* **26**: 181–191.
111. Molbak L, Licht TR, Kvist T, Kroer N, Andersen SR (2003) Plasmid transfer from *Pseudomonas putida* to the indigenous bacteria on alfalfa sprouts: characterization, direct quantification, and in situ location of transconjugant cells. *Appl Environ Microbiol* **69**: 5536–5542.
112. Leung KT, So JS, Kostrzynska M, Lee H, Trevors JT (2001) Using a green fluorescent protein gene-labeled *p*-nitrophenol-degrading *Moraxella* strain to examine the protective effect of alginate encapsulation against protozoan grazing. *J Microbiol Meth* **39**: 205–211.
113. Uchiyama T, Abe T, Ikemura T, Watanabe K (2005) Substrate-induced gene-expression screening of environmental metagenome libraries for isolation of catabolic genes. *Nat Biotechnol* **23**: 88–93.
114. Tresse O, Errampalli D, Lee H, Trevors JD, van Elsas JD (1998) Green fluorescent protein (GFP): a visual marker in Paranitrophenol degrader *Moraxella* sp. *FEMS Microbiol Lett* **164**: 187–193.
115. Møller S, Sternberg C, Andersen JB, Christensen BB, Ramos JL, Givskov M, Molin, S (1998) *In situ* gene expression in mixed-culture biofilms: evidence of metabolic interactions between community members. *Appl Environ Microbiol* **64**: 721–732.
116. Burlage RS, Yang ZK, Mehlhorn T (1996) A transposon for green fluorescent protein transcriptional fusions: application for bacterial transport experiments. Gene **173**: 53–58.
117. Cho JC, Kim SJ (1999) Green fluorescent protein based direct viable count to verify a viable but non-culturable state of *Salmonella typhi* in environmental samples. *J Microbiol Meth* **36**: 227–235.
118. Gage DJ, Bobo T, Long SR (1996) Use of green fluorescent protein to visualize the early events of symbiosis between *Rhizobium meliloti* and alfalfa (*Medicago sativa*). *J Bacteriol* **178**: 7159–7166.
119. So J-S, Lim HT, Oh E-T, Heo T-R, Koh S-C, Leung KT, Lee H, Trevors JT (2002) Visualizing the infection process of *Xanthomonas campestris* in cabbage using green fluorescent protein *W. J Microbiol Biotechnol* **18**: 17–21.
120. Tauriainen S, Karp M, Chang W, Virta M (1997) Recombinant luminescent bacteria for measuring bioavailable arsenite and antimonite. *Appl Environ Microbiol* **63**: 4456–4461.
121. Willardson BM, Wilkins JF, Rand TA, Schupp JM, Hill KK, Keim P, Jackson PJ (1998) Development and testing of a bacterial biosensor for toluene-based environmental contaminants. *Appl Environ Microbiol* **64**: 1006–1012.
122. Tauriainen S, Karp M, Chang W, Virta M (1998) Luminescent bacterial sensor for cadmium and lead. *Biosens Bioelectr* **13**: 931–938.
123. Edmunds JS, Fairey ER, Ramsdell JS (1997) A rapid and sensitive high throughput reporter gene assay for estrogenic effects of environmental contaminants. *Neurotoxicol* **18**: 525–532.
124. Ivask A, Hakkila K, Virta M (2001) Detection of organomercurials with sensor bacteria. *Anal Chem* **73**: 5168–5171.
125. Phoenix P, Keane A, Patel A, Bergeron H, Ghoshal S, Lau PCK (2003) Characterization of a new solvent-responsive gene locus in *Pseudomonas putida* F1 and its functionalization as a versatile biosensor. *Environ Microbiol* **5**: 1309–1327.
126. Ulitzur S, Kuhn J (2000) Construction of *lux* bacteriophages and the determination of specific bacteria and their antibiotic sensitivities. *Methods Enzymol* **305**: 543–557.
127. Batchelor SE, Cooper M, Chhabra SR, Glover LA, Stewart GSAB, Williams P, Prosser JI (1997) Cell density-regulated recovery of starved biofilm populations of ammonia-oxidizing bacteria. *Appl Environ Microbiol* **63**: 2281–2286.

128. Unge A, Tombolini R, Molbak L, Jansson JK (1999) Simultaneous monitoring of cell number and metabolic activity of specific bacterial populations with a dual *gfp-luxAB* marker system. *Appl Environ Microbiol* **65**: 813–821.
129. Ang S, Horng YT, Shu JC, Soo PC, Liu JH, Yi WC, Lai HC, Luh KT, Ho SW, Swift S (2001) The role of *rsmA* in the regulation of swarming motility in *Serratia marcescens*. *J Biomed Sci* **8**: 160–169.
130. Riccillo PM, Muglia CI, de Bruijn FJ, Roe AJ, Booth IR, Aguilar OM (2000) Glutathione is involved in environmental stress responses in *Rhizobium tropici*, including acid tolerance. *J Bacteriol* **182**: 1748–1753.
131. Blackburn NT, Seech AG, Trevors JT (1994) Survival and transport of *lac-lux* marked *Pseudomonas fluorescens* strain in uncontaminated and chemically contaminated soils. *Syst Appl Microbiol* **17**: 574–580.
132. de Lorenzo V, Fernández S, Herrero M, Jakubzik U, Timmis K (1993) Engineering of alkyl- and haloaromatic-responsive gene expression with mini-transposons containing regulated promoters of biodegradative pathways of *Pseudomonas*. *Gene* **130**: 41–46.
133. Guzzo J, DuBow MS (2000) A novel selenite- and tellurite-inducible gene in *Escherichia coli*. *Appl Environ Microbiol* **66**: 4972–4978.
134. Guzzo A, DuBow MS (1994) A *luxAB* transcriptional fusion to the *celF* gene of *E coli* displays increased luminescence in the presence of nickel. *Mol Gen Genet* **242**: 455–460.
135. Tom-Petersen A, Hosbond C, Nybroe O (2001) Identification of copper-induced genes in *Pseudomonas fluorescens* and use of a reporter strain to monitor bioavailable copper in soil. *FEMS Microbiol Ecol* **38**: 59–67.
136. Jaspers MC, Suske WA, Schmid A, Goslings DA, Kohler HP, van der Meer JR (2000) HbpR, a new member of the XylR/DmpR subclass within the NtrC family of bacterial transcriptional activators, regulates expression of 2-Hydroxybiphenyl metabolism in *Pseudomonas azelaica* HBP1. *J Bacteriol* **182**: 405–417.
137. Billi D, Friedmann EI, Helm RF, Potts M (2001) Gene transfer to the desiccation-tolerant cyanobacterium *Chroococcidiopsis*. *J Bacteriol* **183**: 2298–2305.
138. Omura T, Kiyono M, Pan-Hou H (2004) Development of a specific and sensitive bacteria sensor for detection of mercury at picomolar levels in environment. *J Health Sci* **50**: 379–383.
139. Justus T, Thomas SM (1998) Construction of a *umuC'-luxAB* plasmid for the detection of mutagens via luminescence. *Mutat Res* **398**: 131–141.
140. Werlen C, Jaspers MCM, van der Meer JR (2004) Gas-phase end point measurements of bioavialable naphthalene using a *Pseudomonas putida* biosensor. *Appl Environ Microbiol* **70**: 43–51.
141. Riedel K, Ohnesorg T, Krogfelt KA, Hansen TS, Omori K, Givskov M, Eberl L (2001) N-Acyl-L-homoserine lactone-mediated regulation of the Lip secretion system in *Serratia liquefaciens* MG1. *J Bacteriol* **183**: 1805–1809.
142. Reid BJ, Semple KT, Macleod CJ, Weitz HJ, Paton GI (1998) Feasibility of using prokaryotic biosensors to assess toxicity of polycyclic aromatic hydrocarbons. *FEMS Microbiol Lett* **169**: 227–233.
143. Horry H, Durand MJ, Picart P, Bendriaa L, Daniel P, Thouand G (2004) Development of a biosensor for the detection of tributyltin. *Environ Toxicol* **19**: 342–345.
144. Thouand G, Horry H, Durand M-J, Picart P, Bendriaa L, Daniel P, DuBow MS (2003) Development of a biosensor for on line detection of organotin compounds with a recombinant bioluminescent *Escherichia coli* strain. *Appl Microbiol Biotechnol* **62**: 218–225.
145. Van Dyke M, Lee IH, Trevors JT (1996) Survival of *luxAB*-marked *Alcaligenes eutrophus* H850 in PCB-contaminated soil and sediment. *J Chem Technol Biotechnol* **65**: 115–122.

146. Ramos C, Molina L, Molbak L, Ramos JL, Molin S (2000) A bioluminescent derivative of *Pseudomonas putida* KT2440 for deliberate release into the environment. *FEMS Microbiol Ecol* **34**: 91–102.
147. Davidov Y, Smulski D, Van Dyk TK, Vollmer AC, Elsemore DA, LaRossa RA, Belkin S (2000) Improved bacterial SOS promoter::*lux* fusions for genotoxicity detection. *Mut Res* **466**: 97–107.
148. Hay AG, Rice JF, Applegate BM, Bright NG, Sayler GS (2000) A bioluminescent whole-cell reporter for detection of 2,4-dichlorophenoxyacetic acid and 2,4-dichlorophenol in soil. *Appl Environ Microbiol* **66**: 4589–4594.
149. Burlage RS (1998) Organic contaminant detection and biodegradation characteristics. *Methods Mol Biol* **102**: 259–268.
150. Rozen Y, Nejidat A, Gartemann KH, Belkin H (1999) Specific detection of p-chlorobenzoic acid by *Escherichia coli* bearing a plasmid-borne *fcbA'*::*lux* fusion. *Chemosphere* **38**: 633–641.
151. Van Dyk TK, Smulski DR, Elsemore DA, LaRossa RA, Morgan RW (2000) A panel of bioluminescent biosensors for characterization of chemically-induced bacterial stress responses. *ACS Symp Ser 762* (Chemical and Biological Sensors for Environmental Monitoring), pp. 167–184.
152. Wallace WH, Fleming JT, White DC, Sayler GS (1994) An *algD* bioluminescent reporter plasmid to monitor alginate production in biofilms. *Microb Ecol* **27**: 225–239.
153. Simpson ML, Paulus MJ, Jellison GE *et al.* (2000) Bioluminescent bioreporter integrated circuits (BBICs): Whole-cell environmental monitoring devices *Proceedings of the 30th International Conference on Environmental Systems (Society of Automotive Engineers)*, http://www.sae.org/servlets/productDetail?PROD_TYP=PAPER&PROD_CD=2000-01-2420 Toulouse, France.
154. Applegate BM, Kehrmeyer SR, Sayler GS (1998) A chromosomally based *tod-luxCDABE* whole-cell reporter for benzene, toluene, ethylbenzene, and xylene (BTEX) sensing. *Appl Environ Microbiol* **64**: 2730–2735.
155. Corbisier P, Thiry E, Diels L (1996) Bacterial biosensors for the toxicity assessment of solid wastes. *Environ Toxicol Water Qual* **11**: 171–177.
156. Tibazarwa C, Corbisier P, Mench M, Bossus A, Solda P, Mergeay M, Wyns L, van der Lelie D (2001) A microbial biosensor to predict bioavailable nickel in soil and its transfer to plants. *Environ Pollut* **113**: 19–26.
157. Holmes SD, Dubey SK, Gangoli S (1994) Development of biosensors for the detection of mercury and copper ions. *Environ Geochem Health* **16**: 229–233.
158. Belkin S, Smulski DR, Dadon S, Vollmer AC, Van Dyk TK, LaRossa RA (1997) A panel of stress-responsive luminous bacteria for toxicity detection. *Wat Res* **31**: 3009–3016.
159. Vollmer AC, Belkin S, Smulski DR, Van Dyk TK, LaRossa RA (1997) Detection of DNA damage by use of *Escherichia coli* carrying *recA'*::*lux*, *uvrA'*::*lux* or *alkA'*::*lux* reporter plasmids. *Appl Environ Microbiol* **63**: 2566–2571.
160. Min J, Lee CW, Moon SH, LaRossa RA, Gu MB (2000) Detection of radiation effects using recombinant bioluminescent *Escherichia coli* strains. *Radiat Environ Biophys* **39**: 41–45.
161. Rupani S, Konstantin KB, Gu MB, Dhurjati P, Van Dyk T, LaRossa R (1996) Characterization of the stress response of a bioluminescent biological sensor in batch and continuous cultures. *Biotech Progress* **12**: 387–392.
162. Van Dyk TK, Reed TR, Vollmer AC, LaRossa RA (1995) Synergistic induction of the heat shock response in *Escherichia coli* by simultaneous treatment with chemical inducers. *J Bacteriol* **177**: 6001–6004.
163. Belkin S, Smulski DR, Vollmer AC, Van Dyk TK, LaRossa RA (1996) Oxidative stress detection with *Escherichia coli* harboring a *katG'*::*lux* fusion. *Appl Environ Microbiol* **62**: 2252–2256.

164. Khang YH, Yang ZK, Burlage RS (1997) Measurement of iron-dependence of *pupA* promoter activity by a *pup-lux* bioreporter. *J Microbiol Biotechnol* **7**: 352–355.
165. Mioni CE, Howard AM, DeBruyn JM, Bright NG, Applegate BM, Wilhelm SW (2003) Characterization and preliminary field trials of a bioluminescent bacterial reporter of iron bioavailability. *Marine Chemistry* **83**: 31–46.
166. Selifonova OV, Eaton RW (1996) Use of an *ipb-lux* fusion to study regulation of the isopropylbenzene catabolism operon of *Pseudomonas putida* RE204 and to detect hydrophobic pollutants in the environment. *Appl Environ Microbiol* **62**: 778–783.
167. Barkay T, Turner RR, Rasmussen LD, Kelly CA, Rudd JWM (1998) Luminescence facilitated detection of bioavailable mercury in natural waters. In: LaRossa RA (ed) *Methods in Molecular Biology*, Vol. 102: *Bioluminescence Methods and Protocols*, pp. 231–246. Humana Press, Totowa, NJ.
168. Winson MK, Swift S, Fish L, Throup JP, Jorgensen F, Chhabra SR, Bycroft BW, Williams P, Stewart GSAB (1998) Construction and analysis of *luxCDABE*-based plasmid sensors for investigating N-acylhomoserine lactone-mediated quorum sensing. *FEMS Microbiol Lett* **163**: 185–192.
169. Heitzer A, Webb OF, Thonnard JE, Sayler GS (1992) Specific and quantitative assessment of naphthalene and salicylate bioavailability using a bioluminescent catabolic reporter bacterium. *Appl Environ Microbiol* **58**: 1839–1846.
170. Prest AG, Winson MK, Hammond JR, Stewart GS (1997) The construction and application of a *lux*-based nitrate biosensor. *Lett Appl Microbiol* **24**: 355–360.
171. Ahn JM, Mitchell R, Gu MB (2004) Detection and classification of oxidative damaging stesses using recombinant bioluminescent bacteria harboring *sodA*::, *pqi*::, and *katG*::*luxCDABE* fusions. *Enzyme Microb Technol* **35**: 540–544.
172. Layton AC, Muccini M, Ghosh MM, Sayler GS (1998) Construction of a bioluminescent reporter strain to detect polychlorinated biphenyls. *Appl Environ Microbiol* **64**: 5203–5206.
173. Ripp S, Applegate B, Nivens DE, Simpson ML, Sayler GS (2000) Advances in whole-cell bioluminescent bioreporters for environmental monitoring and chemical sensing. In: *AIChE Annual Meeting* Los Angeles, CA, Nov. 12–17. Abstract available online at: http://www.aiche.org/conferences/techprogam/paperdetail.asp?PaperID=3146&DSN=annual2k
174. Wiles S, Whiteley AS, Philp JC, Bailey MJ (2003) Development of bespoke bioluminescent reporters with the potential for *in situ* deployment within a phenolic-remediating wastewater treatment system. *J Microbiol Meth* **55**: 667–677.
175. Riether KB, Dollard MA, Billard P (2001) Assessment of heavy metal bioavailability using *Escherichia coli zntAp*::*lux* and *copAp*::*lux*-based biosensors. *Appl Microbiol Biotechnol* **57**: 712–716.
176. Shingleton JT, Applegate BM, Nagel AC, Bienkowski PR, Sayler GS (1998) Induction of the *tod* operon by trichloroethylene in *Pseudomonas putida* TVA8. *Appl Environ Microbiol* **64**: 5049–5052.
177. Burlage RS, Everman K, Patek D (1999) Method for Detection of Buried Explosives Using a Biosensor. US Patent No. 5,972,638.
178. Vollmer AC, Kwakye S, Halpern M, Everbach EC (1998) Bacterial stress responses to 1-Megahertz pulsed ultrasound in the presence of microbubbles. *Appl Environ Microbiol* **64**: 3927–3931.
179. Elasri MO, Miller RV (1998) A *Pseudomonas aeruginosa* biosensor responds to exposure to ultraviolet radiation. *Appl Microbiol Biotechnol* **50**: 455–458.
180. Erbe JL, Adams AC, Taylor KB, Hall LM (1996) Cyanobacteria carrying an *smt-lux* transcriptional fusion as biosensors for the detection of heavy metal cations. *J Ind Microbiol* **17**: 80–83.
181. De Lorenzo V, Herrero M, Jakubzik U, Timmis KN (1990) Mini-Tn5 transposons

derivatives for insertion mutagenesis, promoter probing, and chromosomal insertion of cloned DNA in Gram-negative eubacteria. *J Bacteriol* **172**: 6568–6572.

182. Winson MK, Swift S, Hill PJ, Sims CM, Griesmayr G, Bycroft BW, Williams P, Stewart GSAB (1998) Engineering the *luxCDABE* genes from *Photorhabdus luminescens* to provide a bioluminescent reporter for constitutive and promoter probe plasmids and mini-Tn*5* constructs. *FEMS Microbiol Lett* **163**: 193–202.
183. Tang X, Lu BF, Pan SQ (1999) A bifunctional transposon mini-Tn*5gfp*-km which can be used to select for promoter fusions and report gene expression levels in *Agrobacterium tumefaciens*. *FEMS Microbiol Lett* **179**: 37–42.
184. McClain MS, Engleberg NC (1996) Construction of an alkaline phosphatase fusion-generating transposon, mTn*10phoA*. *Gene* **170**: 147–148.
185. Xi C, Lambrecht M, Vanderleyden J, Michiels J (1999) Bi-functional *gfp*- and *gusA*-containing mini-Tn*5* transposon derivatives for combined gene expression and bacterial localization studies. *J Microbiol Meth* **35**: 85–92.
186. Wilson KJ, Sessitsch A, Corbo JC, Giller KE, Akkermans ADL, Jefferson RA (1995) β-Glucuronidase (GUS) transposons for ecological and genetic studies of rhizobia and other Gram-negative bacteria. *Microbiology* **141**: 1691–1705.
187. Neuveglise C, Nicaud J-M, Ross-Macdonald P, Gaillardin C (1998) A shuttle mutagenesis system for tagging genes in the yeast *Yarrowia lipolytica*. *Gene* **213**: 37–46.
188. Suarez A, Guttler A, Stratz M, Staender LH, Timmis KN, Guzman CA (1997) Green fluorescent protein-based reporter systems for genetic analysis of bacteria including monocopy applications. *Gene* **196**: 69–74.
189. Weitz HJ, Ritchie JM, Bailey DA, Horsburgh AM, Killham K, Glover LA (2001) Construction of a modified mini-Tn*5* *luxCDABE* transposon for the development of bacterial biosensors for ecotoxicity testing. *FEMS Microbiol Lett* **197**: 159–165.
190. Reidl J, Mekalanos JJ (1995) Characterization of *Vibrio cholerae* bacteriophage K139 and use of a novel mini-transposon to identify a phage-encoded virulence factor. *Mol Microbiol* **18**: 685–701.
191. Kahrs AF, Odenbreit S, Schmitt W, Heuermann D, Meyer TF, Haas R (1995) An improved TnMax mini-transposon system suitable for sequencing, shuttle mutagenesis and gene fusions. *Gene* **167**: 53–57.
192. Baron GS, Nano FE (1999) An erythromycin resistance cassette and mini-transposon for constructing transcriptional fusions to cat. *Gene* **229**: 59–65.
193. Stapleton RD, Sayler GS (1998) Assessment of the microbiological potential for natural attenuation of petroleum hydrocarbons in a shallow aquifer system. *Microb. Ecol.* **36**: 349–361.

Case study

Environmental application of *luxCDABE*-based bioluminescent bioreporters: monitoring environmental contaminants in a groundwater plume

A groundwater research facility at Columbus Air Force Base, Mississipi was contaminated with a simulated jet fuel mixture consisting of naphthalene, toluene, ethylbenzene and *p*-xylene (193). Numerous multi-level sampling wells installed upstream and downstream of the source pollutant allowed for monitoring of the contaminants. Typically water would be pumped up from designated wells and sent to an off-site laboratory for contaminant analysis using gas chromatography/mass spectrometry (GC-MS) techniques. GC-MS analysis is extremely sensitive and accurate and is by far the best method available for detecting chemical contaminants in environmental sources. However, it also requires expensive and bulky instrumentation, trained personnel, the use of hazardous chemicals and a significant allotment of time. As an alternative, *luxCDABE*-based reporters were used as sensors for the groundwater contaminants. Two bioreporters, *Pseudomonas fluorescens* 5RL (154) carrying a bioreporter for naphthalene, and *Pseudomonas putida* TVA8 carrying a bioreporter for toluene, were applied. Analysis occurred on site, where bioreporters were combined with groundwater samples and allowed to incubate for 4 h. Resulting bioluminescence was measured using a Deltatox field portable photomultiplier unit (Azur Environmental, San Diego, CA, USA) interfaced to a laptop computer. Duplicate samples were sent to an off-site laboratory.

Bioluminescent reporters consistently predicted contaminant concentrations within 50% of the GC-MS analytic measurements (*Figure 14.2*) Although in this case not highly quantitative, bioluminescent bioreporters did provide a rapid, general assessment of contaminant presence within the groundwater aquifer, and established an overall snapshot of plume dynamics within a few hours of sampling at a cost of ~10% of that of GC-MS analysis.

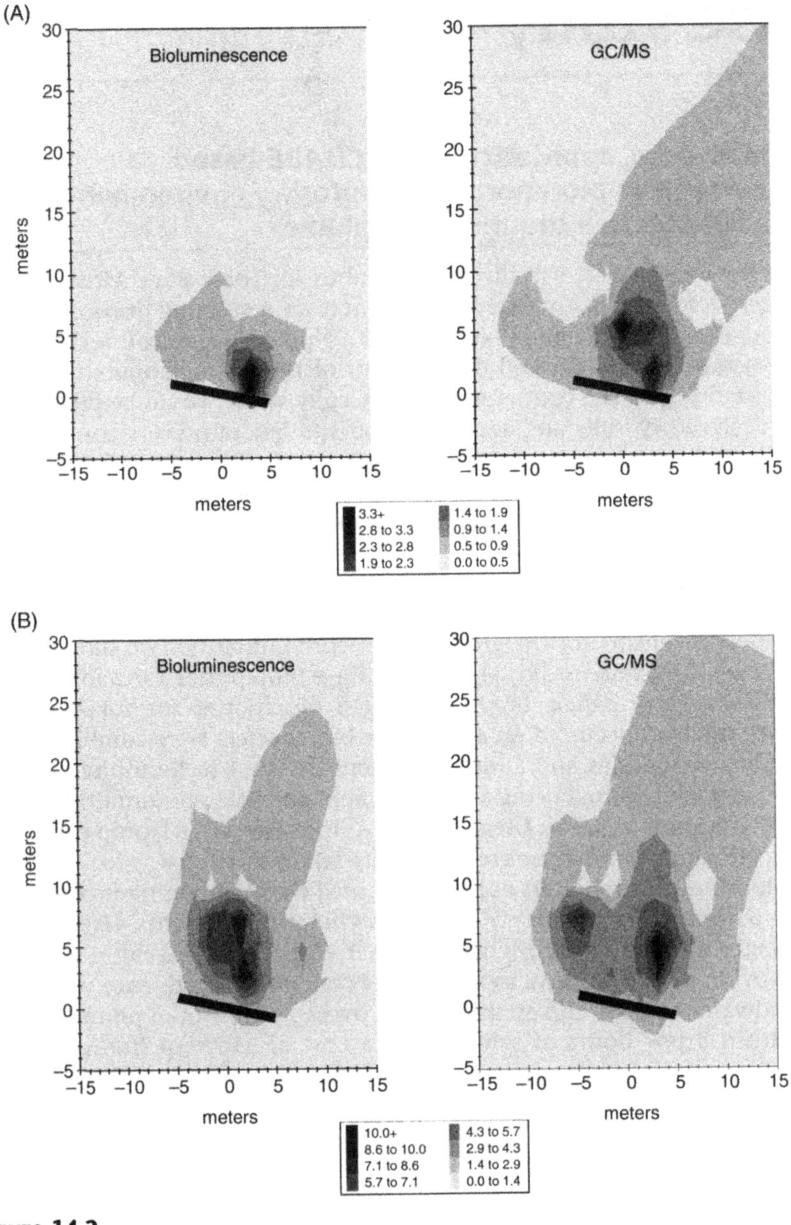

Figure 14.2

Distribution and dispersal of naphthalene and toluene concentrations (ppm) in the Columbus Air Force Base aquifer. (**A**) Naphthalene concentrations were determined using the *luxCDABE*-based bioreporter *P. fluorescens* 5RL and compared to data obtained using standard gas chromatography–mass spectrometry (GC-MS) techniques. (**B**) Plume dynamics in relation to toluene equivalent concentrations were also determined using the bioluminescent bioreporter *P. putida* TVA8, again with comparison to GC-MS data. Black bar represents location of source trench where contaminants were originally introduced.

Bioinformatics and web resources for the microbial ecologist

15

Wolfgang Ludwig

15.1 Introduction

Structure and function analysis of microbial communities is central to our understanding of microbial ecology. Given that the majority of environmental microorganisms cannot be cultured *in vitro* (1), nucleic-acid-based identification methods are commonly used to identify and phylogenetically assign organisms from the environment (Chapters 1 and 2). Identification of uncultured microorganisms is routinely carried out via the so-called rRNA cycle (2), a combination of PCR amplification followed by cloning and sequencing (Chapter 2) or alternatively by specific *in situ* probing (Chapter 9). Direct or indirect analysis of taxonomically informative nucleic acid sequences by comparative sequence analysis or taxon-specific probe hybridization as well as via diagnostic PCR are now standard techniques in microbial ecology studies. In this context the most important molecules are the small subunit rRNA genes. The suitability of the small subunit rRNA genes as phylogenetic markers is discussed in more detail in Chapter 2, but their value is based in no small part on the fact that no other molecule or gene has such a comprehensive database available with respect to the number of entries (>100 000 entries) or with regard to phylogenetic spectrum. Consequently, the current taxonomy of the prokaryotes is based upon phylogenetic conclusions derived from comparative rRNA analyses as outlined in the most recent edition of *Bergey's Manual of Systematic Bacteriology* (3). However, against a background of 3–4 billion years of evolution of cellular life, the information content and taxonomic resolution of any phylogenetic marker is rather limited (4) especially at the lower taxonomic ranks such as the species and strain level. Now in the age of genomics, alternative conserved markers for supplementing and evaluating the higher-level systematics are increasingly considered. For species definition and identification of closely related strains, multi-locus sequencing is increasingly important and will replace genomic DNA–DNA hybridization as the criterion for species definition (5). In order to deal with, analyze and manage the wealth of information from large-scale sequencing methods, databases and software tools for data analysis have to be regarded as essential tools in microbial ecology studies. This not only holds true for identification based upon conserved and housekeeping genes but also for functional analyses. The presence or absence of genetic information on cellular molecules with physiological relevance may allow estimates of the physiological capacities of the organisms in microbial communities. In this context the rapid development of techniques for full genome sequencing of uncultured organisms—environmental genomics (Chapters 10–11)—will provide valuable knowledge on the potential physiological properties of uncultured organisms. This is of even greater importance against the background of continuous improvement in microautoradiography techniques (6) as well as *in situ* mRNA, and plasmid and genomic DNA targeting hybridization methods (Chapter 8). Thus in the near future *in situ* studies on the function of uncultured members of microbial communities will become as routine as rRNA-based identification.

As a consequence of the progress in rapid large-scale sequencing, the number of database and bioinformatics initiatives is continually increasing. Obviously, the experienced users and specialists will easily find data and tools from well-known sources or by following links or using appropriate search engines. However, for the non-experienced user it may be rather difficult to find a path in the jungle of web pages, databases and software facilities. In this chapter, a selection of databases and software tools for routine analysis in microbial ecology will be briefly introduced. Sequence annotation and comparison, phylogenetic analyses and probe-based identification are certainly the major tasks in such studies. The access and utilization of databases and software for similarity searches are described in more detail using the European Bioinformatics Institute (EBI) facilities as an example. Furthermore, the ARB project, which consists of integrated databases and comprehensive software package of directly communicating tools controlled by the user from a common graphical interface, is described in more detail. Many of the software tools for similarity search, alignment, probe design and evaluation described in the context of the EBI databases or the ARB project are also available as standalone versions. The respective internet addresses (URLs) are indicated in the text. The designations of tabs, buttons, windows and boxes in web pages or program windows are indicated in quotation marks ('–') in the text. A list of the major URLs mentioned in the text is given in *Table 15.1*.

Table 15.1 List of important web addresses (URLs) mentioned in the text

Resource	URL
European Bioinformatics Institute	http://www.ebi.ac.uk
GenBank	http://www. ncbi.nih.gov/genbank/
DNA Data Bank of Japan	http://www.ddbj.nig.ac.jp/
Munich Information Center for Protein Sequences	http://mips.gsf.de
Welcome Trust Sanger Institute	http://www.sanger.ac.uk
The Institute for Genomic Research	http://www.tigr.org
Ribosomal Database Project	http://rdp.cme.msu.edu
European Ribosomal RNA Database	http:// rrna.uia.ac.be
ProbeBase	http://www.microbial-ecology.de
The ARB Project	http://www. arb-home.de

15.2 Databases

The most important general sequence databases are those hosted at the European Bioinformatics Institute (EBI) at Hinxton, UK (7), GenBank at Bethesda, MA, USA (8) and the DNA Databank of Japan (DDBJ; 9). The three databases exchange the data submitted to them with each other on a daily basis. Each sequence, when submitted to a database, is assigned a unique identifier (accession number). All three databases will use the same accession number for that sequence. However, database structuring, data access and data formats differ remarkably. Therefore, for the non-experienced user it is recommended (and usually sufficient) initially to focus their activities on only one of the databases. Navigating any of these databases for the first time can be a daunting task. Furthermore, the database projects from time to time change their layout as well modes of data submission and access. In the following sections, procedures for data access, downloading, submission and comparative analysis of sequences will be shown mainly using the EBI databases.

15.2.1 European Bioinformatics Institute (EBI) databases

A variety of general and specialized databases are maintained at EBI (http://www.ebi.ac.uk/). These include databases for primary structure collections, i.e. European Molecular Biology Laboratory (EMBL) Nucleotide Sequence Database, the protein databases SWISS-PROT and TrEMBL, now amalgamated into the Universal Protein Resource UniProt (10) (http://www.ebi.uniprot.org/index.shtml) in addition to higher order structure databases available in the Macromolecular Structure Database (MSD). Information on these and other data collections can be inspected at http://ebi.ac.uk/Databases.

The EMBL nucleotide sequence database incorporates, organizes and distributes nucleotide sequences from all available public sources (7). Complete database releases are available quarterly and can be downloaded as multiple compressed files via anonymous FTP file transfers from ftp://ftp.ebi.ac.uk/pub/databases/embl/release or by browsing (described here for the Netscape browser) the EBI pages. Beginning on the EBI home page: click 'Bioinformatics, Products and Services', select the 'Downloads' tab, press 'embl' in the 'Directory' list, and select 'release'. The individual files can be selected by right mouse click and downloaded to local storage devices selecting 'Save link as ...' from the pop-up menu and defining the desired local path. Daily and weekly updates are accessible at ftp://ftp.ebi.ac.uk/pub/databases/embl/new.

15.2.1.1 Structure of entries

The data are stored in flat file format (*Figure 15.1*). Each entry in the flat file format is composed of different types of tagged lines. The two character line codes on the left-hand side indicate the type of information contained in the lines following. A selection of major line codes is briefly described: 'AC' stands for accession number which is the unique identifier for the entry. The 'SV' line contains the primary accession extended by a version number. The latter facilitates the user's check for updates of the individual database entries. 'DE' marks a line containing a brief description of the entry usually including different types of information such as annotation of the sequences, source and other descriptive data. This field is a good target to start with for database searching. 'OS' means organism species and should contain the correct genus and species designations, and the higher taxa that the organism belongs to are specified in the 'OC' line(s). Lines tagged by 'RA' and 'RT' carry the bibliographical information on authors and title respectively, whereas the journal, volume, pages and year of publication are stored in the 'RL' line. Sequence annotation such as coding regions, signals, and other characteristics or comments are provided in the 'FT' lines. The latter are feature tables structured in a tabular format. The type of information is encoded by a feature key written next to the line code. For example, 'CDS' indicates protein-coding sequence (*Figure 15.1*). At fixed line positions the corresponding sequence positions or regions for the indicated feature are specified. In the following lines starting at the same fixed positions, qualifiers for the respective feature are indicated by a slash and a prefix (*Figure 15.1*). The respective name, value, identifier or other information is enclosed by quotation marks. For the 'CDS' key, one of the possible qualifiers is '/translation' and the equals sign followed by the predicted amino acid sequence in IUPAC one-letter code

```
ID   WS1740TUF  standard; DNA; PRO; 1203 BP.
XX
AC   X76872;
XX
SV   X76872.1
XX
DT   15-SEP-1994 (Rel. 41, Created)
DT   15-SEP-1994 (Rel. 41, Last updated, Version 4)
XX
DE   W.succinogenes (DSM 1740) tuf gene
XX
KW   elongation factor TU; tuf gene.
XX
OS   Wolinella succinogenes
OC   Bacteria; Proteobacteria; epsilon subdivision; Helicobacter group;
OC   Wolinella.
XX
RN   [1]
RX   MEDLINE; 94368062.
RA   Ludwig W., Neumaier J., Klugbauer N., Brockmann E., Roller C.,
RA   Klugbauer S., Reetz K., Schachtner I., Ludvigsen A., Bachleitner M.,
RA   Fischer U., Schleifer K.H.;
RT   "Phylogenetic relationships of Bacteria based on comparative sequence
RT   analysis of elongation factor Tu and ATP-synthase-beta subunit genes";
RL   Antonie Van Leeuwenhoek 64:285-305(1993).
XX
RN   [2]
RP   1-1203
RA   Ludwig W.;
RT   ;
```

```
RL   Submitted (17-DEC-1993) to the EMBL/GenBank/DDBJ databases.
RL   W. Ludwig, Lehrstuhl f. Mikrobiologie, TU Muenchen, 80290 Muenchem, FRG
XX
DR   SWISS-PROT; P42482; EFTU_WOLSU.
XX
FH   Key             Location/Qualifiers
FH
FT   CDS             1..1203
FT                   /db_xref="SWISS-PROT:P42482"
FT                   /transl_table=11
FT                   /gene="tuf"
FT                   /product="elongation factor Tu"
FT                   /protein_id="CAA54199.1"
FT                  /translation="MAKKKFVKYKPHVNIGTIGHVDHGKTTLSAAISAVLATKGLCELK
FT                  DYDAIDNAPEERERGITIATSHIEYETENRHYAHVDCPGHADYVKNMITGAAQMDGAIL
FT                  VVSAADGPMPQTREHILLSRQVGVPYIVVFLNKEDMVDDAELLELVEMEVRELLSNYDF
FT                  PGDDTPIVAGSALKALEEANDQENVGEWGEKVLKLMAEVDRYIPTPERDVDKPFLMPVE
FT                  DVFSIAGRGTVVTGRIERGVVKVGDEVEIVGIRNTQKTTVTGVEMFRKELDKGEAGDNV
FT                  GVLLRGTKKEDVERGMVLCKIGSITPHTNFEGEVYVLSKEEGGRHTPFFNGYRPQFYVR
FT                  TTDVTGSISLPEGVEMVMPGDNVKINVELIAPVALEEGTRFAIREGGRTVGAGVVTKIT
FT                   K"
FT   source          1..1203
FT                   /db_xref="taxon:844"
FT                   /organism="Wolinella succinogenes"
FT                   /strain="DSM 1740"
XX
SQ   Sequence 1203 BP; 331 A; 254 C; 306 G; 312 T; 0 other;
     atggctaaga aaaagtttgt aaaatacaaa ccccacgtta atatcggtac catcggtcac        60
     gttgaccacg gtaaaaccac tcttagtgcc gctatttctg cggtacttgc aaccaaaggt       120
     ctttgcgagc ttaaagatta tgatgcgatc gacaatgctc ctgaagagag agagcgtggt       180
     atcaccatcg ctacttcaca catcgagtat gaaacagaaa atcgacacta cgctcacgtt       240
     gactgccctg gacacgccga ctatgttaaa aacatgatta caggtgctgc tcaaatggat       300
     ggcgcgattc ttgttgtttc tgcggcggat ggccccatgc cccaaactag ggagcacatt       360
     cttctttctc gacaagtagg cgttccttac atcgtggttt tcttgaacaa agaagatatg       420
     gttgatgacg ctgagcttct tgagcttgtt gaaatggaag ttagagaact tcttagcaac       480
     tacgacttcc ctggagatga cactcctatc gttgcaggtt ccgctcttaa agctcttgaa       540
     gaggctaacg accaggaaaa tgttggcgag tggggcgaga aagtattgaa gcttatggct       600
     gaggttgacc gatatattcc tacgcctgag cgagatgtgg ataagccttt ccttatgcct       660
     gttgaagacg tattctccat cgcgggtcgt ggaaccgttg tgacaggaag aattgaaaga       720
     ggcgtggtta aagtcggtga cgaagtagaa atcgttggta tccgaaacac acaaaaaaca       780
     accgtaactg gcgttgagat gttccgaaaa gagctcgaca agggtgaggc gggtgacaac       840
     gttggtgttc ttttgagagg caccaagaaa gaagatgttg agagaggtat ggttctttgt       900
     aaaataggtt ctatcactcc tcacactaac tttgaaggtg aagtttacgt tctttccaaa       960
     gaggaaggcg gacgacacac tccattcttc aatggatacc gacctcagtt ctatgttaga      1020
     actacagacg ttaccggttc tatctctctt cctgagggcg tagagatggt tatgcctggt      1080
     gacaacgtta agatcaatgt tgagcttatc gctcctgtag ccctcgaaga gggaacacga      1140
     ttcgcgatcc gtgaaggtgg tcgaaccgtt ggtgcgggtg tcgttaccaa gatcactaaa      1200
     taa                                                                    1203
//
```

Figure 15.1

Sample DNA sequence accession entry in the European Molecular Biology Laboratory (EMBL) nucleotide database.

enclosed between quotation marks (*Figure 15.1*). Following the feature (FT) lines, the 'SQ' line shows the length of the primary sequence and details of base composition. The sequence is then given in full in rows each consisting of six blocks of 10 base pairs. The detailed description of the line codes is given in the *EMBL Nucleotide Sequence Database User Manual* which can be inspected at http://www.ebi.ac.uk/embl/Documentation/User_manual/usrman.html.

To access the documentation online, select the 'Databases' tab, press the 'EMBL' icon, select 'Documentation' from either the list or the table, press 'User Manual' in the table for text visualization or the icon next to 'User Manual' for downloading the manual to preferred local storage media. A detailed feature table description is directly accessible at http://www.ebi.ac.uk/embl/Documentation/FT_definitions/feature_table.html or by browsing the EBI pages as described for the manual. However, select 'EMBL, Features & Qualifiers' from the table. Some of the fields contain cross-references to other EBI or external databases. Examples are the corresponding SWISS-PROT or UniProt (10) accession number in case of protein encoding DNA, and the MEDLINE number. The former can be found in one of the FT-lines as a qualifier (/db_xref="UniProt/Swiss-Prot:xxxxx") of the 'CDS' feature key. The corresponding UniProt/Swiss-Prot entry, usually comprising detailed information on the protein, can be visualized by clicking the accession number. Clicking on the MEDLINE number which is stored in the RX-line visualizes the respective bibliographic data.

Interactive database searching and browsing is possible by following alternative approaches. Simple searches in the primary and higher structure as well as in the MEDLINE literature databases can be started directly from the home page and most other web pages of EBI (http://www.ebi.ac.uk). Type the search string in the 'Database Search for' text box, select the desired database from the list popping up after clicking 'Nucleotide sequences' and start the search by pressing the 'Go' button. For complex database searching, the Sequence Retrieval System (SRS) should be used. This tool can be directly activated at http://srs.ebi.ac.uk or by browsing the EBI pages. There are several different ways to access data using SRS, a good path to start with is the following: select the 'Toolbox' tab, and press the 'SRS' icon. Detailed description of the tools and operating instructions can be visualized by clicking 'Help' on the SRS start page. Pressing 'Databanks' provides a list of databases available for SRS queries. Further information on the individual database can be visualized by selecting the respective database name. After initiating the SRS tool by clicking the tab 'Library Page' the target database has to be selected by checking the respective boxes (*Figure 15.2*). Multiple selections are allowed. By selecting the 'Query Form' tab, the standard query form appears on the screen (*Figure 15.3*). Complex string searches in all or a selection of up to four different fields (database fields are the lines defined by the line codes described above) can be composed by the user. The fields are selected from a pop-up menu invoked by checking the boxes right to the four 'AllText' windows. Search strings have to be typed into the text boxes. Combinations of strings can be customized by separating them with the following symbols '&' (and), '|' (or), '/' (but not), respectively. A combination of up to four string sets is defined by selecting '&' (and), '|' (or), '/' (but not) from the 'Search Options'

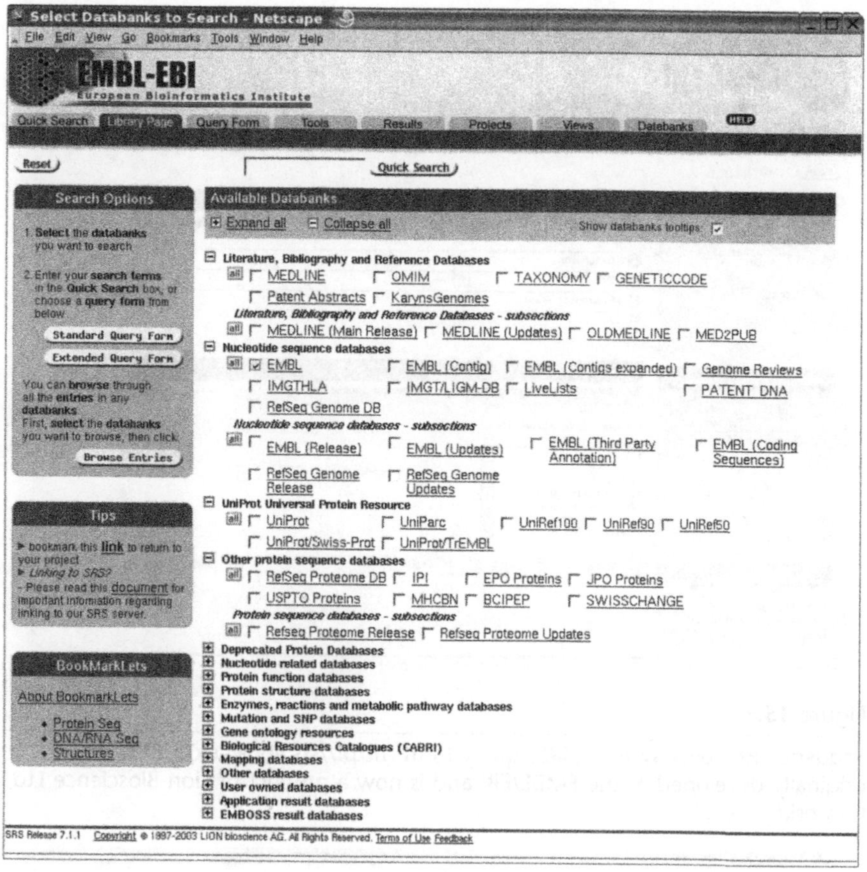

Figure 15.2

Selection of query database for Sequence Retrieval System (SRS) searches (http://srs.ebi.ac.uk/). (SRS was originally developed at the EMBL/EBI and is now a product of Lion Bioscience Ltd, Cambridge, UK.)

box (*Figure 15.3*). The layout of the results output can be customized by selecting the fields (line codes) to be included either by choosing predefined combinations from the 'Results Display Options' box or by selecting from the list shown in the 'Create a View' box. Depending upon the demands of the user it may be necessary to change the output format. A selection of the most common flat file formats is visible by checking the 'sequence format' box. Pressing the 'Search' button starts the database searching. The 'Results' window (*Figure 15.4*) then shows a listing of hits from which individual entries can be selected by checking the respective boxes (*Figure 15.5*). The layout of the output can also be customized using the pop-up menu in the 'Display Options' box. The 'Link' and 'Launch' buttons in the 'Results Options' box give access to other databases or fields linked to one of the entries in the hit list or to further operations such as sequence similarity searches in the selected databases respectively. If the user wants to download the entries to a local storage device, the 'Save' button has to be pressed.

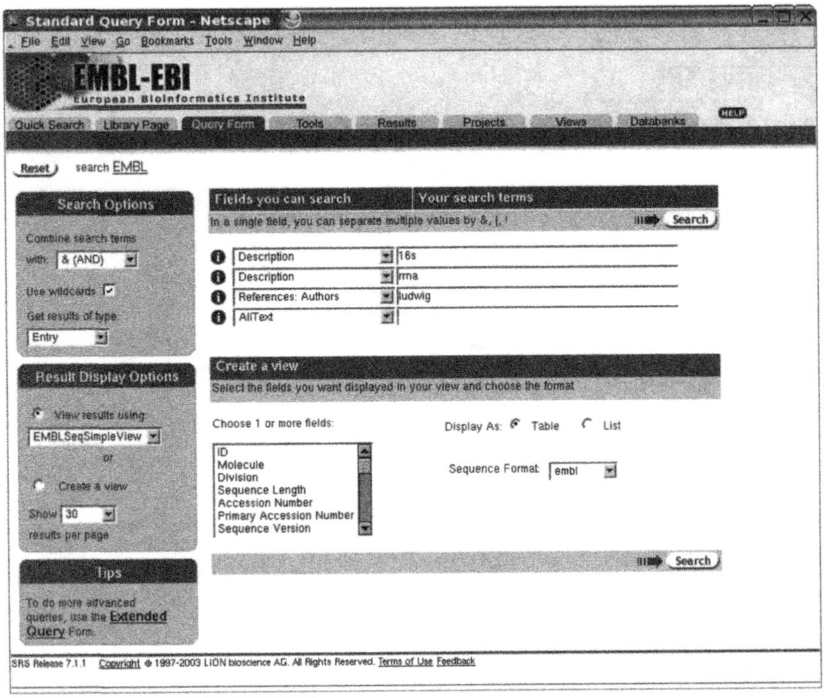

Figure 15.3

Sequence Retrieval System (SRS) query form (http://srs.ebi.ac.uk). (SRS was originally developed at the EMBL/EBI and is now a product of Lion Bioscience Ltd, Cambridge, UK.)

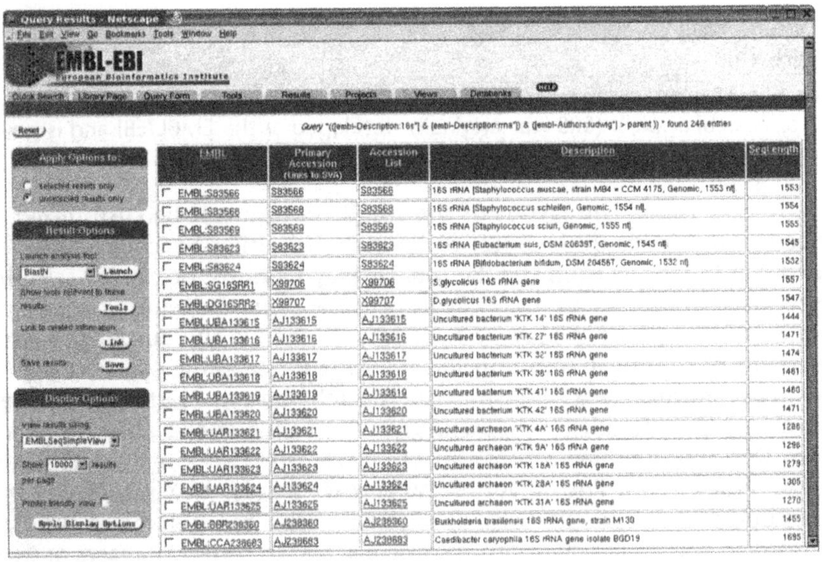

Figure 15.4

Sequence Retrieval System (SRS) results window. (http://srs.ebi.ac.uk). (SRS was originally developed at the EMBL/EBI and is now a product of Lion Bioscience Ltd, Cambridge, UK.)

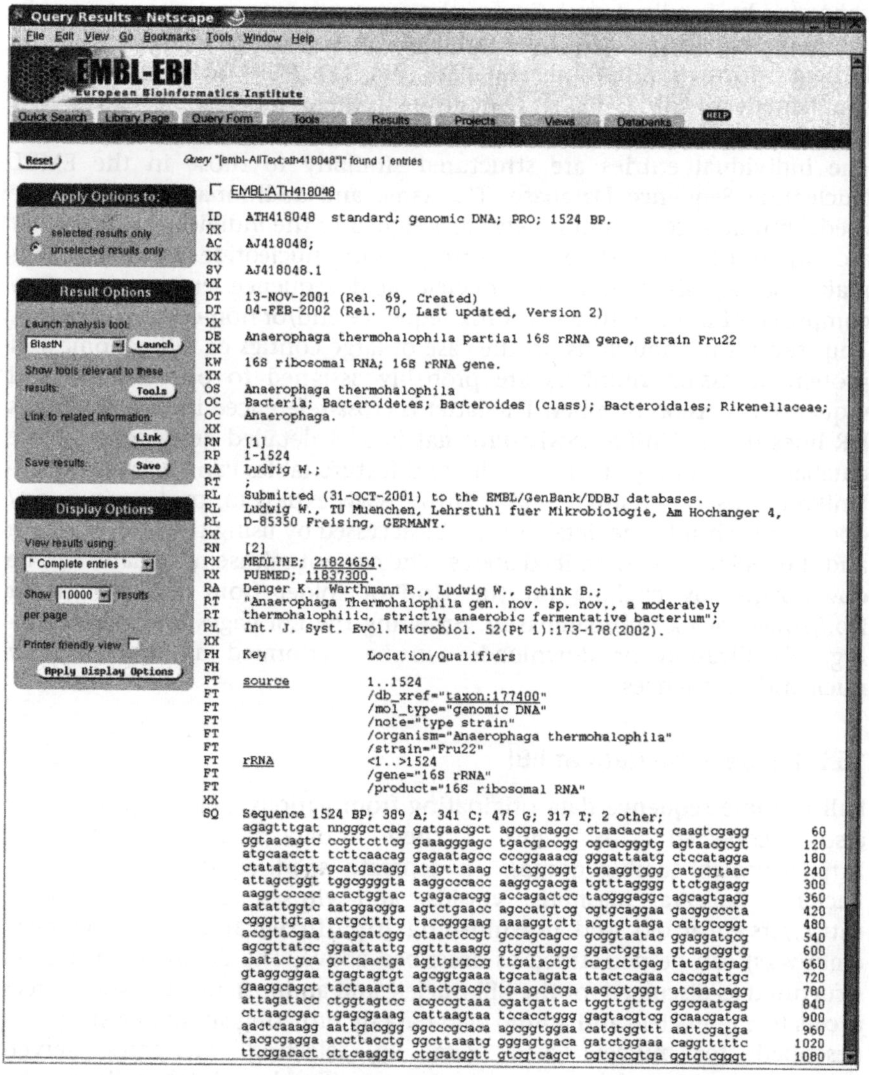

Figure 15.5

Sequence Retrieval System (SRS) results: view of an entry selected from the results hits list (http://srs.ebi.ac.uk). (SRS was originally developed at the EMBL/EBI and is now a product of Lion Bioscience Ltd, Cambridge, UK.)

15.2.1.2 The protein databases at EBI

Protein sequences were for many years deposited in the SWISS-PROT protein database maintained by the Swiss Institute of Bioinformatics (http://us.expasy.org/sprot/) in cooperation with the EBI. In 2002 the UniProt database (10) was created to combine the resources of three existing protein databases: SWISS-PROT, TrEMBL and the Protein Information Resource (PIR) (11). The database contains amino acid sequences mainly generated by *in silico* translation of the corresponding sequence regions in the EMBL Nucleotide Sequence Database (7) and partly adopted from the Protein Information Resource database (PIR; 11). The TrEMBL database is a supplement to SWISS-PROT containing protein primary structures that have not yet been fully annotated according to the SWISS-PROT standards. The individual entries are structured similarly to those in the EMBL Nucleotide Sequence Database. The same and additional line codes are used. Unique accession numbers are assigned to the individual entries that are different from those of the corresponding nucleotide sequences. This makes sense, given that the nucleic acid sequence entries are often comprised of more than one coding sequence and/or non-coding stretches. Thus several to thousands (in the case of large contigs or full genomes) of protein accession numbers are probably assigned to one nucleic acid sequence accession number. The latter is cross-referenced in one of various DR lines in the UniProt/Swiss-Prot flat files. A detailed description of the database structuring, the line codes and feature tables can be found in the UniProt user manual at http://www.expasy.uniprot.org/support/documents.shtml. The database can be accessed by using the simple search and the SRS tools as described above. The current release or updates can be downloaded by clicking on the UniProt folder from the FTP server ftp://ftp.ebi.ac.uk/pub/databases/. Browsing the EBI pages for data searching, visualization or downloads can be performed as described for nucleotide sequences.

15.2.1.3 Genome data at EBI

Full genome sequence data originating from various genome projects are also collected at EBI. These data are stored at http://www.ebi.ac.uk/genomes/ in separate subdirectories for Archaea, Bacteria, Eucarya, plasmids, viroids, and viruses. Database searching (http://srs.ebi.ac.uk) and downloading (ftp://ftp.ebi.ac.uk/pub/databases/embl/wgs) can be performed as described above. The database entries are structured as discussed for the EBI nucleic and amino acid databases. Direct access to the nucleic and amino acid data by browsing should be started as described for nucleic acid data by selecting 'Genomes Server' (http://www.ebi.ac.uk/genomes/) from the 'Databases' tab followed by pressing the desired section (Archaea, Bacteria, Eucarya, plasmids, viroids, viruses). A table containing a list of available genomes is shown on the screen. The flat files containing the full genome data, the individual contigs or the predicted amino acid sequences can be selected by clicking the respective cells of the table.

15.2.1.3.1 Data submission

New sequence data determined and annotated by the user should preferably be submitted to EBI using the web-based tool 'Webin' (http://ebi.ac.uk/submission/). The primary structure and further descriptive information can interactively be imported to the respective fields by typing or pasting to the respective text boxes on the web page. The Instructions for submitters can be inspected or downloaded at http://www.ebi.ac.uk/webin/webin_help.html. If local processing of the data to be submitted is preferred, the 'SEQUIN' standalone tool running on Windows, McIntosh and UNIX operating systems has to be downloaded (ftp:/ftp.ebi.ac.uk/pub/software/sequin) and installed locally. This tool allows submission forms to be completed not only for EBI but also for GenBank and the Japanese DDBJ databanks.

15.2.1.3.2 Data analysis

In many microbial ecology research projects, new sequences are determined for isolates or directly from sample material. Sequence comparison for identification, annotation or predicting function is nowadays a routine task within such studies. However, storage and maintenance of large sequence datasets locally can tie up hardware facilities and prove very time-consuming. The public databases offer several tools for the remote analysis of user-provided sequence data against the background of the full databases of reference sequences.

15.2.1.3.3 FASTA

FASTA (Fast All) is a tool for rapid identification of local sequence similarities (12,13). Several variants of the original software are accessible at EBI for remote database searching. The sequences to be compared are broken into short runs of consecutive characters called words. Segments of nearby word hits are identified first, and scores (relative values indicating the degree of similarity) are assigned to them. Multiple regions of local similarities (segments) are then joined and scores calculated for the ensemble. In the case of protein sequences these comparisons and weightings are based upon a variety of alternative scoring matrices. Such scoring matrices take into account the character of the amino acid changes. Different values are assigned to the 210 possible different character changes (pairs of amino acids). The matrix variants differ with respect to accuracy for comparing proteins of high or lower degrees of sequence similarities (14). Local gapped alignments are generated for the most potential matches around the best initial segment. Finally a full Smith–Waterman (15) alignment search is done. The FASTA program variants handle both nucleic and amino acid sequences. Furthermore, protein sequences can be compared to nucleic acid data by *in silico* translation of the latter. The regions of local similarity can be visualized as a graphic matrix plot or as individual alignments. FASTA-based similarity searches can be performed for user-selected entries from the EBI databases or from user-provided external data. In the former case, FASTA can be started directly from the results window of SRS-mediated database queries. For analysis of user-provided data, FASTA can be started at http://www.ebi.ac.uk/Tools/ and selecting 'FASTA'. The user data has to be typed or pasted into the respective text box on the 'Fasta Submission Form' window. Besides the sequence types (nucleic acid or protein), a number of parameters including scoring matrices and gap penalties can be defined by selecting from the pop-up menus invoked by checking the respective boxes. For the inexperienced user it is recommended to try the default values first. Instructions and description of the procedures can be inspected by pressing 'Fasta Help' on the 'Fasta Submission Form' window (http://www.ebi.ac.uk/fasta33/index.html).

15.2.1.3.4 BLAST

BLAST (Basic Local Alignment Search Tool) (http://www.ebi.ac.uk/blast/index.html) is a simple but extremely fast alternative tool to find similar reference sequences in a database (16) (*Figure 15.6*). Again the highest scoring matches between the query sequence and a database are searched. The local matching regions are then extended in both directions (5' and 3') until the resulting scores pass a default or user defined scoring threshold. The local high scoring alignments are presented in a sorted list of maximal

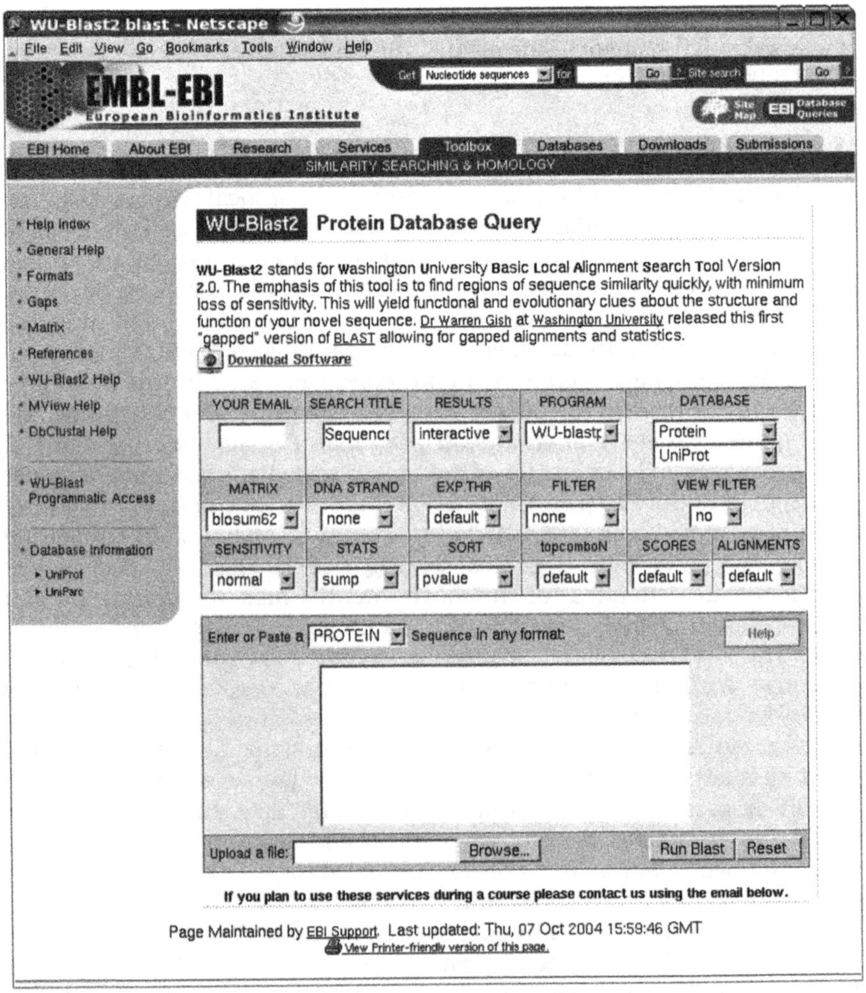

Figure 15.6

European Bioinformatics Institute (EBI) BLAST submission form (http://www.ebi.ac.uk/blast2/nucleotide.html).

sequence pairs (MSPs) (*Figure 15.7*). Almost all public sequence databases including also smaller genome projects provide blast servers based upon one or more of the BLAST programs for online comparisons. At the EBI, BLAST analysis can be done following an approach analogous to that described for FASTA. BLAST programs can also be downloaded for local installation.

15.2.1.3.5 ClustalW

ClustalW (17) is a tool for establishing multiple sequence alignments from scratch. Similar to FASTA and BLAST searches, alignments can be generated for user-defined sets of database entries and again user-provided sequence data can be included. Typically, the database references are selected according to the results of FASTA and/or BLAST searches. The alignment is established in a progressive mode. The program calculates a series of pairwise initial alignments including all sequences of the input set. These binary

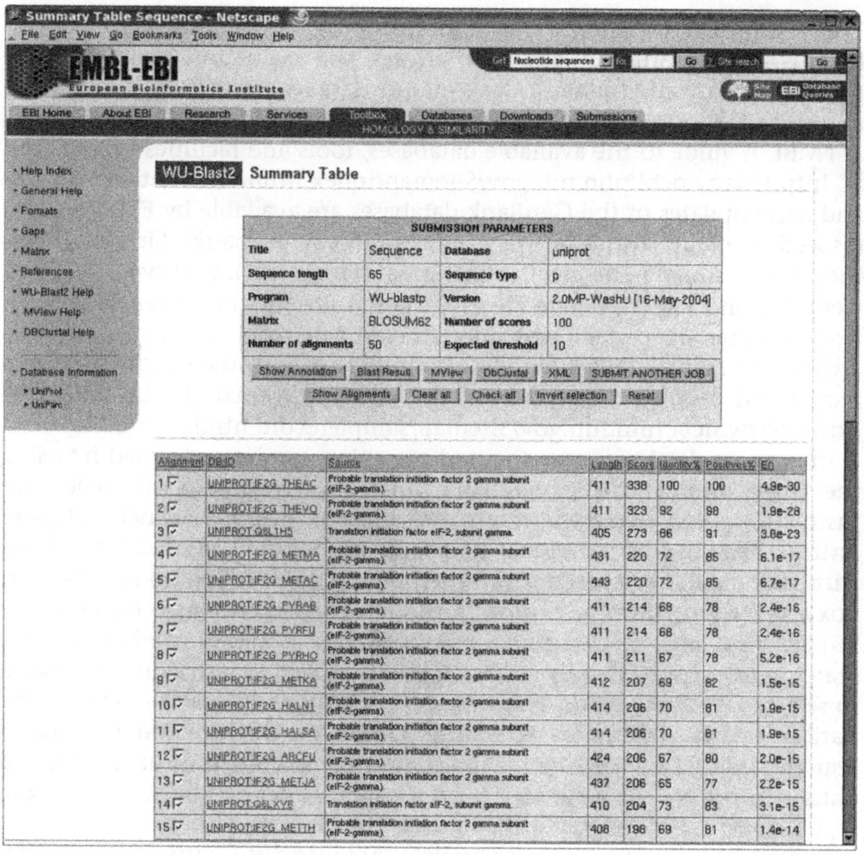

Figure 15.7

European Bioinformatics Institute (EBI) BLAST results: parameters and hit list and initial alignment. (http://www.ebi.ac.uk/blast2)

comparisons are also used to reconstruct a guide tree applying a distance matrix procedure. An overall alignment is generated beginning with the closest neighbors in the guide tree and subsequently adding the other sequences in order of decreasing similarity. Progressive improvement of the alignment is also achieved by subsequently changing the parameters. Firstly, partial alignments of the most obviously similar stretches are used to assign different weights to sequence regions of high and low similarities. The more divergent regions are up-weighted for the next round of alignments. In the case of proteins the substitution matrices are also varied according to sequence divergence. The program also tries to control gap insertion and extension by assigning modified gap penalties according to higher-order structure. Thus in the case of proteins, gaps are preferably inserted in loop regions preventing the potential disruption of regions of more rigid structures such as helices. ClustalW at EBI is accessible via the 'toolbox' tab. Detailed description in 'ClustalWHelp' is available from the 'ClustalW Submission Form' window (http://www.ebi.ac.uk/clustalw/index.html).

15.2.2 GenBank

The National Center for Biotechnology Information (NCBI; http://www.ncbi.nlm.nih.gov/; 18) creates and maintains nucleotide and protein primary and higher-order structure data resources, full genome data as well as taxonomy and literature databases. The latter is a special service at NCBI. A guide to the available databases, tools and facilities can be seen at http://www.ncbi.nlm.nih.gov/Sitemap/index.html. Bimonthly releases and daily updates of the GenBank databases are available by FTP from the respective subdirectories in ftp://ftp.ncbi.nih.gov/genbank/. Similar to EBI, the data are stored in tagged flat file format (*Figure 15.1*). However, although GenBank and EBI cooperate closely, the flat file formats differ. Instead of two character line codes, GenBank uses complete words in a tabulated hierarchy. The feature tables are almost similar in both databases. A sample record and detailed description of the tags and format can be viewed at http://www.ncbi.nlm.nih.gov/Sitemap/samplerecord.html.

Interactive database browsing and searching can be performed by using the 'Entrez' tool. The list of databases which are accessible as well as detailed instructions and descriptions are provided at http://www.ncbi.nih.gov/Entrez/. Searches of individual databases are selected by clicking on the particular database of interest and then typing the search term into the text box and clicking the 'Go' button. Although Entrez is not as versatile as the EBI SRS tool with respect to complex searches, it allows combined query search strings (separated by AND, OR, NOT) as well as restricting the search to selected database fields. These fields or a limited number of field combinations can be selected by activating 'Limits' on the individual databases search pages. The format for visualization or download of all or selected entries in the hit list generated from a search is defined by selecting from a pull-down menu activated by pressing the 'Summary' button.

Sequence similarity searching for user-provided and/or user-defined database sequences can be performed using BLAST variants. These BLAST programs can be operated from http://www.ncbi.nlm.nih.gov/BLAST/. Detailed descriptions and tutorials are accessible at the same web page.

Data submission forms can be completed locally using the stand alone 'Sequin' tool or interactively by typing or pasting the respective information to the text boxes of the 'BankIt' tool (http://www.ncbi.nlm.nih.gov/BankIt/). Instructions on how to complete the online submission as well as submission examples can be inspected at http://www.ncbi.nlm.nih.gov/BankIt/help.html and http://www.ncbi.nlm.nih.gov/BankIt/examples/requirements.html, respectively.

15.3 Genome projects

A listing of the major genome projects, centers and databases can be seen at NCBI (http://www.ncbi.nlm.nih.gov/Genomes/index.html) or from the Genomes on Line Database (GOLD) at http://www.genomesonline.org/. Almost all institutions harboring genome projects allow BLAST searching of the sequence data often before the sequencing and annotation work for the whole genome is finished. A few examples for major centers and smaller genome projects will be briefly introduced.

15.3.1 Munich Information Center for Protein Sequences (MIPS)

MIPS (http://mips.gsf.de) (19) provides genome-related information. This site includes PEDANT (20), the database for the comprehensive analysis of protein sequences from genomes, several genome databases for Eukarya and additional services and tools for genome analysis. The PEDANT database could be of major interest in microbial ecology and environmental genomics studies. The gene products are automatically characterized with respect to various criteria using a large variety of bioinformatics tools. The user can navigate using the genome browser (http://pedant.gsf.de/) through a number of pre-computed catalogues and select proteins according to structural and functional classes as well as their interaction in physiological pathways. Text, pattern and BLAST searches can be performed in all genomes and predicted proteins.

15.3.2 Sanger Institute

The Wellcome Trust Sanger Institute (http://www.sanger.ac.uk) is a genome research center set up in 1992 to further our knowledge of genomes, and in particular to play a substantial role in the sequencing and interpretation of the human genome. A number of microbial genomes have been completed or are under work at the center. The data can be accessed at http://www.sanger.ac.uk/Projects/Microbes/. The Sanger center also provides a free genome viewer and annotation software tool (Artemis). It is available for the most commonly used operating system platforms and can be downloaded from http://www.sanger.ac.uk/Software/Artemis/. Comprehensive documentation is provided at http://www.sanger.ac.uk/Software/Artemis/stable/manual/.

15.3.3 The Institute for Genomic Research (TIGR)

Also founded in 1992, TIGR (http://www.tigr.org/) is a research center with primary interests in structural, functional and comparative analysis of genomes and gene products from a wide variety of organisms including viruses, Bacteria, Archaea, and Eucarya. TIGR's Genome Projects are a collection of curated databases containing DNA and protein sequence, gene expression, cellular role, protein family, and taxonomic data for microbes, plants and humans. Anonymous FTP as well as Blast (http://tigrblast.tigr.org/cmr-blast/) access to sequence data is provided. The Comprehensive Microbial Resource (CMR) is a tool that allows access to all of the bacterial genome sequences completed to date (21). The CMR is fully described in: http://www.tigr.org/tigr-scripts/CMR2/CMRHomePage.spl. MUMmer (22) is a freely available suffix-tree-based system for rapid alignment of whole genome sequences.

15.3.4 Mobile genetic elements (MGE) databases

In addition to whole genome databases, there is considerable interest in having access to specific sequence information on the group of MGE that comprise the horizontal gene pool (see Chapter 10). In recent years a number of database sites have been established to meet this interest. The first site established was the IS Database (http://www-is.biotoul.fr/) providing a comprehensive resource on insertion sequence (IS) elements. More recently, the Plasmid Genome Database (http://www.genomics.ceh.ac.uk/plasmiddb/index.html) has provided comprehensive datasets and a BLAST search facility to interrogate >600 complete plasmid sequences. The third and most ambitious of these MGE database sites is the ACLAME database (http://aclame.ulb.ac.be/) that aims to provide a comprehensive MGE based on the functional modules of which individual MGE are comprised (23).

15.4 Ribosomal RNA databases

Comparative rRNA sequence analysis is still a central tool for eliciting phylogenetic relationships, amending microbial taxonomy and identification of microorganisms. There are three major projects maintaining and providing databases of processed rRNA primary structure data.

15.4.1 Ribosomal Database Project

The Ribosomal Database Project II (RDP; 24) was initiated by Carl Woese and co-workers. The first release of a specialist database for small subunit rRNA sequences was published in 1992. Initially hosted at Argonne National Laboratory, then later at the University of Illinois, Champaign, Urbana, it is now maintained at the Center for Microbial Ecology at Michigan State University (http://rdp.cme.msu.edu). Currently two versions exist:

(i) RDPII version 9 (http://rdp.cme.msu.edu/index.jsp) which includes the latest updates of rRNA gene sequences;
(ii) the older version (release 8.1) that can be accessed from the link on the version 9 site.

Version 8.1 is still very much of interest because of an excellent collection of online analysis tools. The RDP maintains databases of aligned large and small subunit rRNA sequences containing higher-order structure information. Small subunit rRNA data are stored in separate databases for Prokaryotes, Eukaryotes and mitochondria. Based upon phylogenetic analyses, it has been shown that the alignment of rRNA primary structures cannot simply be done by maximizing sequence similarity, but higher-order structure data or prediction have to be taken into account (4). Consequently, the alignments in RDP databases are optimized by specialists and can be used as guides for the alignment of user-provided data.

The databases can be downloaded as compressed files in GenBank (*.gb.bz2) or FASTA (*.fasta.bx2) format, and either as aligned or non-aligned datasets on http://rdp.cme.msu.edu/misc/resources.jsp. A new facility at RDB is hierarchy browsing of databases (http://rdp.cme.msu.edu/hierarchy/ hb_intro.jsp) according to RDP's or NCBI's (8) taxonomic hierarchy. The new RDP hierarchy is based upon that proposed in the recent edition of *Bergey's Manual of Systematic Bacteriology* (3) and should be used preferentially. Browsing and searching can be optionally performed on all entries or confined to almost complete sequences and/or type strain data. The latter is an important facility with respect to identification and new description of taxonomic entities. After selecting entries, visualization and downloading of the data is possible in a small selection of formats.

The older RDPII project version 8.1 release also provides several online functions for comparative analysis of user-provided sequences and database entries. Interactive data selection and parameter definition can be done by pressing the 'run' button for the respective program on the 'Online Analyses' page. To access this page click onto release 8.1 from the RDPII home page, and then click onto 'Online Analyses'. This then shows a menu page consisting of a Table of programs. The 'info' and 'arrowhead' buttons in the table on this page give access to documentation and starting of the program, respectively. The online functions include a tool ('Sequence match') for finding the database primary structures sharing the highest sequence similarity with the user-submitted raw sequence. The data are first broken up into short oligonucleotides. The ratio of common and unique 'words' roughly reflects the degree of overall similarity. The most similar reference sequences are used as a guide for aligning the user-entered sequence data via the 'Sequence Aligner' tool. It has to be taken into account that this procedure only works correctly if the imported user sequences are highly similar to the database references. The alignment of more divergent sequences has to be optimized manually according to secondary structure prediction (4). This is not possible online. Preliminary phylogenetic analyses can be performed for user-defined data selections by calculating sequence similarity matrices ('Similarity Matrix') or applying the neighbor-joining treeing method of the PHYLIP (25) package ('Phylip interface').

Another facility at RDP is chimera check. This tool is especially important in environmental studies applying the so called rRNA cycle (2) where there is a risk of producing chimeric rDNA fragments (Chapter 2) during PCR amplification. The 'Chimera check' tool splits the sequence into two parts at every tenth position and for every split compares the sum of the highest number of word hits (most similar database sequence) for both parts and the full sequence. A substantial difference between the overall sequence sections indicates the presence of a chimeric structure. The user has to be aware that this procedure only works properly if close relatives are in the database for both sources of the chimeric sequence. Furthermore, fragments originating from multiple sources cannot easily be identified as such.

The RDP also provides a facility to evaluate the specificity of hybridization probes defined by the user (Probe Match; http://rdp.cme.msu.edu/probematch/search.jsp). The output shows the alignment of the probe to regions of sequences including mismatched targets from the database. The maximum number of changes, insertion, and deletions allowed in the results list is predefined by the submitting user. A list of potential target organisms according to the taxonomic hierarchy in *Bergey's Manual of Systematic Bacteriology* (3) is provided in addition. This tool does not allow probe design but does help to evaluate the specificity of new and established probes against the most recent RDP database version.

15.4.2 European Ribosomal RNA database

The European Ribosomal RNA database initially located at the University of Antwerp (Belgium) has now moved to the University of Ghent (Belgium; http://www.psb.ugent.be/rRNA/). Small (26) and large (27) subunit rRNA sequences are compiled and provided in aligned format. The alignment takes into account secondary structure prediction based upon comparative analysis of the database sequences. The data are stored in a special distribution format, a tagged flat file with three-letter line codes separated by a colon, from the respective information such as accession numbers, bibliography and taxonomy. The predicted secondary structure is indicated by bracketed expression. An example file can be seen at http://www.psb.ugent.be/rRNA/help/formats/distributionformat.html. Other file formats for downloading are EMBL and DCSE. The latter is needed if the DCSE sequence editor standalone software (http://rrna.uia.ac.be/dcse/) is used. The most convenient way for downloading is to select data with respect to file format and organisms using the query interface (http://www.psb.ugent.be/rRNA/ssu/query/index.html; note that for large subunit RNA, 'ssu' may be replaced by 'lsu'). String searches can be performed in the database fields containing annotation. Alternatively, the organisms' respective sequences can be selected by browsing through a list of names, structured according to taxonomy or phylogenetic affiliations.

Some standalone software tools are available for downloading. Besides the already mentioned DCSE editor for sequence alignment and visualization, RnaViz (http://rrna.uia.ac.be/rnaviz/), a graphical user interface for producing publication-quality secondary structure drawings of rRNA molecules, is available.

An rRNA BLAST server (http://www.psb.ugent.be/rRNA/blastrrna.html) allows the user to submit sequence data for sequence similarity searches. A new functionality is 'quick phylogeny' which allows the user to submit sequences and find the approximate phylogenetic position for them. The program uses BLAST (16) for finding the most similar database reference sequences and ClustalW (17) to establish an alignment and a tree. Given that no alignment or tree evaluations are performed, the results obviously have to be regarded as preliminary.

15.5 rRNA-targeted probes

15.5.1 ProbeBase

ProbeBase (http://www.microbial-ecology.de/probebase/index.html; 28) is an initiative of the microbial ecology group at the Lehrstuhl für Mikrobiologie of the Technical University of Munich (http://www.mikro.biologie.tu-muenchen.de) and is now located at the University of Vienna. It is a comprehensive database containing published rRNA-targeted oligonucleotide probe sequences, DNA microarray layouts and associated information. All probes contained in the database were designed and evaluated *in silico* using the ARB program package and databases (http://www.arb-home.de). Furthermore, probe specificity was experimentally tested applying fluorescent *in situ* hybridization (FISH; 2) (Chapter 9), microarray (29) or other hybridization formats. ProbeBase can be searched for target organisms or probe names. Online analysis of user-provided sequences for the occurrence of targets for the probes maintained in the database is also possible.

15.5.2 Roscoff database

Another database for experimentally evaluated rRNA targeted probes specific for eukaryotes, protists (mostly photosynthetic) and cyanobacteria is maintained at the Station Biologique de Roscoff. The data on probe sequence, target organisms and bibliography are accessible at http://www.sb-roscoff.fr/Phyto/Databases/RNA_probes_introduction.php.

15.5.3 PRIMROSE

In the context of rRNA-targeted probes, PRIMROSE (30), a program for identifying potentially useful oligonucleotides for use as probes or PCR primers, might be of interest for the microbial ecologist. It is freely available from http://www.cf.ac.uk/biosi/research/biosoft/. The program is designed to work on the RDP (24) sequence databases for finding potentially useful oligonucleotides with up to two degenerate positions. The RDP databases have to be downloaded by the user.

15.6 ARB project

The ARB project (arbor, latin: tree; http://www.arb-home.de) (31) was initiated over 10 years ago. The two major tasks according to the ARB concept are:

(i) the maintenance of a structured integrative database combining processed primary structures and any type of additional data assigned to the individual sequence entries;

(ii) a comprehensive selection of software tools directly interacting with one another and the central database which are controlled via a common graphical interface.

Initially designed for rRNA data, the ARB software package can also be applied to DNA and protein sequences. ARB provides a versatile workbench to establish and maintain databases and to perform various types of data analysis. Given that the database is under the control of the user and can be freely modified, the database and software have to be installed locally at the user's site. Software and rRNA databases are accessible to the public (http://www.arb-home.de) and are in use worldwide. System and hardware requirements can be seen at the ARB home page. A selection of the major modules of ARB are briefly described here.

15.6.1 ARB databases

The primary structures representing organisms, genes or gene products are stored in individual database fields. The primary data are processed with respect to alignment, higher-order structure prediction, conservation profiles and other criteria. According to the ARB concept of an integrative database, any type of additional data can be stored within the user-defined database fields assigned to the respective sequence. They are either stored in the local database or linked to it via local networks or the internet. The database is hierarchically structured and highly compressed. The ARB project provides databases for large and small subunit rRNA and a selection of evolutionarily conserved protein genes. The data can be imported and exported in a variety of commonly used formats.

15.6.1.1 Data visualization and access

Any original or online processed database entry can be visualized in the ARB main window at the terminal nodes of a phylogenetic tree (*Figure 15.8*). These trees can also be used as guide for database browsing. Sets of visualization windows can be customized with respect to layout and selection of data (fields) to be analyzed and shown. A versatile search tool allows complex queries in all or selected combinations of database fields.

15.6.1.2 Sequence editors

The sequence data are accessible for visualization and modification via a powerful editor (*Figure 15.9*). The original or virtually transformed sequence data are shown in user-defined layout and color codes. Different editing modes (optionally allowing character or alignment changes) as well as protection levels (to prevent unintended modification or loss of data) that can be assigned to the individual sequences are valuable tools to ensure data consistency. Groups of sequences can be optionally represented by its consensus which is calculated online according to user-defined parameters. Sets of search strings can be defined, and perfect or partly mismatched targets highlighted in the particular sequences. In the case of rRNA data, an online higher-order structure check is performed and the presence and character of base pairing is indicated by user-defined symbols. The ARB secondary structure editor (*Figure 15.10*) fits any sequence into the common consensus model. Any of the search strings activated in the primary structure editor can be indicated in the secondary structure model.

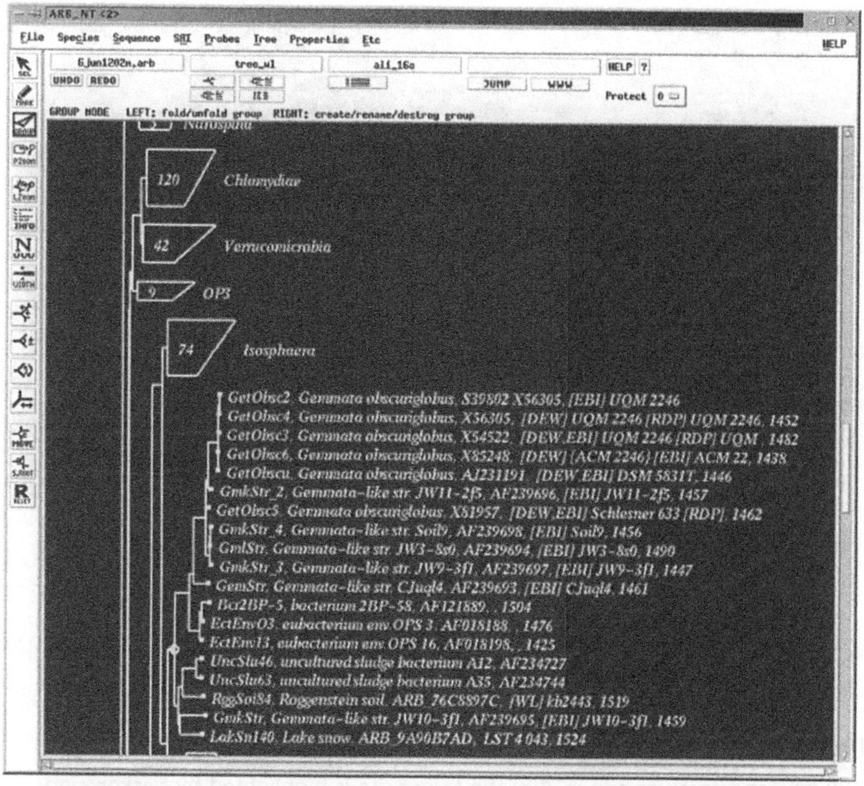

Figure 15.8

ARB main window. Phylogenetic groups (condensed) in the tree are indicated by a box with a number showing number of sequences. Visualized database field entries at the terminal nodes: 'name' (ARB specific unique ID); 'full_name' genus and species (other designations in case of clones or not validly described organisms); accession number (EMBL, GenBank, NNBJ); strain designation (tags EBI, RDP, DEW indicate source databases).

15.6.1.3 Sequence alignment

For *de novo* sequence alignment, ClustalW (17) is implemented in the ARB package. However, in most cases new sequences have to be inserted into existing alignments. ARB uses the PT server (positional tree) to rapidly find the most similar database reference sequences. The ARB fastaligner then includes these references as guide.

15.6.1.4 Filters, masks and profiles

The ARB package provides tools for determining conservation or base composition profiles, higher-order structure masks as well as filters to either include or exclude particular alignment positions for subsequent operations such as phylogenetic treeing. These profiles can be based upon the full database or user-defined subsets.

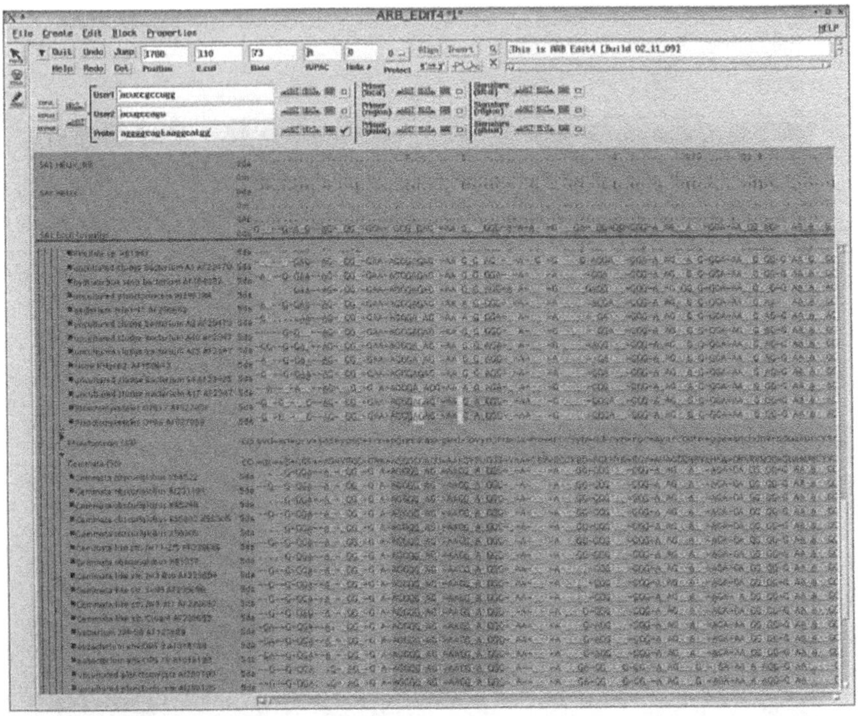

Figure 15.9

ARB primary structure editor. A probe target site is highlighted by background shading (dark).

15.6.2 Phylogenetic treeing

15.6.2.1 ARB parsimony

The central tree-making tool – ARB_parsimony – is specially developed for handling of several thousand sequences. An intrinsic software component superimposes branch length on the parsimony-generated tree topology reflecting primarily the significance of the tree topology and secondly indicating the number of estimated character changes. A prominent feature of the ARB_parsimony tool is the possibility of adding sequences to an existing tree without permitting any changes of the initial tree. This allows for the inclusion of partial sequences without destroying an existing optimized tree topology (4). The second feature allows tree optimization, cycles of NNI (nearest neighbor interchange) and KL (Kernigham and Lin analysis) (32) tree modifications to be performed. This optimization can be confined to subtrees. Thus tree optimization is possible applying the appropriate filters for the respective phylogenetic levels and groups.

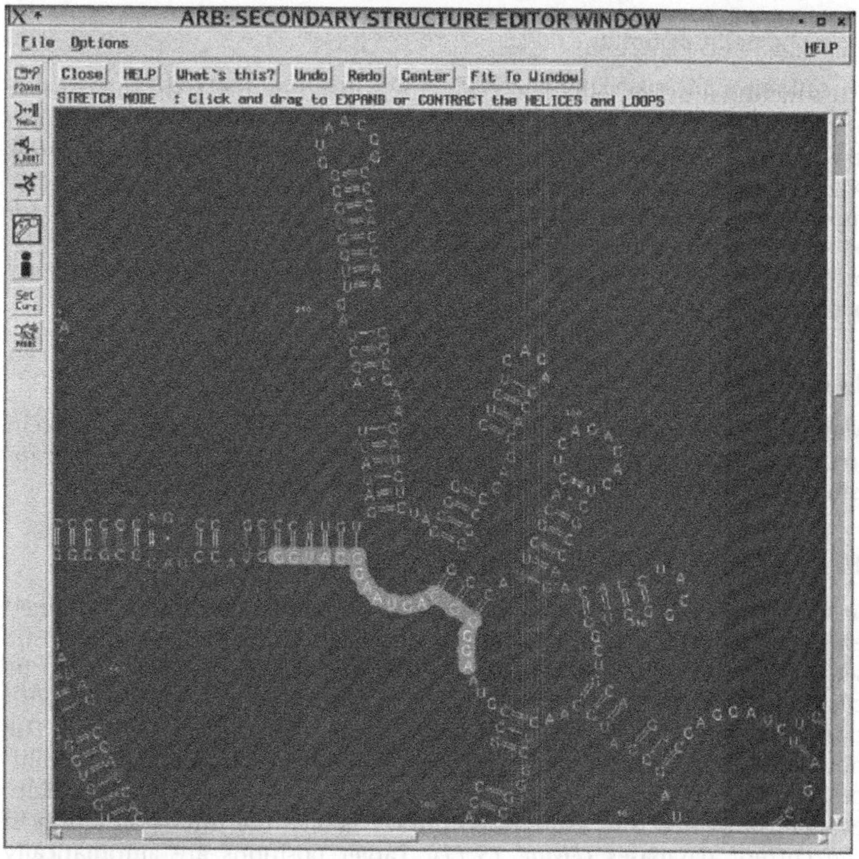

Figure 15.10

ARB secondary structure editor. A probe target site is highlighted by shading.

15.6.2.2 PHYLIP

Alternative treeing methods are incorporated in the ARB package; these include part of the PHYLIP programs. Felsenstein's (25) PHYLIP package is comprised of a variety of standalone programs for phylogeny among which are some of most commonly used non-commercial treeing programs. Precompiled versions for the major operating systems can be downloaded from http://evolution.genetics.washington.edu/phylip/getme.html. Detailed documentation is available at http://evolution.genetics.washington.edu/phylip/phylipweb.html.

15.6.2.3 FastDNAml

Olsen's (33) version of the maximum likelihood approach fastDNAml is also positioned in the ARB hierarchy of interacting software tools. A standalone version can be downloaded from ftp://ftp.bio.indiana.edu/molbio/evolve/.

15.6.2.4 TREE-PUZZLE

The functionalities of TREE-PUZZLE (34,35) are accessible in the full context of ARB databases and tools. It reconstructs phylogenetic trees by maximum likelihood. It implements a fast tree search algorithm known as quartet puzzling that allows analysis of large datasets and assigns estimations of support to internal tree branches. The current version of TREE-PUZZLE can be downloaded from http://www.tree-puzzle.de. A complete list of the program features is given in the manual (http://www.tree-puzzle.de/manual.html).

15.6.2.5 MOLPHY

Mainly for maximum-likelihood-based phylogenetic analyses of protein sequence data, part of the MOLPHY package (http://www.ism.ac.jp/ismlib/softother.e.html; 36) is included in the ARB facilities.

15.6.2.6 Probe design and evaluation

In silico taxon- or gene-specific probe design and evaluation for user-defined entities is based upon the PT-server-mediated search for diagnostic sequence stretches. Further parameters for searching, which can be customized by the user, are: length (up to 100 mers), G + C content, position and character of diagnostic nucleotides and target position. The multi-probe tool helps to construct sets of probes according to the multiple probe concept (37–39). The probe match tool allows specificity evaluation of new and established probes against the full information in the current databases (*Figure 15.11*). Target positions are automatically shown in the primary and secondary structure editors. This facility is of importance when designing probes for *in situ* cell hybridization.

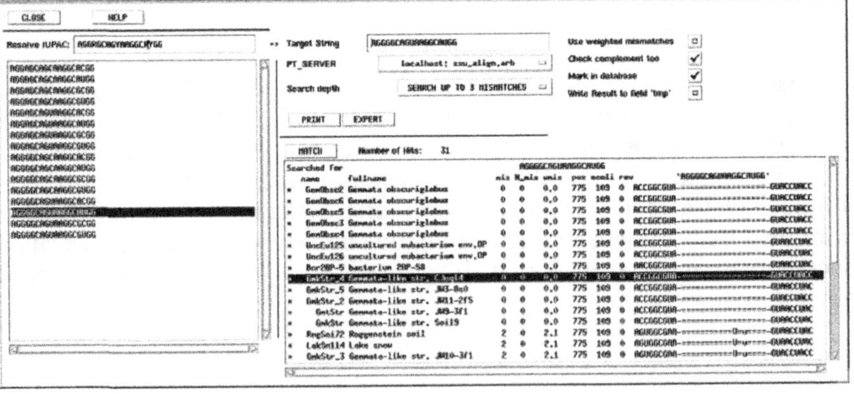

Figure 15.11

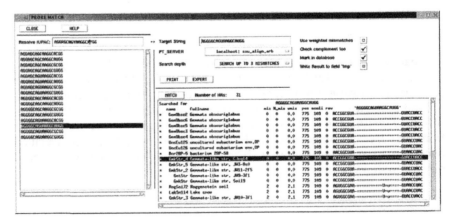

ARB probe match results window. The highlighted version of a degenerated probe (sequence in the 'Resolve IUPAC:' text box) is analyzed against >80 000 reference sequences. Alignment of the target region is shown with the hit list.

15.6.2.7 ARB probe library

The ARB probe library gives access to all possible probes which phylogenetically or taxonomically make sense according to the multiple probe concept. These probes were designed on the basis of the current small subunit rRNA ARB database of almost complete sequences. A client can be downloaded from http://www.mikro.biologie.tu-muenchen.de. This platform-independent client visualizes a phylogenetic tree. Numbers at the nodes indicate the number of potential probe targets common to the members of the corresponding subtree. The target sequences and additional descriptive information are visualized by mouse-clicking on the respective node. Users have to be aware that the potential probe targets are designed *in silico* but have not been tested *in vitro*.

15.6.3 Future developments

ARB-Genome is under development and will allow the ARB functionalities to be applied to fully annotated genomes as well as to individual genes. Visualization tools for genome maps are included, providing similar functionality to that of the trees in conventional ARB. In addition, according to the concept of integrative databases, experimental parameters and data (proteomics, transcriptomics) can be assigned to the organisms or genes.

15.7 Concluding remarks

Despite the progress in bioinformatics and genomics as well as the collaboration of some of the databases, the concept of integrative databases—the assignment of any information, be it primary data of expression studies, physiological pathways or conventional descriptive data such as a picture of the habitats or recordings of the voices of higher organisms to the sequence data—is not yet optimally implemented. As with databases, a similar situation is seen with respect to interoperability of the vast number of software tools. The ARB concept is an attempt to combine direct interaction of various tools and data(bases) with a maximum of accepted interventions by the individual user.

References

1. Staley JT, Konopka A (1985) Measurement of *in situ* activities of nonphotosynthetic microorganisms in aquatic and terrestrial habitats. *Annu Rev Microbiol* 39: 321–346.
2. Amann R, Ludwig W, Schleifer KH (1995) Phylogentic identification and *in situ* detection of individual microbial cells without cultivation. *Microbiol Rev* 59: 143–169
3. Garrity G (ed.) (2001) *Bergey's Manual of Systematic Bacteriology*, 2nd edn. Springer-Verlag, New York.
4. Ludwig W, Klenk HP (2001) Overview: a phylogenetic backbone and taxonomic framework for prokaryotic systematics. In: Garrity G (ed) *Bergey's Manual of Systematic Bacteriology*, 2nd edn, pp. 49–65. Springer, New York.
5. Stackebrandt E, Frederiksen W, Garrity GM *et al.* (2002) Report of the ad hoc committee for the re-evaluation of the species definition in bacteriology. *Int J Syst Evol Microbiol* 52: 1043–1047.

6. Lee N, Juretschko S, Daims H, Schleifer KH, Wagner M, Andreasen K, Nielsen J, Nielsen P (1999) Combination of fluorescent *in situ* hybridisation and microautoradiography—a new tool for structure function analyses in microbial ecology. *Appl Environ Microbiol* **65**: 1289–1297.
7. Stoesser G, Baker W, van den Broek A *et al.* (2002) The EMBL nucleotide sequence database. *Nucleic Acids Res* **30**: 27–30.
8. Benson DA, Karsch-Mizrachi I, Lipman DJ, Ostell J, Rapp BA, Wheeler DL (2002) GenBank. *Nucleic Acids Res* **30**: 17–20.
9. Tateno Y, Imanishi T, Miyazaki S, Fukami-Kobayashu K, Saitou N, Sugawara H, Gojobori T (2001) DNA Data Bank of Japan (DDBJ) for genome scale research in life science. *Nucleic Acids Res* **30**: 27–30.
10. Apweiler R, Bairoch A, Wu CH *et al.* (2004) UniProt: the Universal Protein Knowledgebase. *Nucl Acids Res* **32**: D115–D119.
11. Barker, WC, Garavelli, JS, Huang, H *et al.* (2000) The protein information resource. *Nucleic Acids Res* **28**: 41–44.
12. Lipman DC, Pearson WR (1985) Rapid and sensitive protein similarity searches. *Science* **227**: 1435-1441.
13. Pearson WR, Lipman DC (1988) Improved tools for biological sequence comparison. *Proc Natl Acad, Sci USA* **85**: 2444–2448.
14. Sansom C (2000) Database searching with DNA and protein sequences: an introduction. *Briefings Bioinform* **1**: 22–32.
15. Smith TF, Waterman MS (1981) Identification of common molecular subsequences. *J Mol Biol* **147**: 195–197.
16. Altschul SF, Madden TL, Schaffer AA, Znang J, Znang Z, Miller W, Lipman DJ (1997) Gapped BLAST and PSI-Blast: a new generation of protein-specific gap penalties and weight matrix choice. *Nucleic Acids Res* **25**: 3389–3402.
17. Thompson JD, Higgins DG, Gibson DJ (1994) CLUSTAL W: improving the sensitivity of progressive multiple sequence alignment through sequence weighting, position-specific gap penalties and weight matrix choice. *Nucleic Acids Res* **22**: 4673–4680.
18. Benson DA, Karsch-Mizrachi I, Lipman DJ, Rapp BA, Wheeler DL (2001) GenBank. *Nucleic Acids Res* **30**: 17–20.
19. Mewes HW, Frishman D, Güldener U *et al.* (2001) MIPS: a database for genomes and protein sequences. *Nucleic Acids Res* **30**: 31–34.
20. Frishman D, Albermann K, Hani J, Heumann K, Metanomski A, Zollner A, Mewes HW, (2001) Functional and structural genomics using PEDANT. *Bioinformatics* **17**: 144–157.
21. Peterson JD, Umayam LA, Dickinson TM, Hickey EK, White O (2001) The Comprehensive Microbial Resource. *Nucleic Acids Res* **29**: 123–125.
22. Delcher AL, Kasif S, Fleischmann RD, Peterson J, White O, Salzberg SL (1999) Alignment of whole genomes. *Nucleic Acids Res* **17**: 2369–2376.
23. Leplae R, Hebrant A, Wodak SJ, Toussaint A (2004) ACLAME: A classification of mobile genetic elements. *Nucl Acids Res* **32**: D45–D49.
24. Maidak BL, Cole JR, Lilburn TG *et al.* (2001) The RDP-II (ribosomal database project). *Nucl Acids Res* **29**: 173–174.
25. Felsenstein J (1989) PHYLIP—phylogeny inference package (version 32). *Cladistics* **5**: 164–166.
26. Wuyts J, Van de Peer Y, Winkelmans T, De Wachter R (2002) The European database on small subunit ribosomal RNA. *Nucl Acids Res* **30**: 183–185.
27. Wuyts J, De Rijk P, Van de Peer Y, Winkelmans T, De Wachter R (2002) The European large subunit ribosomal RNA database. *Nucl Acids Res* **30**: 175–177.
28. Loy A, Horn M, Wagner M (2003) ProbeBase: an online resource for rRNA-targeted oligonucleotide probes. *Nucl Acids Res* **31**: 514–516.
29. Loy A, Lehner A, Lee N, Adamczyk J, Meier H, Ernst J, Schleifer KH, Wagner M

(2002) An oligonucleotide microarray for 16S rRNA gene-based detection of all recognised lineages of sulfate-reducing prokaryotes in the environment. *Appl Environ Microbiol* **68**: 5064–5081.
30. Ashelford KE, Weightman AJ, Fry JC (2002) PRIMROSE: a computer program for generating, and estimating the phylogenetic range of 16S rRNA oligonucleotide probes and primers in conjunction with the RDP-II database. *Nucl Acids Res* **30**: 3481–3489.
31. Ludwig W, Strunk O, Westram R *et al.* (2004) ARB, a software environment for sequence data. *Nucl Acids Res* **32**: 1363–1371.
32. Kernigham BW, Lin S (1970) An efficient heuristic procedure for partitioning graphs. *Bell System Tech J* **49**: 291–307.
33. Olsen GJ, Matsuda H, Hagstrom R, Overbeek R (1994) FastDNAml: a tool for construction of phylogenetic trees of DNA sequences using maximum likelihood. *Comput Appl Biosci* **10**: 41–48.
34. Strimmer K, von Haeseler A (1996) Quartet Puzzling: a quartet maximum likelihood method for reconstructing tree topologies. *Mol Biol Evol* **13**: 964–969.
35. Schmidt HA, Strimmer K, Vingron M, von Haeseler A (2002) TREE_PUZZLE E: maximum likelihood phylogentic analysis using quartets and parallel computing. *Bioinformatics* **18**: 502–504.
36. Adachi J, Hasegawa M (1996) Molphy version 2.3, programs for molecular phylogenetics based on maximum likelihood. Computer Science Monograph No 28, The Institute of Statistical Mathematics, Tokyo, Japan.
37. Ludwig W, Amann R, Martinez-Romero E, Schönhuber W, Bauer S, Neef A, Schleifer KH (1998) rRNA based identification systems for rhizobia and other bacteria. *Plant and Soil* **204**: 1–9.
38. Amann R, Ludwig W (2000) Ribosomal RNA-targeted nucleic acid probes for studies in microbial ecology. *FEMS Microbiol Rev* **24**: 555–565.
39. Behr T, Koob C, Schedl M *et al.* (2000) A nested array or rRNA targeted probes for the detection and identification of *enterococci* by reverse hybridization. *Syst Appl Microbiol* **23**: 563–572.

Index

For Product Safety Concerns and Information please contact our EU representative GPSR@taylorandfrancis.com Taylor & Francis Verlag GmbH, Kaufingerstraße 24, 80331 München, Germany

Printed and bound by CPI Group (UK) Ltd, Croydon, CR0 4YY
08/06/2025
01897009-0016